XXXVIII

THE
EARLY AND LATER
HISTORY OF PETROLEUM

THE

EARLY AND LATER

HISTORY OF PETROLEUM

WITH

AUTHENTIC FACTS

IN REGARD TO ITS DEVELOPMENT IN

WESTERN PENNSYLVANIA

BY

J. T. HENRY

[1873]

REPRINTS OF ECONOMIC CLASSICS

AUGUSTUS M. KELLEY · PUBLISHERS

NEW YORK 1970

First Edition 1873

(Philadelphia: Jas. B. Rodgers Co., Printers, *52 & 54
North Sixth Street*, 1873)

Reprinted 1970 by
AUGUSTUS M. KELLEY · PUBLISHERS
REPRINTS OF ECONOMIC CLASSICS
New York New York 10001

.   .   .   .

S B N   678 00622 9
L C N   76 107917

.   .   .   .

PRINTED IN THE UNITED STATES OF AMERICA
by SENTRY PRESS, NEW YORK, N. Y. 10019

Woodburytype. A. P. R. Co., Phila.

From a Photograph by MATHER, of Titusville, in Dec. 1861.

THE PHILLIPS AND WOODFORD WELLS. TARR FARM.

(SEE HISTORY OF FLOWING WELLS.)

Woodford Well. Struck in Dec. 1861.

Phillips Well. Struck in Sept. 1861.

THE

EARLY AND LATER

HISTORY OF PETROLEUM,

WITH

AUTHENTIC FACTS IN REGARD

TO ITS

DEVELOPMENT IN WESTERN PENNSYLVANIA,

The Oil Fields of Europe and America. Gas Wells. Spiritual Wells. Oil Well Shafts. Petroleum Products.
Oil Companies. Pipe Line Statistics. Early Modes of Transportation. Flowing Wells of 1861,
to 1864. Pit Hole in 1865. The Lubricating Oil District, &c. Also, Statistics of Product,
Export, and Consumption, with prices of Oil from 1859, to 1872, &c., &c.

THE PARKERS' AND BUTLER COUNTY OIL FIELDS.

ALSO, LIFE SKETCHES OF

PIONEER AND PROMINENT OPERATORS,

WITH THE

REFINING CAPACITY OF THE UNITED STATES.

BY J. T. HENRY.

PHILADELPHIA:
JAS. B. RODGERS CO., PRINTERS, 52 & 54 NORTH SIXTH ST.
1873.

Dedicatory.

—

TO

WILLIAM H. ABBOTT,

Titusville, Penn'a.

—

In submitting to a discerning, yet indulgent public, this volume, relating to the History of Petroleum, with Life Sketches, of many of the prominent men in the Pennsylvania Oil Region, identified therewith, I beg to associate with my unpretending labors, your own honored name, known and acknowledged as that of an enterprising business man, a public-spirited citizen, a high-toned Christian gentleman. Your characteristic reserve, I know, would prompt you to shrink from such conspicuous mention; but the public voice pronounces you one of the worthiest members of the community, and I cannot deny myself the privilege of paying this dedicatory tribute to one who has sustained my enterprise with never-failing sympathy and encouragement.

J. T. HENRY.

PREFACE.

———

In the preparation of this work, the end and aim has been, to supply some needed historical data of the discovery of Petroleum Oil, in Western Pennsylvania, West Virginia, Ohio, and at other points in the United States and Canadas. The Pennsylvania Oil Region is by far the largest field, and produces eight-tenths of all the Petroleum, now so largely entering into the commerce of the commonwealth, and of the nation, and has surely come to be an indispensable benefaction to millions of households, in every civilized nation upon the globe. To the early and later developments in this section, therefore, this work is mainly devoted. While this portion of its pages is deemed to be full and complete, and, we may add, reliably so, a sufficiently elaborate account of other Oil Fields upon this continent, and throughout the old world are given, and from the best sources attainable. In this respect no pains have been spared to render it complete and authentic in all the details treated of.

The Statistical pages of the work, though not elaborate, are regarded as full enough for the purposes of an intelligent understanding of the magnitude of the industry. Such as we give, are known to be accurate. The Refining capacity of the more important refining centres of the United States—except Boston, Mass.,—may be relied upon as very nearly authentic.

The BIOGRAPHICAL SKETCHES, which make up so important a part of the work, are from among the representative men of the Region,—and they furnish, individually and collectively, very many interesting and valuable facts, with reference to developments, early and late, in the Oil Region of Pennsylvania, never before given to the public in an authentic form.

And here we may as well refer to the fact that no "sketches" are given of operators at Petroleum Centre, Columbia, Tarr Farm, Rouseville, Oil City, President, and but one at Franklin. As "brevity is the soul of wit," we may add—*the neglect is not ours.*

While we have, in the list of biographies given, made a fair showing of the representative men of the Oil Region of Western Pennsylvania, we have not exhausted this portion of our subject. There are scores of men connected, in one way and another, with this great mining industry, whose history, and whose developments as producers, would add interest to our effort, and lustre to their names,—but we have been unable, in many cases, to obtain permission thus to do justice to them. We may do this in future editions of the work.

Our acknowledgments are due to JAMES McCARTY, of Oil City, and to Mr. RICHARD LINN, formerly of *The Petroleum Monthly,* for valuable aid in the preparation of the work,—and to Mr. CARBUTT, of Philadelphia, for the beautiful photographic illustrations that accompany its pages, especially in the department of Biographies,—and to JAS. B. RODGERS Co., for the superior excellence of the mechanical and typographical part of the work.

TITUSVILLE, October 20th, 1873.

CONTENTS.

BIOGRAPHICAL SKETCHES.

EARLY AND LATER

HISTORY OF PETROLEUM.

CHAPTER I.

EARLY NOTES OF PETROLEUM.—PETROLEUM SPRINGS.

WHILE the history of Petroleum in America prior to developments brought about by artesian boring, will probably be accounted of little practical value, it is apprehended that a work of this sort overlooking it would be incomplete.

In Europe and Asia it has been an object of some commercial value for centuries, and there is good reason to believe that it has been known and used since the earliest ages of the world.

It is impossible to go back to the time when petroleum was first discovered. From its frequent occurrence in the form of springs in many parts of the world, it is evident that it has always been known—certainly more than four thousand years.

Layard and Botta, in their discoveries at Nineveh, adduce positive evidence that the inhabitants of this ancient city had knowledge of the existence and use of petroleum. In building the city, an asphaltic mortar ("slime" according to the Old Testament,) was employed, the asphalt for which was a partially evaporated petroleum. That used at Babylon was obtained from the Springs of Is, on the Euphrates, which, at a later date, attracted the attention of Alexander, of Trajan, and of Julian; they, even to this day, supply the neighboring villages with oil.

9

Herodotus, 500 years before Christ, spoke of the oil wells of Zante ; and Pliny and Dioscorides described the oil of Agrigentum, which was used in lamps under the name of " Sicilian Oil."

The wells of Amiano, on the banks of the Taro, were formerly used for lighting the City of Genoa.*

There is reason to believe that at some former period in the history of the American continent, the existence and uses of petroleum had been better understood than they were for some centuries before the recent artesian developments. The numerous pits, until recently, and perhaps even still to be seen along the valley of Oil Creek, cribbed with roughly hewn timber, but nearly hidden by the rubbish of ages, indicate a development comparatively extensive. Trees were found growing in the centre of some of these pits, which, we are told, on the evidence of the concentric circles in the wood, were shown to be the growth of centuries. Many circumstances concur in referring these excavations to a period of time, and to a race of people, who occupied the country prior to the advent of those aborigines, found here by our Latin or Saxon ancestors. They were probably the work of that mysterious people who left the traces of their rude civilization in the copper mines about Lake Superior and the mounds of the South-West.

When we consider how easily, partially nomadic races, of which they probably were, degenerate, and how suddenly they are sometimes extinguished, this disposition of the matter seems plausible; but as all that is more within the sphere of the archæologist than the historian of a modern industry, we have passed it by without research.

There is a tradition in Venango Co., Pa., that the oil springs on Oil Creek formed a part of the religious cere-

* In the body of this work we devote a number of chapters to foreign oil fields.

mony of the Seneca Indians, who formerly lived on these wild hills.   The Aborigines dipped it from their wells and mixed it with their war-paint, which is said to have given them a hideous appearance, varnishing their faces, as it were, and enabled them to retain the paint for a long time, and to keep their skin entirely impervious to water. The uses of this oil for their religious worship is spoken of by the French commander of Fort Duquesne, in the year 1750.   " I would desire," writes the commandant to his Excellency, General Montcalm, " to assure you that this is a most delightful land.   Some of the most astonishing natural wonders have been discovered by our people."

" While descending the Allegany, fifteen leagues below the mouth of the Connewango, and three above the Venango, we were invited by the chief of the Senecas to attend a religious ceremony of his tribe.   We landed, and drew up our canoes on a point where a small stream entered the river.   The tribe appeared unusually solemn. We marched up the stream about half a league, where the company, a large band it appeared, had arrived some days before us.   Gigantic hills begirt us on every side.   The scene was really sublime.   The great chief then recited the conquests and heroism of their ancestors.   The surface of the stream was covered with a thick scum, which, upon applying a torch at a given signal, burst into a complete conflagration.   At the sight of the flames, the Indians gave forth the triumphant shout that made the hills and valleys re-echo again.   Here, then, is revived· the ancient fire worship of the East; here, then, are the children of the Sun."   Tracing the course of the French commander down the Allegany river on our present maps, we find the spring spoken of, as evidently upon Oil Creek, and on marching half a league above that stream we will probably reach Rouseville, where Cherry Run flows into Oil Creek.   The " gigantic hills " are still here, and the "thick scum" which the Indians gathered, and which careful, prudent men, now guard against conflagration, flows into peaceable tanks, and, instead of lighting up the wilderness for exhibitions of uncouth savages, sends joy and comfort into thousands of distant homes.

Later again we find a most interesting account of a Petroleum Spring in the southwestern part of the state of New York in the "American Journal of Science" for 1833, written by Prof. Silliman, Sr., a man who rendered early and valuable services to the cause of scientific investigation in this country. As it contains some interesting reflections on the origin of petroleum, nothing better can be done than to give it in full:

Notice of a Fountain of Petroleum called the OIL SPRING.

" The Oil Spring, as it is called, is situated in the western part of the County of Allegany, in the State of New York. This county is the third from Lake Erie on the south line of the State, the counties of Cattaraugus and Chautauqua lying west, and forming the southwestern termination of the State of New York. The Spring is very near the line which divides Allegany and Cattaraugus.

Being in the county of Allegany, I was indebted to the kindness of a friend, who on the 6th of September took me from Angelica to the Spring. After crossing the Genesee River, our ride was to the town of Friendship, six miles; then to Cuba, eight miles; and thence into the township of Hinsdale, three and a half miles, making seventeen and a half miles from Belvidere, the country-seat of Philip Church, Esq., and twenty-one miles from Angelica village. The place will be found without difficulty by taking a guide at Hicks' tavern, which is on the corner of the road to Cuba where it is intersected by the road to Warsaw, two miles west of Cuba.

The last half mile is in the forest: a road is cut, for

the greater part of the way, through the woods; but the path becomes finally an obscure foot-track in which a stranger without a guide might easily lose his way, or at least fail of finding the object of his search. The country is rather mountainous; but the road running between the ridges is very good, and leads through a cultivated region rich in soil and picturesque in scenery. Its geological character is the same with that which is known to prevail in this western region; a silicious sandstone, with shale, and in some places limestone is the immediate basis of the country. The sandstone and shale (the limestone I did not see) lie in nearly horizontal strata. The sandstone is usually of a light gray color, and both it and the shale abound with entrocites, encrinites, corallines, terebratula, and other reliquæ characteristic of the secondary or transition formation. The Oil Spring or fountain rises in the midst of a marshy ground. It is a muddy and dirty pool of about eighteen feet in diameter, and is nearly circular in form.

There is no outlet above ground, no stream flowing from it; and it is of course a stagnant water, with no other circulation than that which springs from the changes of temperature and from the gas and petroleum that are constantly rising on the surface of the pool.

The water is covered with a thin layer of petroleum or mineral oil, giving it a foul appearance as if coated with dirty molasses, having a yellowish-brown color. Every part of the water was covered by this film, but it had nowhere the iridescence which I recollect to have observed at St. Catharine's well, a petroleum fountain near Edinburgh in Scotland. There the water was pellucid, and the hues produced by the oil were brilliant, giving the whole a beautiful appearance. The difference is, however, easily accounted for. St. Catharine's well is a lively, flowing fountain, and the quantity of petroleum is only sufficient to cover it partially, while there is nothing to soil the stream; in the present instance, the stagnation of the water, the comparative abundance of the petroleum and the mixture of leaves and sticks and other productions of a dense forest preclude any beautiful features. There are, however, upon this water here and there spots of what seems to be a purer petroleum probably recently risen, which is free from mixture, and which has a bright brownish-yellow appearance — lively and sparkling. Were the fountain covered entirely with this purer production, it would be beautiful.

We were informed that when the fountain is frozen, there are always some air holes left open, and that in these the petroleum collects in unusual abundance and purity, having distinctly the beautiful appearance which has just been mentioned as now occurring here and there upon the water. The cause of this is easily understood. The petroleum being protected by the ice from the impurities which at other times fall into it, escapes contamination, and being directed to the air holes both by its lightness and by the gas which mixes with it collects there in greater quantity and purity. All the sticks and leaves, and the ground itself around the fountain, are rendered more or less adhesive by the petroleum.

They collect the petroleum by skimming it like cream from a milk-pan. For this purpose they use a broad, flat board, made thin at one edge like a knife. It is moved flat upon and just under the surface of the water, and is soon covered by a coating of petroleum which is so thick and adhesive that it does not fall off, but is removed by scraping the instrument upon the lip of a cup. It has then a very foul appearance like very dirty tar or molasses; but it is purified by heating it, and straining it while hot through flannel or other woolen stuff. It is used by the people of the vicinity for sprains and rheumatism and for sores upon their horses. It is not monopolized by any one, but is carried away freely by all who care to collect it, and for this purpose the spring is frequently visited. I could not ascertain how much is annually obtained. But the quantity is considerable. It is said to rise more abundantly in hot weather than in cold. Gas is constantly escaping through the water, and appears in bubbles upon its surface. It becomes much more abundant, and rises in large volumes whenever the mud at the bottom is stirred by a pole. We had no means of collecting or of firing it; but there can be no doubt that it is the carburetted hydrogen—probably of the lighter kind, but rendered heavier and more odorous by holding a large portion of the petroleum in solution. Whenever it is examined we should expect, of course, to find carbonic acid gas mingled with it, and not improbably ozate or nitrogen. We could not learn that any one had attempted to fire the gas as it rises, or to kindle the film of petroleum upon the water. We were told that an intoxicated Indian had fallen into the pool and been drowned many years ago, but that his body had never been recovered. The story may be true, and if true, it would be a curious inquiry whether the antiseptic properties of petroleum so well exemplified in the Egyptian mummies may not have preserved his body from putrefaction.

The history of this spring is not distinctly known. The Indians were well acquainted with it, and a square mile around it is still reserved for the Senecas. As to the geological origin of the spring, it can scarcely admit of a doubt that it rises from beds of bituminous coal below. At what depth we know not, but probably far down. The formation is doubtless connected with the bituminous coal of the neighboring counties of Pennsylvania and of the west rather than with the anthracite beds of the central parts of Pennsylvania.

A Branch of the Oil Creek (not the same with Oil
Creek in Venango Co., Pa.—ED.) which flows into the
Allegany River, a principal tributary of the Ohio, passes
near this spring, and we crossed the rivulet in going to it.
There we had the pleasure of seeing water that was on its
way to New Orleans and the Gulf of Mexico. We had
just passed the Genesee which flows into Lake Ontario,
and is thus seeking the Atlantic through the St. Law-
rence, and a little to the east, rise waters which flow to
the Susquehanna and the Chesapeake Bay; and thus this
elevated land, said to be one thousand four hundred feet
above the ocean level, is a grand rain shed for the supply
of rivers, seeking their exit through very remote and
opposite parts of the continent.

I cannot learn that any considerable part of the large
quantity of petroleum used in the Eastern states under
the name of Seneca Oil comes from the spring now described. I am assured that its source is about one hundred
miles from Pittsburgh on the Oil Creek, which empties into
the Allegany River, in the township and county of Venango. It exists there in great abundance, and rises in
purity to the surface of the water. By dams enclosing
certain parts of the river or creek it is prevented from
flowing away, and is absorbed in blankets from which it
is wrung. Although I have this statement from an eye-
witness, (he mentions in a footline that this eye-witness
was a stage driver at Rochester—ED.), still it would be an
interesting service, claiming a grateful acknowledgment,
if some gentleman in the vicinity of the petroleum, or at
Pittsburgh would furnish an account of it for this or some
similar journal. And as there are numerous springs of
this mineral oil in various regions of the West and South-
West, connected especially with the saline and bituminous
coal formations, it would promote the cause of science if
notices of any of them were forwarded for publication.

The petroleum sold in the Eastern states under the name of Seneca Oil is of a dark brown color, between that of tar and molasses; and its degree of consistency is not dissimilar according to temperature. Its odor is strong, and too well known to need description. I have frequently distilled it in a glass retort, and the naphtha which collects in the receiver is of a light straw color, and much lighter, more odorous and inflammable than petroleum. In the first distillation a little water usually rests in the receiver at the bottom of the naphtha. From this it is easily decanted, and a second distillation prepares it perfectly for preserving potassium and sodium, the object which led me to distil it. And these metals I have kept under it, as others have done for years. Eventually they acquire some oxygen from or through the naphtha, and the exterior portion of the metal returns slowly to the condition of alkali—more rapidly if the stopper is not tight. The petroleum remaining from distillation is thick like pitch. If the distillation has been pushed far the residuum will flow only languidly into the retort, and in cold weather it becomes a soft solid, resembling much the multha or mineral pitch. The famous lake of multha and petroleum in the island of Trinidad is well known. I have specimens from that place in all the conditions between fluid petroleum and firm pitch. It is unnecessary to repeat that the English use it on their ships of war as a substitute for tar and pitch, and that the bituminous mass in the natural lake, which covers several square miles, is sufficiently tenacious to support a man during the colder part of the year, but at the opposite season is too soft to sustain any considerable weight.

In alluding to the probable connection with bituminous coal of the oil spring named at the head of this notice, I did not mean to imply that petroleum and other bituminous substances *necessarily prove* that there is coal beneath, for it has been ascertained that bitumen exists in a limited degree in many minerals, as appears from some of the phenomena of volcanoes, and was proved experimentally by the late Hon. George Knox, in an extensive series of researches published in the philosophical transactions of London.

As regards the probability of finding coal the opinion should be thus modified: If the country on whose waters, or in whose rocks petroleum or other varieties of bitumen appear, is such a one as in its geological structure is consistent with the usual associations of coal, then the existence of bitumen, especially if it be abundant, and more especially if the rocks themselves are impregnated with it, affords a strong presumption in favor of the existence of coal beneath. Such is the fact in this part of the State of New York. The shale at Genesee is highly bituminous, and burns readily, with abundant flame. I cannot answer for the rocks in the immediate vicinity of the Oil Spring, as they are not in view.

The people have dug a few feet for coal at the distance of a few yards from the spring; the excavation is too shallow to decide anything except that petroleum rose in this place also as at the spring, thus proving that the bituminous impregnation is not peculiar to that spot.

If these remarks should excite any interest in the minds of landed proprietors in that vicinity, I would venture to suggest to them that it would not be wise without some more evidence to proceed to sink shafts, for they would be very expensive and might be fruitless. It would be much wiser *to bore*, which would enable them at a comparatively moderate expense to ascertain the existence, depth and thickness of the coal should it exist. But even this should not be done without a previous diligent examination of water courses, banks, precipices, excavations for wells, cellars, roads, &c., which might perhaps materially aid the inquiry. The well-known existence of bituminous coal beds at the distance of a few miles in Pennsylvania renders it highly probable that they may pass under this region, but perhaps at too great a depth to admit of profitable extraction; for the abundance of coal in other parts of Pennsylvania and the west, the magnitude and easy accessibleness of the beds and the excellence of the coal will long render it impossible that thin beds in other parts of the country, especially if lying deep in the ground, should be wrought without ruinous expenditure. It is worthy of remark that the cattle drink freely of the waters of the oil springs—a fact that we should hardly expect since they are so foul, and since there is abundance of pure water near, and also because we should expect that the petroleum would render the water very disgusting to animals. Perhaps they may find in this something of the reputed virtues of tar-water. I could not learn that the birds ever light upon or near the spring. The mephitic gases might perhaps make it a *real avernus* to them."

For such as take a lively interest in discovering the origin and learning the history of Petroleum, we can hardly think of a paper more useful than the above. In the first place, the writer was one of the ablest scientists of our country in his day, and had manifestly looked into the subject before, and in whose educated mind the phenomenon was referred to natural and plausible, if not true, causes. Then the evidence of what was known on the subject of its existence in the country *previous* to the development, which cannot fail to interest the reader, is not liable to the suspicion of invention which clings to some of the more modern publications, even when written over professional titles.

The readiness with which the eminent author connects its origin with bituminous coal, if it answered no better purpose, would serve to disembarrass some would-be scho-larly individuals who are inclined to deny that they ever held the same opinion in a cruder form, because it is not the theory prevailing among the practical oil operators of the day, who indeed, as far as we can judge, for the most part persistently refuse to entertain any theory at all. No doubt they will hasten to take back all their denials when they find their first and very natural suppo-sitions shared by such respectable authority.

Indeed, though the theories of the origin of Petroleum are numerous and all of them liable to some apparently insuperable objections, we think the one indicated by Prof. Silliman in the above paper is the one, with some modifi-cations, which still obtains among the *best* minds that have given the matter consideration. Formerly it was held to be a distillation of bituminous strata at *high tem-peratures*, and the work, we believe, of a very inconsidera-ble period of time; whereas now it is thought to be a result of heat applied at a very low temperature, but for infinite ages. The first theory supposed the production to have ceased completely with the cause which produced it; while the second favors the belief that it is being slowly but constantly generated in the carboniferous formations. In the manufacture from bituminous coal of kerosene oil, a substance much resembling petroleum, there is a consi-derable amount of gas given off that does not condense, which, supposing the petroleum to be the result on simi-lar constitutions of subterranean distillation, accounts for the gas which invariably accompanies it, though the pro-duction of gas does not surely indicate the production of oil, as would probably be the case if the distillation took place at a very low temperature. The village of Fredonia, Chautauqua Co., N. Y., near the shores of Lake Erie, was lighted by natural carburetted-hydrogen gas in the year 1828, which was supposed to be the production of bituminous coal beneath, but instrumental surveys have we believe, since shown, that there is no coal beneath, and all borings for oil in that vicinity have proved fruitless, though they have been prosecuted to a great depth. The supply of gas was more than sufficient to light the town, though the hole bored in the fetid limestone rock in the edge of the small stream in which the gas was first noticed to escape, was only an inch and a half in diameter. Great quantities of gas escaped within a few miles and further up the same stream. It would be interesting to know what connection these have with petroleum, or whether or not they have been affected by the enormous discharge of gas, which has gone on for years now in the oil region of Pennsylvania. It is not likely; but we are unable to say. While their composition is the same, it is not probable that their sources are identical.

In a paper communicated to the " American Journal of Science," for July, 1833, on " *The Saliferous Rock Formation in the Valley of the Ohio*," by Dr. S. P. Hildreth, of Marietta, touching incidentally on the subject of Carburetted-Hydrogen Gas and Petroleum, he says: " All salt wells afford more or less of this interesting gas, an agent intimately concerned in the free rise of the water, and universally present where salt water is found. Indeed so strong is the evidence afforded by the rising of this gas to the surface of the existence of the salt rock below, that many wells are sunk on this evidence alone. It is without doubt a product of the saliferous formation as it rises in many wells without any appearance of petroleum, which latter product is probably generated by bituminous coal, and in all wells, from a depth far below where coal has been discovered in sufficient quantity to furnish such an immense and constant supply as is continually rushing from the earth in these saliferous regions. In many wells, salt water and inflammable gas rise in company with a steady uniform flow. In others, the gas rises at intervals of ten or twelve hours, or perhaps as many days, in vast quantity, and with overwhelming force, throwing the water from the well to the height of fifty or one hundred feet in the air, and again retiring within the bowels of the earth to acquire fresh power for a new effort. This phenomenon is called " blowing," and is very troublesome and vexatious to the manufacturer. The explosion is sometimes so powerful as to cause the copper tube which lines the upper part of the well to collapse, and to entirely misplace and derange the fixtures about it. By constant use this difficulty is sometimes overcome by the exhaustion of the gas, and in others the wells have been abandoned as hopeless of amendment.

A well on the Muskingum, ten miles above McConnelsville, at six hundred feet in depth, afforded such an immense quantity of gas, and in such a constant stream, that while they were boring, it several times took fire from the friction of the iron on the poles against the sides of the wall, or from the scintillations from the auger, driving the workmen away, and communicating the flame to the shed which covered the works. It spread itself along the surface of the earth, and ignited several combustible bodies at the distance of several rods. It became so troublesome and difficult to extinguish whenever ignited, being in this respect a little like the Greek fire so celebrated by Gibbon, that from this cause only the well has been abandoned. In the days of superstition and ignorance this would doubtless have been attributed to the anger of the genius who presided over the spot, and thus protected it from the unhallowed approaches of man.

At A. P. Stone's well, on the opposite side of the river, a little below McConnelsville, the gas rises in small regular puffs or discharges, averaging one for every minute or two, causing the water to flow in jets from the spout as it falls into a large cistern below.   The water rises in the head through a bored log to the height of twenty-five feet above the surface of the earth.   Through a hole in the top of a small receiver or cup, the gas rises in a constant stream, and when a candle or torch is applied, kindles into a beautiful flame, burning steadily until extinguished by closing the hole—affording in the stillness and darkness of midnight a striking and interesting phenomenon.   It is supposed that this well alone furnishes sufficient gas, if properly applied, to light the town very handsomely. No petroleum rises with it, and very little in any of the other wells of this locality.   The quantity of gas in different wells varies very considerably;   all, however, afford sufficient to keep the water in constant agitation over the mouth of the well.   The supply of water depends very much on the quantity of gas discharged.   A few miles above Charleston, on the Big Kanawha, great quantities of the carburetted hydrogen are slowly emitted through the earth.   A tract of several rods in extent, near the river bank, is so charged with it that on making shallow cavities in the sand, and applying a fire-brand, it immediately becomes ignited, and burns with a steady flame for an indefinite period, or until extinguished by covering it with sand.   The boatmen, a rude but jolly race, often amuse themselves by tracing a circle in the sand around some one of the company unacquainted with the mystery, and applying fire, a flame immediately springs up as if by magic around the astonished wight, which being entirely confined to the circle traced, adds much to his terror, and increases the delight of his boisterous companions.   In a short time the sand beneath the burning gas becomes red hot.   The neighboring women sometimes make use of it to boil their water when washing clothes on the bank of the river, and boatmen occasionally cook their food in the same easy and cheap manner.   This spot would afford a

fine site for the temple of the fire-worshippers of ancient
Persia. In low stages of the water, gas and oil are seen
oozing from the bed of the river at various points. On
the Little Muskingum River, a few miles from Marietta,
this gas is discharged in many places—often through a
pool or sink-hole filled with water—in which case it is
called a burning spring. Petroleum is often found rising
from the earth near the spring. Throughout the whole
saliferous region, so far as I have any knowledge, on penetrating the salt rock a greater or less quantity of carburetted-hydrogen gas is discharged through the opening;
in some places accompanied by petroleum, and in others
without this co-existent production."

Continuing, he says of

Petroleum or Fossil Oil:

Since the first settlement of the country west of the
Apallachian range, the hunters and early pioneers have
been acquainted with this oil. Rising in a hidden and
mysterious manner from the bowels of the earth, it soon
arrested their attention, and acquired great value in the
eyes of these simple sons of the forest. Like some mira-
culous gift from Heaven, it was thought to be a sovereign
remedy for nearly all the diseases common to those pri-
meval days; and from its success in rheumatism, burns,
coughs, sprains, &c., was justly entitled to all its celebrity.
It acquired the name of Seneca Oil—that by which it is
generally known—from having first been found in the
vicinity of Seneca Lake, New York. From its being
found in limited quantities, and its great and extensive
demand, a small vial of it would sell for forty or fifty
cents. It is at this time in general use among the inhabitants of the country for saddle bruises and that complaint called the scratches in horses. It seems to be
peculiarly adapted to the flesh of horses, and cures many

of their ailments with wonderful certainty and celerity. Flies and other insects have a natural antipathy to its effluvia, and it is used with much effect in preventing the deposit of eggs by the "blowing fly" in the wounds of domestic animals during the summer months. In neighborhoods where it is abundant it is burned in lamps in place of spermaceti oil, affording a brilliant light, but filling the room with its own peculiar odor. By filtering it through charcoal much of this empyreumatic smell is destroyed, and the oil greatly improved in quality and appearance. It is also well adapted to prevent friction in machinery; for, being free of gluten, so common to animal and vegetable oils, it preserves the parts to which it is applied for a long time in free motion. Where a heavy vertical shaft runs in a socket it is preferable to all or any other articles. This oil rises in greater or less abundance in most of the salt wells on the Kanawha, and collecting as it rises in the head on the top of the water, is removed from time to time with a ladle, and put by for sale or use.

The greater abundance of stone coal in this locality, than in that of the Muskingum, gives it a decided advantage in the elaboration of petroleum. On the latter river the wells afford but little oil, and that only during the time the process of boring is going on. It ceases soon after the wells are completed, and yet all of them abound more or less in gas.

A well on Duck Creek, about thirty miles north of Marietta, owned by Mr. McKee, furnishes the greatest quantity of any in this region. It was dug in the year 1814, and is four hundred and seventy-five feet in depth. Salt water was reached at one hundred and eighty-five feet, but not in sufficient quantity. However, no more water was found below this depth. The rocks passed were similar to those on the Muskingum River, above the flint stratum, or like those between the flint and salt deposits at McConnelsville. A bed of coal two yards in thickness was found at the depth of one hundred feet, and gas at one hundred and forty-four feet, or forty-one feet above the salt rock. The hills are sandstone, based on lime, one hundred and fifty or two hundred feet in height, with abundant beds of stone coal near their feet. The oil from this well is discharged periodically, at intervals of from two to four days, and from three to six hours duration at each period. Great quantities of gas accompany the discharges of oil, which for the first few years amounted to from thirty to sixty gallons at each eruption. The discharges at this time, are less frequent and diminished in amount, affording only about a barrel per week, which is worth at the well from fifty to seventy-five cents a gallon. A few years ago, when oil was most abundant, a large quantity had been collected in a cistern holding thirty or forty barrels. At night some one engaged about the works approached the well-head with a lighted candle. The gas instantly became ignited, and communicated the flames to the contents of the cistern, which, giving way, suffered the oil to be discharged down a short declivity into the creek, where the water passes with a rapid current close to the well. The oil still contined to burn most furiously, and spreading itself along the surface of the stream for half a mile in extent, shot its flames to the tops of the highest trees, exhibiting the novel and perhaps never-before witnessed spectacle of a river actually on fire."

Here we find Petroleum obtained more than thirty years before its final development, yet attended with all the accidents, and presenting all the phenomena that characterize its production in Pennsylvania. These are not accounts open to the suspicion of exaggeration. They were written and published more than a generation before the *philosophy* of Petroleum broke upon the understanding of man. Here we find it repeatedly forced upon his attention in the very way it was finally developed, and still the idea of artesian boring was never designedly applied to the production of Petroleum till at last suggested, as is often the case with great ideas, by a most trivial occurrence.

Here was a well bored for salt to the depth of six or eight hundred feet—to the average depth of Petroleum wells in Pennsylvania—producing scarcely anything but Petroleum, every well in the region throwing up more or less of the oil, and yet it seems never to have occurred to any one that if bored for expressly, it could be found in paying quantities.

What a comment on the narrowness of a mind preoccupied! How have we overlooked all the great truths of philosophy, until at last they secured a sprouting place in some mind unprejudiced by practice and unbiased by theory. And then we call it inspiration, when the germ that has fallen in good ground, and sprouted, and blossomed, and borne fruit, has proved to be such a very wonderful little seed, and one that all saw, and none comprehended, when it was lying exposed in the stony places. And then we wonder we had not comprehended it before, especially as it was easier to comprehend it—the simple, little idea—than to shut it out from our understanding as the thing quite incomprehensible.

Its value as a lubricator was indisputably established; its medicinal properties were appreciated; very little stood in the way of its adaptation to purposes of illuminating, and so much of even that objection was removed by the simple process of filtering through charcoal—on which process a patent has since been obtained—that in the light of present events, it is impossible to understand how its importance could have been overlooked, could have failed to suggest, if not the philosophy of its existence in the earth, the manner in which it is held among the rocks, at least the hope of making a fortune by developing it after the manner of raising brine.

Even twenty years later we find Mr. Kier of Pittsburgh, profitably engaged in bottling and selling it as a great natural panacea; consuming in this way regularly about three barrels a day, obtained from his father's salt well at Tarentum, a few miles above Pittsburgh, on the Allegany River, labelling his bottles to the effect that this most wonderful remedy was obtained four hundred feet below the surface of the earth—distilling it, even, so that nearly every objection to it as an illuminator in the crude state was removed, and yet pocketing the returns without giving a thought to its origin or extent.

With Mr. Kier naturally enough the consideration was to utilize what he had, to make the most of it, rather than by research and development to bring forth that which would have been in every sense a drug in the market. He grasped one idea—its utility, and suggested the next—its development.

Thus link by link, was forged slowly, the chain of events which united thought and action, effecting what is known as the "discovery" of Petroleum.

CHAPTER II.

THE DISCOVERY OF THE VALUE OF PETROLEUM.

THE reader will hardly need to be informed that the circumstances related in the last chapter were not what led to the discovery of the economic value of Petroleum. There can hardly be said to be any merit in witnessing and describing circumstances purely accidental in their occurrence, when to do so effects no impression useful to the cause of human progress. However accidental, the events referred to must be allowed to have been very suggestive; but inasmuch as they did *not* suggest anything of practical importance to the very intelligent gentlemen who beheld them, it would seem to increase our obligations to the person who finally did grasp the simple idea of the philosophy of Petroleum, and that without any knowledge of the circumstances mentioned. But it must be conceded that, in their cases, a most important incentive was wanting—the article had no certain market, no determinate value. The fact that it sold for fifty and twenty-five cents a gallon proves nothing. It sold for that in 1859, but the first day's productions of the first well "broke the market." There was no demand for it, because its uses were unknown— or at least not definitely understood. If any one thinks that a larger and more constant supply would have earlier brought about a knowledge of its importance (had that supply been accidentally obtained, and had its continuance been altogether conjectural,) let him reflect how comparatively slow was its introduction to general use, even when the supply was so large that its cost was merely nominal, and its continuance an established fact. No! The world was not yet ready for it. And though the long course of scientific research in other directions, which prepared for its final reception, has escaped public attention, it is easy to show that the way was not prepared in a moment; but it is difficult, looking back from the light of the present, to excuse a stupidity which cannot now be understood, because it has been out-grown.

No science has been more active, progressive, and useful in the last forty years than Chemistry. But its strides have been as silent as they have been rapid; and though as a science it has almost grown up within the age of living men, and while it has done more, perhaps, than all other sciences to enable us to understand physical relations, there are still plenty of intelligent people who know no more of it than the name. It is mainly to this science that we owe those elaborate experimental researches which demonstrated the practical utility of Petroleum to the domestic comforts of refined civilization.

About the year 1830 a German chemist named Reichenbach, while experimenting with the bitumen found in wood, discovered a white, tasteless, inodorous, waxy substance which he called Paraffine, because of its antipathy to unite with other substances. Like Selligne of France, Reichenbach had devoted much attention to the production of illuminating oils from the coals and bituminous shales in his own country, as well as various other portions of the Continent of Europe, and like him experimented for years without producing anything of value.

The small quantity of paraffine obtained, was hoarded as a curiosity of the laboratory, and for many years, it is said, was the only bit of that substance in existence. It does not appear that Reichenbach himself at the time, clearly understood the chemical change by which it was produced, but his researches were continued, and together with Selligne's and several others, his name now appears as the inventor or discover of kerosene, or coal oil. It seems likely that either of them is entitled to all he claims, for they appear to have worked separately; and aside from the meagre details of their progress given to the scientific journals of the day, to have known little of each other's experiments.

It is true, no vast credit can attach to the invention or discovery on account of originality, for the way was very clearly pointed out by facts already known. But its importance can hardly be over-estimated. The refining influence—we might say the civilizing influence—of a good cheap illuminator, could easily be shown if this were the place for an extended essay on the subject.

The introduction of gas in the cities of Asia and Europe, furnish statistics on the subject most interesting from the political, as well as the humanitarian point of view. But the want of an illuminator for the habitations of the poor, and for such places as could not be economically lighted with gas was manifest, and in obedience to a common want we find it invented *almost* simultaneously in three countries—England, France and Germany—occupying the same plane in civilization. Such coincidences frequently occur; but it is a remarkable fact that they never occur unless the invention has become a manifest necessity to the comfort of the society, and the progress of the civilization which the inventors have enjoyed.

Reichenbach's ounce or two of Paraffine was preserved for many years as a great curiosity to his scientific friends. A small quantity of it was also discovered by Selligne, a few years later; but it was reserved for Mr. James Young of Scotland to complete the value of the discovery by showing how to produce it at will, and in quantity, and by utilizing it to the comforts of mankind. In 1850 he procured a patent in England for the manufacture of " paraffine oil, or oil containing paraffine, and paraffine from bituminous coals." Some years later a similar patent was issued to him by the government of the United States.

Before this oil had been manufactured from bituminous shales and coals on a small scale in France—by Selligne, we think—in Germany, and in the Austrian Empire. Even on the American Continent, Dr. Abraham Gesner manufactured oil from coal as early as 1846, and exhibited the same in the course of his lectures in the British Provinces. Uniting himself with capitalists, Mr. Young promptly began the manufacture of paraffine oil on a large scale. No lamp had yet been invented in which it would burn without a most offensive smoke, and while the heaviest of this manufacture was used for lubricating machinery, the lightest was reduced to paraffine wax, manufactured into candles, and sold as spermaceti, to which it bears a striking resemblance. It is a product obtained by destructive distillation of the oil—that is, one substance is destroyed before the other is produced. The change which takes place is purely chemical, and not mechanical, as it would be if the paraffine was separated from the oil by which it was merely held in solution. Mr. Young's process consists in breaking the coals into pieces about the size of a hen's egg, which are then distilled in the common gas retort, with worm pipes and the ordinary refrigerators of stills, the water in them being kept at a temperature of about 55° Fahr., by a stream of cold water entering the worm cistern. The retort is kept at a low red heat, and heated up gradually. The product is an oil containing the paraffine.

The crude oil is put into a cistern, and steam heat applied up to about 156°. This separates some of the impurities, and the oil is run off into another vessel, leaving the impurities behind.   The oil is then distilled in an iron still, with a worm pipe and refrigerator, the water in the latter being kept at 55° Fahr.   The oil thus distilled is then agitated with 10 per cent. of sulphuric acid for an hour. It is then allowed to settle twelve hours, when it is drawn off from the acid and impurities into an iron vessel, where it is again agitated with a solution of caustic soda of specific gravity 1.300.   Six hours are again allowed for the alkali and impurities to settle when the oil is again drawn off and distilled with half its bulk of water, which is run into the still from time to time to supply the quantity distilled off.   The light oil comes over with the steam, and is employed for illumination.   The oil left in the still is carefully separated from all water, and put into a leaden vessel, and then agitated with two per cent. of sulphuric acid.   It is then allowed to settle twenty-four hours. This oil is then run into another vessel, and for every one hundred gallons there are added twenty-eight pounds of chalk ground up with water into a paste.   The oil and chalk are agitated together until the oil is freed of acid. After it has remained a week at rest, it is used for lubricating machinery, and may be mixed with animal or vegetable oils for that purpose.

To obtain the Paraffine the oil containing it is brought down to a temperature of 30° Fah., when paraffine will crystallize and separate itself from the oil; or it may be filtered and finally submitted to pressure.   Again it is agitated with its bulk of sulphuric acid, and the operation repeated until the acid ceases to be colored by the paraffine, which is kept melted during the operation."

Mr. Gerker's method, differed from this not very materially, but had for its object not the production of paraffine, and therefore.the similarity of the treatment ceased just where the production of paraffine began.

His purpose was merely to procure an illuminating oil,

which he called "Kerosene," a name almost identical in its meaning with "Paraffine oil." The patents granted him by our government, known as the "Kerosene Patents," were sold to the North American Kerosene Gas light Company of New York, which in 1854 began the manufacture of Kerosene oil at their works on Newtown Creek, Long Island, New York. Its introduction was discouragingly slow. The refining process was not thoroughly understood, and nothing had then been found to overcome the odor which was most offensive. Men interested in the manufacture of camphene, and burning fluids of all sorts spread the belief that it was very explosive. But the beauty of its light commended it in spite of the odor and the fear of explosion, though in fact when first manufactured it was no more explosive than ordinary sperm oil. One great apparent need was a lamp which would burn it without a smoke, and admit of its being moved around, which could not be done when it was burned in the camphene lamp. This difficulty was afterwards remedied by the introduction of the Vienna burner by Mr. Anstin.

Mr. Young's patents specified a paraffine oil from "coals." The great profit of the business induced many to embark in the manufacture of the oil, and he soon found that his patent-right was being invaded both in this country and England. Proceedings were instituted at once to compel the payment of a royalty of three cents per gallon, and also for damages by infraction of the patent-right. In most cases these were strenuously resisted; the defendants in many cases claim that their oils were not from coals, but from bituminous shales and asphaltum. These gave the proprietors of the lands with whom for the most part sharp bargains had been driven, a pretext for checking the despoliation of their property, under contracts for the removal of coal, and Mr. Young in common with many of the other English manufacturers found himself involved in litigation that threatened to prove interminable, and was only settled at last by compromise. One of the best effects of these costly suits was a most exhaustive scrutiny into all the varieties of coals and bitumens, by the best scientific authorities in all countries, and though in a few instances carboniferous deposits, which were clearly shown to be asphalts or of that nature, by their solution in benzine and naphtha, were pronounced " coals " by ignorant jurors, it has not affected the value of those researches to the industries of the world, nor changed the opinions of educated men. Mr. Young's patent, after having been the source of a princely fortune, expired in England in 1864, and four years later in the United States, but it had brought him no revenue from this country since the discovery of Petroleum in Pennsylvania.

Before that event took place, however, a market was being prepared for it on the continent of Europe, by the manufacture on an extensive scale of oils from coals, schists and bituminous shales.

In Germany, on the Rhine, and in various parts of France and Switzerland, large manufactories were erected, and it soon became not a luxury of life, but a necessity. And in the Empire of Austria, in some of the Northern provinces of which it was extensively manufactured, a lamp had been invented in which the oil burned with a beautiful clear flame, and without smoke. The light could also be carried about indifferently without extinguishing. Here was the great desideratum at last, the greatest obstacle overcome. This lamp was promptly introduced into the United States. Our government has issued innumerable patents for alterations in this lamp, but not more than half a dozen have really been any improvement on the first one brought here, and many of them are greatly inferior. The quality of our light of course has been greatly improved, but it is less the result of any improvement in the lamp, than a more perfect method of refining the oil.

When Mr. Young began the manufacture of mineral oil, the success of his efforts sent consternation into some branches of industry on this side of the Atlantic.

The great manufacturers of animal oils along our Eastern seaboard were first to take the alarm. For years they had almost monopolized the whale fisheries, and large amounts of capital were invested in the production of the smaller fish oils and lard oil. They saw in his success the breaking down of their monopoly, the destruction of their trade; and determined to preserve their importance, they commenced manufacturing mineral oils themselves. Casting about for the means of self-preservation, they very soon discovered that our own country afforded even greater facilities for the production of these commodities than either Great Britain or any other part of Europe.

Our bituminous coal measures, were found to be the most extensive and accessible in the world. Upon examination, valuable oil-producing shales were discovered. Mines could be obtained on the most reasonable terms; could often be purchased in fee for a few dollars per acre. Everything was favorable, and it seemed as if our prestige in oil was not only to be maintained, but vastly increased.

The great difficulty was the cost of labor, which was four times as great as in England, and nearly six times greater than in Germany, where much of the work was done by women. This, to be sure, was in some measure, compensated by the difference in the cost of lands; but it was a compensation which must soon have been overcome.

Eastern capitalists invested largely in the coal lands of Virginia, Kentucky and Missouri, and to obviate the expense of transporting the coal by rail, began the erection of oil works at the mines.

Near Boston, Mass., Saml. Downer had erected works on a most extensive scale, which cost about half a million dollars, while at Portland he had other works put up at an expense of $250,000 for the manufacture of oil from imported coal; and they continued to increase, till at the time oil was struck, there were not less than fifty or sixty of these establishments in the United States, one of which was in Portland, one in New Bedford, four in Boston, one in Hartford, five in the environs of New York, eight or ten in western Pennsylvania, twenty-five in Ohio, eight in Virginia, six in Kentucky, and one in St. Louis. Many, if not most, of these were of small capacity, however, and the greater part of them were not more than fairly started when the discovery of petroleum prostrated the whole business, and threatened its projectors with overwhelming loss, from which they were happily rescued by converting their oil factories into refineries, which was done with very little trouble.

While the object of this chapter has been to show the gradual steps by which the economic value of petroleum was discovered, or rather demonstrated, and while the reader will, we presume, believe with us, that had its value not been thus conclusively determined, and had not the way for its reception at home and abroad been opened by the previous extensive introduction of coal oil both as a lubricator and an illuminator, its development must have been indefinitely delayed; for it was a belief in its identity—for practical purposes—with coal oil that prompted the series of investigations which resulted in its most wonderful development.

The event which finally determined its economic value, which proved its identities with, and differences from, coal oil, which showed that while for many purposes it was about the same—for most purposes it was superior, was the exhaustive analysis procured and paid for by

George H. Bissel and others, the report of which is
appended by permission of Professor Silliman to this
chapter rather than place it in the chronological order to
be observed throughout this work :

MESSRS. EVELETH, BISSELL & REED,

 Gentlemen,—

I herewith offer you the results of my somewhat extended researches upon the Rock Oil, or Petroleum, from
Venango County, Pennsylvania, which you have requested me to examine with reference to its value for
economical purposes.

Numerous localities, well known in different parts of
the world, furnish an oily fluid exuding from the surface
of the earth, sometimes alone in " tar springs," as they
are called in the western United States ; frequently it is
found floating upon the surface of water in a thin film,
with rainbow colors, or in dark globules, that may, by
mechanical means, be separated from the fluid on which
it swims.

In some places wells are sunk for the purpose of accumulating the product in a situation convenient for collection by pumping the water out.   The oil exudes on
the shores of lakes and lagoons, or rises from springs
beneath the beds of rivers.   Such are the springs of
Baku, in Persia, and the wells of Amiano, in the duchy
of Parma, in Italy.   The usual geological position of the
rocks furnishing this natural product, is in the coal measures—but it is by no means confined to this group of
rocks, since it has been found in deposits much more
recent, and also in those that are older—but in whatever
deposits it may occur, it is uniformly regarded as a product of vegetable decomposition.   Whether this decomposition has been effected by fermentation only, or by the

aid of an elevated temperature, and distilled by heated vapor, is perhaps hardly settled.

It is interesting, however, in this connection to remember, that the distillation, at an elevated temperature, of certain black bituminous shales in England and France, has furnished large quantities of an oil having many points of resemblance with Naphtha, the name given to this colorless oil, which is the usual product of distilling Petroleum. The very high boiling point of most of the products of the distillation of the Rock Oil from Venango County, Pa., would seem to indicate that it was a pyrogenic (fire-produced) product.

Bitumen, Asphaltum, Mineral Pitch, Chapapote, &c., &c., are names variously given to the more or less hard, black resinous substance which is produced usually from the exposure of Petroleum to the air, and is found either with or without the fluid Naphtha or Petroleum. The most remarkable examples of the occurrence of these substances, so intimately connected with the history of Rock Oil, are the Lake Asphaltites of the Dead Sea, so memorable in history, the well-known Bitumen Lake of Trinidad, and the deposits of mineral pitch or Chapapote in Cuba. In one of the provinces of India, vast quantities of Petroleum are annually produced, the chief consumption being local, for fuel and lights, but a portion is also exported to Europe for the production of Naphtha. In the United States, many points on the Ohio and its tributaries, are noted as producing this oil; nearly all of them within the coal measures. A detailed history of these various localities can be found recorded in books of science, and their repetition here would be out of place.

General Character of the Crude Product.

The crude oil, as it is gathered on your lands, has a dark brown color, which, by reflected light, is greenish or bluish. It is thick even in warm weather—about as thick as thin molasses. In very cold weather it is somewhat more stiff, but can always be poured from a bottle even at 15° below zero. Its odor is strong and peculiar, and recalls to those who are familiar with it, the smell of Bitumen and Naphtha. Exposed for a long time to the air, it does not thicken or form a skin on its surface, and, in no sense, can it be called a drying oil. The density of the crude oil is .882, water being 1·000. It boils only at a very high temperature, and yet it begins to give off a vapor at a temperature not greatly above that of boiling water. It takes fire with some difficulty, and burns with an abundant smoky flame. It stains paper with the appearance of ordinary fat oils, and feels smooth and greasy between the fingers. It is frequently used in its crude state to lubricate coarse machinery. In chemical characters, it is entirely unlike the fat oils. Most of these characters are common to Petroleum from various places. In one important respect, however, the product of your lands differs from that obtained in other situations, that is, it does not, by continued exposure to the air, become hard and resinous like mineral pitch or bitumen. I have been informed by those who have visited the locality, that on the surface of the earth above the springs which furnish your oil, there is no crust or deposit of this sort such as I have seen in other situations where Petroleum or mineral tar is flowing. This difference will be seen to be of considerable importance, as it is understood and represented that this product exists in great abundance upon your property, that it can be gathered wherever a well is sunk in the soil, over a great number of acres, and that it is unfailing in its yield from year to year. The question naturally arises, of what value is it in the arts, and for what uses can it be employed? These researches answer these inquiries.

Examination of the Oil.

To determine what products might be obtained in the oil, a portion of it was submitted to fractional distillation.* The temperature of the fluid was constantly regulated by a thermometer, the heat being applied first by a water bath, and then by a bath of linseed oil. This experiment was founded upon the belief that the crude product contained several distinct oils, having different boiling points. The quantity of material used in this experiment, was 304 grammes. The thermometer indicated the degrees of the Centigrade scale, but, for convenience, the corresponding degrees of Fahrenheit's scale are added. The water bath failed to distil any portion of the oil at 100° C. (=212°Fah.) only a small quantity of acid water came over. An oil bath, linseed oil, was then substituted, and the temperature was regularly raised by slow degrees until distillation commenced. From that point the heat was successively raised by stages of ten degrees, allowing full time at each stage for complete distillation of all that would rise at that temperature before advancing to the next stage. The results of this tedious process are given in the annexed table— 304 Grammes of crude oil, submitted to fractional distillation, gave :

		Temperature		Quantity.
1st Prod.	at 100° C.=213° Fah.	(acid water,)	5 Gms.
2d "	at 140° C. to 150° C.=284° to 302° Fah.	26 "
3d "	at 150° C. to 160° C.=302° to 320° Fah.	29 "
4th "	at 160° C. to 170° C.=320° to 388° Fah.	38 "
5th "	at 170° C. to 180° C.=338° to 367° Fah.	17 "
6th "	at 180° C. to 200° C.=356° to 392° Fah.	16 "
7th "	at 200° C. to 220° C.=392° to 428° Fah.	17 "
8th "	at 220° C. to 270° C.=428° to 518° Fah.	12 "

Whole quantity distilled by this method . . 160
Leaving residue in the retort 144

Original quantity, ... 304

---

*Fractional distillation is a process intended to separate various products in mixture, and having unlike boiling points, by keeping the mixture contained in an alembic at regulated successive stages of temperature as long as there is any distillate at a given point, and then raising the heat to another degree, &c.

Product No 1, as above remarked, was almost entirely water, with a few drops of colorless oil, having an odor similar to the original fluid, but less intense.

Product No. 2 was an oil perfectly colorless, very thin and limpid, and having an exceedingly persistent odor, similar to the crude oil, but less intense.

Product No. 3 was tinged slightly yellow, perfectly transparent, and apparently as limpid as the 2d product, with the same odor.

Product No. 4 was more decidedly yellowish than the last, but was in no other respect distinguishable from it.

Product No. 5 was more highly colored, thicker in consistence, and had a decided empyreumatic odor.

Product No. 6. This and the two subsequent products were each more highly colored and denser than the preceding. The last product had the color and consistency of honey, and the odor was less penetrating than that of the preceding oils. The mass of crude product remaining in the retort (equal 47.4 per cent.,) was a dark, thick, resinous-looking varnish, which was so stiff when cold, that it could be inverted without spilling. This showed no disposition to harden or skin over by exposure to the air. The distillation was arrested at this point in glass, by our having reached the limit of temperature for a bath of linseed oil. The *density* of the several products of this distillation, shows a progressive increase, thus:

No. 2,	density,	733
No. 3	"	752
No. 4	"	766
No. 5	"	776
No. 6	"	800
No. 7	"	848
No. 8	"	854

To form an idea of the comparative density of these several products, it may be well to state, that Sulphuric Ether, which is one of the lightest fluids known, has a density of .736, and Alcohol, when absolutely pure, .800.

The *boiling points* of these several fluids present some anomalies, but are usually progressive, thus, No. 2 gave signs of boiling at 115° C. (=239° Fah.) and boiled vigorously and remained constant at 225° C. to 228° C., (=437° to 442° Fah.)   No. 3 began to boil 120°, (=248° Fah.,) rose to 270° (=518° Fah.,) where it remained constant.   No. 4 began to vaporize at 140°, (=284° Fah.,) rose to 290°, (=554° Fah.,) where it remained constant. On a second heating the temperature continued to rise, and passed 305°, (=581° Fah.)   No. 5 gave appearance of boiling at 160°, (=320° Fah.,) boiling more vigorously as the heat was raised, and was still rising at 308°, (=581° Fah.)   No. 6 commenced boiling at 135°, (=275° Fah.), boiled violently at 160°, (=320° Fah.,) and continued rising above the range of the mercurial thermometer.   No. 7 commenced ebullition at the same temperature as No. 6, and rose to 305°, (=581° Fah.,) where the ebullition was not very active.   Much time was consumed in obtaining these results.   We infer from them that the Rock Oil is a mixture of numerous compounds, all having essentially the same chemical constitution, but differing in density and boiling points, and capable of separation from each other, by a well-regulated heat.

The uncertainty of the boiling points indicates that the products obtained at the temperatures named above, were still mixtures of others, and the question forces itself upon us, whether these several oils are to be regarded as *educts* (*i. e.*, bodies previously existing, and simply separated in the process of distillation,) or whether they are not rather produced by the heat and chemical change in the process of distillation.   The continued application of an elevated temperature alone is sufficient to effect changes in the constitution of many organic products, evolving new bodies not before existing in the original substance.

Properties of the Distilled Oils.

Exposed to the severest cold of the past winter, all the oils obtained in this distillation remained fluid. Only the last two or three appeared at all stiffened by a cold of 15° below zero, while the first three or four products of distillation retained a perfect degree of fluidity. Exposed to air, as I have said, they suffer no change. The chemical examination of these oils showed that they were all composed of Carbon and Hydrogen, and probably have these elements in the same numerical relation. When first distilled, they all had an acid reaction, due to the presence of a small quantity of free sulphuric acid, derived from the crude oil. This was entirely removed by a weak alkaline water, and even by boiling on pure water. Clean copper remained untarnished in the oil which had thus been prepared, showing its fitness for lubrication, so far as absence of corrosive quality is concerned. The oils contain no oxygen, as is clearly shown by the fact that clean potassium remains bright in them. Strong *Sulphuric Acid* decomposes and destroys the oil entirely. *Nitric Acid* changes it to a yellow, oily fluid, similar to the changes produced by Nitric Acid on other oils. *Hydrochloric, Chromic* and *Acetic Acids,* do not affect it. *Litharge* and other metallic oxyds do not change it, or convert it in any degree to a drying oil. *Potassium* remains in it unaffected, even at a high temperature. *Hydrates of Potash, Soda,* and *Lime,* are also without action upon it. *Chloride of Calcium* and many other salts manifest an equal indifference to it. Distilled with *Bleaching Powders* (chloride of lime) and water, in the manner of producing chloroform, the oil is changed into a product having an odor and taste resembling chloroform. Exposed for many days in an open vessel, at a regulated heat below 212°, the oil gradually rises in vapor, as may be seen by its staining the paper used to cover the vessel from dust, and also by its sensible diminution. Six or eight fluid ounces, exposed in this manner in a metallic vessel for six weeks or more, the heat never exceeding 200°, gradully and slowly diminished, grew yellow, and finally left a small residue of dark brown lustrous-looking resin, or pitchy substance, which in the cold was hard and brittle. The samples of oil employed were very nearly colorless. This is remarkable when we remember that the temperature of the distillation was above 500° Fah. The oil is nearly insoluble in pure alcohol, not more than 4 or 5 per centum being dissolved by this agent. In ether the oil dissolves completely, and on gentle heating is left unchanged by the evaporization of the ether. India Rubber is dissolved by the distilled oil to a pasty mass, forming a thick black fluid which, after a short time, deposits the india rubber. It dissolved a little amber, but only sufficient to color the oil red. It also dissolves a small portion of copal in its natural state, but after roasting, the copal dissolves in it as it does in other oils.

Use for Gas Making.

The Crude Oil was tried as a means of illumination. For this purpose, a weighed quantity was decomposed, by passing it through a wrought iron retort filled with carbon, and ignited to full redness. The products of this decomposition were received in a suitable apparatus. It produced nearly pure carburetted hydrogen gas, the most highly illuminating of all the carbon gases. In fact, the oil may be regarded as chemically identical with illuminating gas in a liquid form. The gas produced equalled ten cubic feet to the pound of oil. It burned with an intense flame, smoking in the ordinary gas jet, but furnishing the most perfect flame with the Argand burner.

These experiments were not prosecuted further, because it was assumed that other products, now known and in use, for gas making, might be employed at less expense for this purpose, than your oil. Nevertheless, this branch of inquiry may be worthy of further attention.

Distillation at a higher Temperature.

The results of the distillation at a regulated temperature in glass led us to believe, that in a metallic vessel, capable of enduring a high degree of heat, we might obtain a much larger proportion of valuable products. A copper still, holding five or six gallons, was therefore provided, and furnished with an opening, through which a thermometer could be introduced into the interior of the vessel. Fourteen imperial quarts (or, by weight, 560 ounces) of the crude product were placed in this vessel, and the heat raised rapidly to about 280° C. (=536° Fah.), somewhat higher than the last temperature reached in the first distillation. At this high temperature, the distillation was somewhat rapid, and the product was easily condensed without a worm. The product of the first stage was 130 ounces (or over 28 per cent.), of a very light-colored thin oil, having a density of .792. This product was also acid, and, as before, the acid was easily removed by boiling with fresh water. The temperature was now raised to somewhat above 300° C. (=572° Fah.), and 123 ounces more distilled, of a more viscid and yellowish oil, having a density of .865. This accounts for over 43 per cent. of the whole quantity taken. The temperature being raised now above the boiling point of mercury, was continued at that until 170 ounces, or over 31 per cent., of a dark brown oil had been distilled, having a strong empyreumatic odor. Upon standing still for some time, a dark blackish sediment was seen to settle from this portion, and on boiling it with water, the unpleasant odor was in a great degree removed, and the fluid became more light-colored and perfectly bright. (It was on a sample of this that the photometric experiments were made.) The next portion, distilled at about 700° Fah., gave but about 17 ounces, and this product was both lighter in color and more fluid than the last. It now became necessary to employ dry hickory wood as a fuel, to obtain flame and sufficient heat to drive over any further portions of the residue remaining in the alembic.

It will be seen that we have already accounted for over 75 per cent. of the whole quantity taken. There was a loss on the whole process of about 10 per cent., made up, in part, of a coaly residue that remained in the alembic, and partly of the unavoidable loss resulting from the necessity of removing the oil twice from the alembic, during the process of distillation, in order to change the arrangements of the thermometer, and provide means of measuring a heat higher than that originally contemplated.

About 15 per cent. of a very thick, dark oil completed this experiment. This last product, which came off slowly at about 750° Fah., is thicker and darker than the original oil, and when cold is filled with a dense mass of pearly crystals. These are Paraffine, a peculiar product of the destructive distillation of many bodies in the organic kingdom. This substance may be separated, and obtained as a white body, resembling fine spermaceti, and from it beautiful candles have been made. The oil in which the crystals float is of a very dark color, and by reflected light is blackish green, like the original crude product. Although it distills at so high a temperature, it boils at a point not very different from the denser products of the first distillation. The Paraffine, with which this portion of the oil abounds, does not exist ready-formed in the original crude product; but it is a result of the high temperature employed in the process of distillation, by which the elements are newly arranged.

I am not prepared to say, without further investigation, that it would be desirable for the Company to manufacture this product in a pure state, fit for producing candles (a somewhat elaborate chemical process); but I may add that, should it be desirable to do so, the quantity of this substance produced may probably be very largely increased by means which it is now unnecessary to mention.

Paraffine derives its name from the unalterable nature of the substance, under the most powerful chemical agents. It is white, in brilliant scales of a greasy lustre; it melts at about 116°, and boils at over 700° Fah.; it dissolves in boiling alcohol and ether, and burns in the air with a brilliant flame. Associated with Paraffine are portions of a very volatile oil, *Eupione,* which boils at a lower temperature, and by its presence renders the boiling point of the mixture difficult to determine. I consider this point worthy of further examination than I have been able at present to give it, *i. e.* whether the last third, and possibly the last half, of the Petroleum, may not be advantageously so treated as to produce from it the largest amount of Paraffine which it is able to produce.

The result of this graduated distillation, at a high temperature, is that we have obtained over 90 per cent. of the whole crude product in a series of oils, having valuable properties, although not all equally fitted for illumination and lubrication.

A second distillation of a portion of the product which came over in the latter stages of the process, (a portion distilled at about 650° Fah., and having a high color), gave us a thin oil of density about .750, of light yellow color and faint odor.

It is safe to add that, by the original distillation, about

50 per cent. of the crude oil is obtained in a state fit for use as an illuminator without further preparation than simple clarification by boiling a short time with water.

Distillation by high Steam.

Bearing in mind that by aid of high steam, at an elevated temperature, many distillations in the arts are effected which cannot be so well accomplished by dry heat, I thought to apply this method in case of the present research. Instances of this mode of distillation are in the new process for Stearine candles, and in the preparation of Rosin Oil. I accordingly arranged my retort in such a manner that I could admit a jet of high steam into the boiler, and almost at the bottom of the contained Petroleum. I was, however, unable to command a jet of steam above 275° to 290° Fah., and, although this produced abundant distillation, it did not effect a separation of the several products, and the fluid distilled had much the same appearance as the Petroleum itself, thick and turbid. As this trial was made late in the investigation, I have been unable to give it a satisfactory issue, chiefly for want of steam of a proper temperature. But I suggest, for the consideration of the Company, the propriety of availing themselves of the experience already existing on this subject, and particularly among those who are concerned in the distillation of Rosin Oil—a product having many analogies with Petroleum in respect to its manufacture.

Use of the Naphtha for Illumination.

Many fruitless experiments have been made in the course of this investigation which it is needless to recount. I will, therefore, only state those results which are of value.

1. I have found that the only lamp in which this oil can be successfully burned, is the Camphene lamp, or one

having a button to form the flame, and an external cone
to direct the current of air, as is now usual in all lamps
designed to burn either Camphene, Rosin Oil, Sylvic Oil,
or any other similar product.

2. As the distilled products of Petroleum are nearly or
quite insoluble in alcohol, burning fluid (*i. e.*, a solution
of the oil in alcohol) cannot be manufactured from it.

3. As a consequence, the oil cannot be burned in a
hand lamp, since, with an unprotected wick, it smokes
badly. Neither can it be burned in a Carcel's mechanical
lamp, because a portion of the oil being more volatile
than the rest, rises in vapor on the elevated wick required
in that lamp, and so causes it to smoke.

I have found all the products of distillation from the
copper still capable of burning well in the Camphene
lamp, except the last third or fourth part (*i. e.*, that portion which came off at 700° Fah. and rising, and which
was thick with the crystals of Paraffine). Freed from
acidity by boiling on water, the oils of this distillation
burned for twelve hours without injuriously coating the
wick, and without smoke. The wick may be elevated
considerably above the level required for Camphene, without any danger of smoking, and the oil shows no signs of
crusting the wick tubes with a coating of Rosin, such as
happens in the case of Camphene, and occasions so much
inconvenience. The light from the rectified Naphtha is
pure and white without odor. The rate of consumption
is less than half that of Camphene, or Rosin Oil. The
Imperial pint, of 20 fluid ounces, was the one employed
—a gallon contains 160 such ounces. A Camphene lamp,
with a wick one inch thick, consumed of rectified Naphtha in one hour 1¾ ounces of fluid. A Carcel's mechanical lamp of ⅝ inch wick, consumed of best Sperm Oil, per
hour 2 ounces. A "Diamond Light" lamp, with "Sylvic
Oil," and a wick 1½ inch diameter consumed, per hour,
4 ounces.

I have submitted the lamp burning Petroleum to the inspection of the most experienced lampists who were accessible to me, and their testimony was, that the lamp burning this fluid gave as much light as any which they had seen, that the oil spent more economically, and the uniformity of the light was greater than in Camphene, burning for twelve hours without a sensible diminution, and without smoke.   I was, however, anxious to test the amount of light given, more accurately than could be done by a comparison of opinions.   With your approbation I proceeded therefore to have constructed a *photometer*, or apparatus for the measurement of light, upon an improved plan.   Messrs. Grunow, scientific artists of this city, undertook to construct this apparatus, and have done so to my entire satisfaction.   This apparatus I shall describe elsewhere—its results only are interesting here.   By its means I have brought the Petroleum light into rigid comparison with the most important means of artificial illumination.   Let us briefly recapitulate the results of these

Photometric Experiments.

The *unit* adopted for comparison of intensities of illumination is Judd's Patent Sixes Sperm Candle.

The Sperm Oil used was from Edward Mott Robinson, of New Bedford—the best winter Sperm remaining fluid at 32° Fah.   The Colza Oil and Carcel's lamps were furnished by Dardonville, lampist, Broadway, New York. The Gas used was that of the New Haven Gas Light Co., made from best Newcastle coal, and of fair average quality.

The distance between the standard candle, and the illuminator sought to be determined, was constantly 150 inches—the Photometer traversed the graduated bar in such a manner as to read, at any point where equality of illumination was produced, the ratio between the two lights. I quote only single examples of the average results, and with as little detail as possible, but I should state that the operation of the Photometer was so satisfactory that we obtained constantly the same figures when operating in the same way, evening after evening, and the sensitiveness of the instrument was such that a difference of one half inch in its position was immediately detected in the comparative illumination of the two equal discs of light in the dark chamber. This is, I believe, a degree of accuracy not before obtained by a Photometer.

Table of illuminating power of various artificial lights compared with Judd's patent candles as a unit.

Source of Light.	Ratio to Candle.—1.
Gas burning in Scotch fish-tail tips, 4 feet to the hour	1 : 5.4
" " " " " 6 " "	1 : 7.55
" " Cornelius " " 6 " "	1 : 6.3
" " English Argand burner 10 " "	1 : 16.
Rock Oil, burning in 1 inch wick Camphene Lamp, consuming 1 3-4 ounces of fluid to the hour	1 : 8 1
Carcel's Mechanical Lamp, burning best Sperm Oil, 2 ounces of fluid to the hour, wick 7-8 of an inch	1 : 7.5
Carcel's " " " " " " Colza Oil,	1 : 7.5
Camphene Lamp, (same size as Rock Oil above,) burning best Camphene, 4 fluid ounces per hour	1 : 11.
"Diamond Light" by "Sylvic Oil," in 1 1-2 inch wick, 4 ounces per hour	1 : 8.1

From this table it will be seen that the Rock Oil Lamp was somewhat superior in illuminating power to Carcel's Lamp of the same size, burning the most costly of all oils. It was also equal to the "Diamond Light" from a lamp of one half greater power, and consequently is superior to it in the same ratio in lamps of equal power. The camphene lamp appears to be about one-fifth superior to it, but, on the other hand, the Rock Oil surpasses the Camphene by more than one half in economy of consumption, (i. e., it does not consume one half so much fluid by measure), and it burns more constantly. Compared with the Sylvic Oil and the Sperm, the Rock Oil gave on the ground glass diaphragm, the whitest disc of illumination, while in turn the Camphene was whiter than the Rock Oil light.   By the use of screens of different colored glass, all inequalities of *color* were compensated in the use of the photometer, so that the intensity of light could be more accurately compared.   Compared with Gas, the Rock Oil gave more light than any burner used except the costly Argand consuming ten feet of gas per hour.   To compare the *cost* of these several fluids with each other, we know the price of the several articles, and this varies very much in different places.   Thus, gas in New Haven costs $4 per 1,000 feet, and in New York $3.50 per 1,000, in Philadelphia $2.00 per 1,000 and in Boston about the same amount.

Such Sperm Oil as was used costs $2.50 per gallon, the Colza about $2, the Sylvic Oil 50 cents, and the Camphene 68 cents—no price has been fixed upon for the rectified Rock Oil.

I cannot refrain from expressing my satisfaction at the results of these photometric experiments, since they have given the Oil of your Company a much higher value as an illuminator than I had dared to hope.

Use of the Rock Oil as a Lubricator for Machinery.

A portion of the rectified oil was sent to Boston to be tested upon a trial apparatus there, but I regret to say that the results have not been communicated to me yet. As this oil does not gum or become acid or rancid by exposure, it possesses in that, as well as in its wonderful resistance to extreme cold, important qualities for a lubricator.

Conclusion.

In conclusion, gentlemen, it appears to me that there is much ground for encouragement in the belief that your

Company have in their possession a raw material from which, by simple and not expensive process, they may manufacture very valuable products.

It is worthy of note that my experiments prove that nearly the *whole* of the raw product may be manufactured without waste, and this solely by a well directed process which is in practice one of the most simple of all chemical processes.

There are suggestions of a practical nature, as to .the economy of your manufacture, when you are ready to begin operations, which I shall be happy to make, should the company require it—meanwhile, I remain, gentlemen,

Your obedient servant,

B. SILLIMAN, JR.,

Professor of Chemistry in Yale College.

NEW HAVEN, APRIL 16, 1855.

CHAPTER III.

EARLY AND INTERESTING FACTS.

WE now approach that interesting period in the history of Petroleum in America, which witnessed the first movement toward a practical development of its astounding resources.

The reader who has carefully scanned the report to Mr. Bissell and others, submitted in the last chapter, will have observed that, however it may have been with himself, the existence of Petroleum was not a novelty to scientific minds.

While he will perceive with admiration, the completeness and comprehensiveness with which every phase of the subject was examined, and reflect with astonishment upon the manner in which every mode of treatment was foreshadowed, it cannot fail to strike the reader as remarkable, that notwithstanding the value of this product to our country, has been about nine hundred millions of dollars; notwithstanding, (thirteen years have elapsed since the first well was sunk) and the total number of wells since sunk to obtain it must reach twenty thousand, its origin has not yet been absolutely determined. Indeed very little more is known than was conjectured by the rash pioneers, who only just failed of achieving the development nearly a generation earlier, in the valleys of the Ohio and Kanawha. We shall follow up the history of that development, which has since proved to be of so great importance to the prosperity of the country, and the comfort and convenience of mankind.

Bearing in mind the frequent appearance of Petroleum

in the salt wells of the Kanawha valley in Virginia, and along the valley of the Ohio near the mouth of the Muskingum, the reader will not be surprised to hear of its appearance in the salt wells of Tarentum, on the Allegany river, twenty miles above Pittsburgh, Pa. But its appearance at this place was singular, in so far that instead of appearing where the wells were first sunk, many of them were successfully pumped for brine for a long time before the manifestations of oil disturbed operations. Mr. Kier, who together with his father was a large owner in these salt works, states that one well on the left bank of the river, after having been pumped constantly for twenty years without a show of Petroleum, passed to new owners, who rigged it with a new engine of greater power, and in a few days it began to yield four or five barrels per day.

The wells on the other side of the river owned by S. M. Kier and his father, had for years previous yielded a small quantity of Petroleum, which being suffered to waste for a long time, spread itself over the surface of the old canal, and became accidentally ignited, when it came so near causing the destruction of a large amount of property, as well as endangering human life, that it was afterwards poured upon the ground.

About the year 1849, Mr. Kier, Jr., conceived the thought of putting it up in bottles and selling it as a specific remedy for all the ills of life. He opened an establishment in Pittsburgh, where it was put up in half pint bottles, which were wrapped in the following descriptive sheet, and sold for a half dollar apiece:

KIER'S

PETROLEUM, OR ROCK OIL, CELEBRATED FOR ITS WONDERFUL CURATIVE POWERS. A NATURAL REMEDY! PROCURED FROM A WELL IN ALLEGANY CO., PA., FOUR HUNDRED FEET BELOW THE EARTH'S SURFACE. PUT UP AND SOLD BY SAMUEL M. KIER, 363 LIBERTY STREET, PITTSBURGH, PENN'A.

The healthful balm, from Nature's secret spring,
The bloom of health and life, to man will bring;
As from her depths the magic liquid flows,
To calm our sufferings, and assuage our woes.

THE PETROLEUM HAS BEEN FULLY TESTED ! It was placed before the public as a REMEDY OF WONDERFUL EFFICACY. Every one not acquainted with its virtues, doubted its healing properties. The cry of humbug was raised against it. It had some friends—those that were cured through its wonderful agency. These spoke out in its favor. The lame, through its instrumentality were made to walk—the blind, to see. Those who had suffered for years under the torturing pains of RHEUMATISM, GOUT AND NEURALGIA, were restored to health and usefulness. Several who were blind, have been made to see, the evidence of which will be placed before you. If you still have doubts, go and ask those who have been cured ! Some of them live in our midst, and can answer for themselves. In writing about a medicine, we are aware that we should write TRUTH—that we should make no statements that cannot be proved. We have the witnesses: crowds of them, who will testify in terms stronger than we can write them, to the efficacy of this remedy; who will testify that the Petroleum has done for them what no medicine ever could before : cases that were pronounced hopeless, and beyond the reach of remedial means; cases abandoned by Physicians of unquestionable celebrity, have been made to exclaim, " THIS IS THE MOST WONDERFUL REMEDY YET DISCOVERED !" We will lay before you the certificates of some of the most remarkable cases : to give them all, would require more space than would be allowed by this circular. Since the introduction of the Petroleum, many Physicians have been convinced of its efficacy, and now recommend it in their practice; and we have no doubt that it will stand at the head of the list of valuable Remedies. If the Physicians do not recommend it the people will have it of themselves; for its transcendent power to heal WILL and MUST become known and appreciated; when the voices of the cured speak out; when the cures themselves stand out in bold relief, and when he who for years has suffered with the tortures and pangs of an immedicable legion, that has been shortening his days and hastening him " to the narrow house appointed for all the living," when he speaks out in its praise, who will doubt it ? The Petroleum is a Natural Remedy; it is put up as it flows from the bosom of the earth, without anything being added to or taken from it.

It gets its ingredients from the beds of substances which it passes over in its secret channel. They are blended together in such a form as to defy all human competition. The Petroleum, in this respect, is like Mineral Water, whose virtues in most chronic diseases, are acknowledged, not only by Physicians, but by the community at large. These singular fluids flowing out of the earth, impregnated with medicinal substances of different properties, and holding them in such complete solution as to require the aid of Chemistry in order to detect them, bear ample proof to the fact that they are compounded by the master hand of Nature, for the alleviation of human suffering and disease. If Petroleum is medicine at all, it is a good one, for Nature never half does her work; and that it is a medicine of unequalled power we have the most abundant testimony. It will be used when many of the new remedies now in vogue will have been forgotten forever. It will continue to be used and applied as a Remedy as long as man continues to be afflicted with disease. That it will cure every disease to which we are liable, we do not pretend ; but that it will cure a great many diseases hitherto incurable, is a fact which is proven by the evidence in its favor. Its discovery is a new era in medicine, and will inure to the health and happiness of man.

All of which was followed by about a hundred testi-

monials of wonderful cures of hopelessly incurable diseases.

In fact, after the manner of patent medicines in our own
day, it was declared perfectly capable of doing or undoing
whatsoever anybody could wish done or undone. It was
trundled around the country by agents who traveled in
vehicles decorated in gilt, with pictures of the good Sa-
maritan ministering to a sufferer, writhing in inhuman
contortions under a palm tree.

Although the oil cost him next to nothing, as it was
obtained from his own wells which were pumped for salt,
and for a long time he could not dispose of even the whole
of the two or three barrels a day produced, yet the expense of introducing it as a medicine in this way consumed the profits. As the stuff, however, possessed
considerable medicinal virtue, the demand continued to
increase until quite a valuable trade was established,
when he withdrew his agents and furnished it exclusively
through the drug stores.

This at first left quite a quantity on hand, for his sales
sensibly fell off for a while, after the agents were withdrawn in 1852, and having previously burned the crude
oils at the wells, it occurred to him that he might utilize
this surplus if he could only devise some way of rendering it less offensive, in the way of smoke and odor.

The most obvious suggestion was to distil it. This he
accomplished by fitting a caldron kettle with a cover and
a worm. The first result was a dark distillate, little
better than the crude itself; but after he learned to manage
his fires so as not to send it over too rapidly, he produced
by twice distilling, an article about the color of clear cider,
which, like all distillates, had an odor infinitely more
offensive than the crude Petroleum, and as he knew nothing of treating it with acids, as is done at the present
time—as indeed was extensively done very shortly after

that time with coal oil—he seemed to be progressing very slowly toward the production of an illuminator.

After some improvement on the camphene lamp, however, he perceived with joy that his distillate would burn without smoking, provided the flame was kept low enough and the lamp left perfectly quiet. From this rude beginning, he went on improving both the quality of his fluid and the adaptability of his lamp, thus manufacturing and selling for a dollar and a half a gallon all the Petroleum he could not dispose of as a medicine, for burning, till at last by the introduction of the "Virna burner" and the treatment of his distillate with acids he had brought the matter nearly to its present state of perfection, when the first Petroleum well in Venango County broke his monopoly, and put an end to the manufacture of coal oil in the United States.

Up to the time when his first attempts to utilize Petroleum for a burning fluid, a very little of which had been collected on Oil Creek by absorption in blankets, from which it was wrung, amounting in all perhaps to a couple of barrels per month, the principal part of which was gathered from a spring which bubbled up in the middle of the creek on the M'Clintock Farm, three miles above Oil City.

Many writers have given very exaggerated accounts of the quantity of oil exuding from these springs, and convey to the reader the impression that the surface of the creek was an unbroken sheet of Petroleum, while the truth is, only in high water, when the freshets brought down that which had collected in the bays, was it at all noticeable.

The spring next in importance was near the northern line of the county on the lands of Brewer, Watson & Co. It was beside this spring the first artesian well was sunk for Petroleum, and this also seems to have been the principal scene of development in early times, for here are still to be traced many pits, cribbed with roughly hewn timbers preserved beneath the accumulation of centuries.

The first written document looking to a mechanical development is the following between J. D. Angier, still a resident of Titusville, and the firm of Brewer, Watson & Co., consisting of Ebenezer Brewer and James Rynd, of Pittsburgh, and Jonathan Watson, Rexford Pierce and Elijah Newberry of Titusville, associated in an extensive lumbering business on Oil Creek:

The Agreement.

"Agreed this fourth day of July, A. D. 1853, with J. D. Angier of Cherrytree Township, in the county of Venango, Pa., that he shall repair up and keep in order the old oil spring on land in said Cherrytree Township, or dig and make new springs, and the expenses to be deducted out of the proceeds of the oil, and the balance, if any, to be equally divided, the one half to J. D. Angier and the other half to Brewer, Watson & Co., for the full term of five years from this date.   If profitable."

{ BREWER, WATSON & Co.
{ J. D. ANGIER.

Following out the spirit of his agreement, Mr. Angier proceeded at once to erect some slight works for collecting the oil. A few rude trenches were dug, centering in a common basin from which the water was raised by a pump, connected with the saw mill of Brewer, Watson & Co., into a series of broad shallow troughs, shelving off to the ground. Where the water passed from each trough into the next, was rigged an ingenious little skimmer adjusted just under the surface of the water, so as to collect the oil.

The water passing under was again agitated by the fall

which favored a further separation of the oil, which was collected as before by the skimmer at the end of the trough. In this manner three or four gallons a day were collected, and even as high as six gallons, where the ground had been recently agitated by digging, but the expense consumed the profit, and after a few months, the experiment was suffered to drop.

Mr. Angier describes a remarkable phenomenon observed in this method of obtaining Petroleum. While digging in the gravelly clay, three or four feet beneath the surface, the workmen frequently struck "pockets" of oil often containing a quart.

In the summer of the year 1854, Dr. F. B. Brewer, whose father was at the head of the firm of Brewer, Watson & Co., visited relatives at Hanover, New Hampshire, and carried a bottle of Petroleum to Professor Crosby, of Dartmouth College, where he had graduated some ten years before.

A few weeks later George H. Bissell, a native of the town and a graduate of the same College, but then practicing law in New York city, while on a visit to his mother called to spend an evening with his old tutor, Prof. Crosby, and was shown the Petroleum, upon the wonderful properties of which the Professor expatiated with great enthusiasm.

Coal oil was then just being introduced in the eastern states for illuminating and lubricating, and the similarity of the products, naturally suggested the question why Petroleum might not be used for the same purpose. Of Mr. Kier's attempts in that direction, nothing was of course known. They were upon too limited a scale to attract attention. The only doubt was as to the supply; and that was of course a serious doubt.

Coal oil was selling for a dollar a gallon, and from the glowing description, which had been given of the spring by Dr. Brewer, it seemed reasonable to hope that many thousand gallons might be collected annually.

Professor Crosby had a son, who was ready for any enterprise that promised a chance of making money. He seems to have been persuaded from the first, that the oil spring was a humbug, but he had the penetration to see that it was a humbug of the "taking" sort; and dilating on the representations of Dr. Brewer, he induced Mr. Bissell, on certain conditions to pay the expenses of a trip to Titusville, for the purpose of inspecting the spring. The most obvious method of handling such a piece of property for the purpose of making money, was to throw it into a joint-stock company.

If he brought back a favorable report of the spring, Mr. Bissell pledged himself to organize the company and launch the enterprise on the New York stock market.

Mr. Bissell authorized him to propose to the firm of Brewer, Watson & Co., the formation of a joint-stock company, with a capital stock of $250,000, divided into ten thousand shares of twenty-five dollars each—Brewer, Watson & Co., to receive one fifth of the whole stock; and five thousand dollars for the tract containing the oil spring, to be paid out of the first money realized from the sale of treasury stock, which was also to be one fifth of the whole. Mr. Crosby was to take one fifth, and assume one fifth of the expense of getting up the company, while the other two fifths were to remunerate Mr. Bissell, and his associate in Law, Mr. Eveleth, for the trouble of organization.

To this Dr. Brewer, as the agent of Brewer, Watson & Co., replied in the following letter, a copy of which has been preserved and is furnished by Dr. Brewer:

Gentlemen,—

We have received through Mr. A. H. Crosby, your

proposals to put in market in a joint-stock company, certain springs yielding a peculiar oil surpassing in value any other oil now in use for burning, for lubricating machinery, and as a medicinal agent.

The springs yielding this oil, are situated on Oil Creek in Venango county, near the corner of Warren and Crawford counties, and cover a large surface of territory.

The yield is abundant, and is believed to be inexhaustible. We have some simple machinery constructed at an expense of, say two hundred dollars, that yields on an average to each spring worked three gallons per day, requiring perhaps one day in a week the attention of one man, which when estimated with regard to the percentage, will show as follows :

Capital invested $200, int., at 10 per cent....$20
Two months of run $20 per month............ 40

Total. $60

The cost of raising 1.095 gallons of oil, worth here seventy-five cents per gallon, making $821.25. Deducting expenses $60 leaves $761.25.

Now this is only one spring, and worked very imperfectly, but actually paying an interest on $10,000.

I make these figures as they are, and have been whenever the spring has been worked, and this is no fancy thing for a stock, but an exceedingly large paying stock, and one that with proper machinery would afford a much larger percentage.

Now your proposition, as far as it goes, is satisfactory ; but it does not go far enough to guarantee to us a certain *quid pro quo* for what we have paying us now. And in asking us who will represent only one fifth of the company to furnish the actual capital gratuitously, to the other four fifths, for what we expect to realize on one fifth, is not perhaps asking too much, but it would in our opinion be granting too much.

There are other parties in Pittsburgh who were very so-licitous to put the thing in market a year ago by purchas-ing our interest, but we prefer the plan you suggest if you will warrant us a certain amount for our premises; and we will propose as follows: Pay to Brewer, Watson & Co. $5,000, to be reimbursed to the stock company from the first sale of stock, or as you may deem proper in any way, and we will assign or deed to the company, the right to go on and erect such machinery as the company may think proper, to procure oil on a certain one hundred acres of land known here as the Willard Farm, and embracing most of the oil territory as yet discovered, and further, all springs on our other lands adjoining, not interfering with our lumbering and farming interests.

This will give us an equivalent, or partially so, for what we furnish the company; and we wish the company to pay from the sale of stock its current expenses, whatever they may be, and by sale of stock to provide for a dividend, if thought best—such stock to have preference over all other except for the reimbursement of the purchase money. The other stock should be sold as you propose. These, gen-tlemen, are our views, hastily thrown together. If the general outlines meet your favor the minutia can be ar-ranged with you in New York."   *   *   *

After spending a few weeks with Dr. Brewer, Mr. Crosby hurried back to Hanover to report, but finding that Mr. Bissell had returned to New York he forwarded the letter, of which the above is a copy, to Mr. Eveleth, who was in Maine, and apparently not comprehending its terms he hastened after Mr. Bissell, and reported his proposition ac-cepted. Mr. Bissell announced himself satisfied, and at once began the preliminary arrangement for organizing the company. In a gush of innocent satisfaction with the success of his negotiations Mr. Crosby telegraphed to Dr. Brewer that *his* proposition was accepted. He returned again to New Hampshire, and a few days later he wrote the following:

HANOVER, N. H., *September* 11, 1854.

MY DEAR DOCTOR,

I intended to have written you again before leaving the city, but as I was very busy, and as the main question was settled by my telegram of the Monday previous, I concluded to wait until my return home.

I cannot now tell you exactly when we shall be ready to meet you in New York, but will write next Monday again, and shall then be able to tell you when you had better start.

The oil I suppose you can take with you to Erie, and ship it so that it will be in New York nearly as soon as you are, and that will be in sufficient season to offer it for exhibition, as we shall then have circulars, stock-books, and everything else ready to issue to a gullible public."

But this gulling of the public, is not an enterprise of unmitigated interest with men who have everything to gain, and Eveleth & Bissell having considerable to lose, objected to this scheme in so far that ten days later this embryo broker, curbing his wild ambition to surge into the stock exchange, and get up a panic with his "fancy," writes Dr. Brewer to the effect that after a "long talk" with Eveleth and Bissell it was decided to "put the thing through by daylight."

Thus vanished his bright dream of oil spring "fancy" at a premium of five hundred per cent., and his enthusiasm thenceforward continued to wane.

Shortly afterward, Dr. Brewer, empowered as the attorney of the lumbering firm, visited New York city to ratify the terms of sale and contract, but the letter which Crosby had forwarded to Mr. Eveleth having been lost in the mails, the result was a general misunderstanding.

Eveleth & Bissell suspecting that Crosby had intentionally deceived them—which was surely not the case—now refused to credit his statements as to the value of the springs, dismissed Dr. Brewer, and peremptorily dropped the whole matter. But they had already incurred obligations to the extent of several hundred dollars for seals, certificates of stock, stock books, etc. On the eve of his departure for Titusville, they sent a line to his hotel, saying they would reconsider the matter, and inviting him to call. He did so, and an arrangement was effected on substantially the same basis as before proposed; but Crosby, who was unable to meet his portion of the expenses, was left entirely out of the bargain. The agreement to sell ratified, for their better information, it was decided that one of them should visit the locality at once, and examine the spring, and bring away a draft from which a map could be made.

The oil which had been sent on to Mr. Bissell was distributed for examination among several prominent chemists, and a week or two later he wrote the following letter, which may convey some idea of what it cost, both in time and money, to bring about the organization of the company and to procure the analysis of the Petroleum, which must be regarded as the most important step in all these negotiations, if we accept only the birth of the great fact which made development possible. Messrs. Eveleth & Bissell were young men, and though possessed of considerable means, did not rank among the "heavy" of New York, and the whole expense of the organization and the analysis was advanced by them in one of the most stringent seasons that has ever marked the financial history of our country. It was done, too, at a sacrifice of personal convenience, which could only have been prompted by an earnest faith in the ultimate success of their enterprise— but the letter:

NEW YORK, *Nov.* 6, 1854.

F. B. BREWER, ESQ.:

Dear Sir: We have had to encounter many obstacles in the way of organizing our joint-stock company, and shall be unable to get out our papers at the time originally proposed.

Mr. Eveleth will go on at the earliest possible period, and will then be prepared to arrange everything to our mutual satisfaction. I do not think, however, that it will be possible for Mr. Eveleth to arrive in Titusville before the 18th or 20th inst.

We have obtained our stock-books, certificates of stock, signs, &c., &c., and have done everything to insure success when we fairly get under way. We have forwarded several gallons of the oil to Mr. Atwood of Boston, an eminent chemist, and his report of the qualities of the oil and the uses to which it may be applied are very favorable. Professor Silliman of Yale College is giving it a thorough analysis, and he informs us that so far as he has yet tested it, he is of opinion that it contains a large proportion of benzole and naphtha, and that it will be found more valuable for purposes of application to the arts than as a medicinal, burning or lubricating fluid.

Our expense of a thorough analysis will be very heavy; but we think the money will be well spent. We send you a proof-sheet of our certificate of stock. The book will be printed of course on bank-note paper.

Let us hear from you at your earliest convenience, and believe us, Very truly yours,

EVELETH & BISSELL.

The whole cost of the analysis, including the photometrical comparison, for which a new and improved instrument was especially provided, was between eleven and twelve hundred dollars, every cent of which was advanced by these young men.

The above letter is the first that bears the seal of The Pennsylvania Rock Oil Company; but as the company came not into legal existence till nearly two months later, it was probably applied in obedience to a whim, or perhaps to give their correspondent an idea of its impression.

CHAPTER IV.

REAL ESTATE TRANSACTIONS ON OIL CREEK.

THE first deed from Brewer, Watson & Co., bears date four days later, and conveyed in fee simple to George H. Bissell and Jonathan G. Eveleth, of New York city, one hundred and five acres of land in Cherrytree Township, Venango county, Pennsylvania, embracing the island at the junction of Pine Creek and Oil Creek, on which a part of the works of the lumbering firm were situated. It was on this island that Mr. Angier's trenches were dug, and the first artesian well bored for Petroleum five years afterwards. The consideration for the property mentioned in the deed was twenty-five thousand dollars, though the real consideration was but five thousand. It was thought that if the consideration should appear to be such an insignificant fraction of the capital stock it would be more difficult to dispose of the shares, and therefore, as is usually done in the formation of joint stock companies, the land was put in at a figure far above its cost. The deed, though dated on the 10th of November, was not formally executed till the first of January following, for the reason that Messrs. Eveleth and Bissell had opened negotiations with a party of gentlemen in New Haven, under whose notice the matter had been brought by Prof. Silliman, who evinced an inclination to subscribe for a large portion of the stock, and in case they did, it was proposed to place one or more of their number on the board of directors, and have the property conveyed directly from Brewer, Watson and Co., to The Pennsylvania Rock Oil Company. But this failing, Eveleth and Bissell gave their joint and several notes for the purchase money, save five hundred dollars paid in cash, and on the first day of January the deed was executed by the members of the firm living in Titusville, and four days afterwards by the remaining members in Pittsburgh. It was asserted in a paper on this subject which appeared in the *Atlantic Monthly,* for 1869, that Dr. Brewer never received pay for his land, which is quite untrue; for though Dr. Brewer never had anything to do with the land except in the capacity of agent for his father and the other members of the firm, the notes he received were certainly paid for; they were all found cancelled, and with payment indorsed, among papers submitted by Mr. Bissell. This may seem a matter of trifling interest in the history of the vast industry born of these transactions, but as well as being a piece of personal injustice, it is a palpable absurdity, for Mr. Bissell afterwards acquired, and perhaps still retains, an immense amount of property in the county, that would have been liable for those debts.

On the 30th of December, 1854, the following certificate of incorporation was filed, as by law required, with the Recorder of the city of New York, and also at Albany with the Secretary of State :

{ *" Certificate of Incorporation of the*
{ *Pennsylvania Rock Oil Company.*

{ STATE OF NEW YORK,
{   CITY AND COUNTY OF NEW YORK, SS.

Be it known that we, the undersigned, do hereby associate ourselves as a body politic and corporate, pursuant to the N. Y. Revised Statutes, 4th edition, vol. 1st, Chap. 18th, Art. 2d, and also Laws of New York, 1853, chap. 333, in relation to the formation and management, powers and responsibilities of corporations.

And the following are the articles of our agreement and association :

ART. 1. The name of the corporation shall be the Pennsylvania Rock Oil Company.

ART. 2. The objects for which said Company is formed, are to raise, procure, manufacture and sell Rock Oil.

ART. 3. The capital stock of the said Company shall be two hundred and fifty thousand dollars, and shall be divided into ten thousand shares of twenty-five dollars each.

ART. 4. The business of said Company shall commence on the 1st day of January, 1855, and continue fifty years.

ART. 5. The business of said Company shall be under the management of seven trustees, and the board of trustees for the first year shall consist of the following persons, viz :

George H. Bissell, of New York; J. G. Eveleth, of New York; Franklin Reed, of New York; Francis B. Brewer, of Titusville, Pennsylvania; Anson Sheldon, of New Haven, Connecticut; James H. Salisbury, of New York; and Dexter A. Hawkins, of New York.

ART. 6. The principal place of business shall be in the city and county and State of New York.

In witness whereof we have hereunto set our hands and affixed our seals, this thirtieth day of December, Anno Domini, one thousand eight hundred and fifty-four."

Here follow the signatures of the above-mentioned seven trustees, of whom, all but Dr. Brewer, who represented the stock of Brewer, Watson & Co., were mere lay-figures, occupying positions it was necessary for appearance' sake, that some one should fill. Not more than one of them at most, represented stock held in his own right, stock for which he had paid.

On the 16th of January 1855, Eveleth Bissell conveyed to the Trustees of The Pennsylvania Rock Oil Company all their right and title to the lands, but the deed fortunately was not recorded, and the estate continued ostensibly in them till the following autumn, when it was conveyed to other parties for the benefit of the *new* Pennsylvania Rock Oil Company.

After the organization of the Company in January, an effort was made to get the stock taken at some price, but the great stringency in the money market, not less than the unusual character of the enterprise, placed the stock in the ever dangerous category of " fancies," and prevented its being taken to any great extent in the city of New York.

Yet every effort was made ; and even Crosby, then engaged as a reporter on one of the newspapers of the city, became again an agent in the enterprise. He received a few shares in acknowledgment of former services, and a few more to engage some influence he was supposed to possess by having at all times the ear of the public ; but no sooner had he got his couple of hundred shares transferred to his own name on the books of the Company, than, figuratively speaking, he dropped the ear of the public— which had never been a profitable ear to him—and gave his exclusive attention to the disposal of his own stock.

Selling stock may be pleasant enough when one has stocks that sell, but unfortunately for his hopes, Mr. Crosby's were not of that sort.

With him, as with others, the times were hard—in fact, as is usually the case with such jovial characters, the times were especially hard in his case. He was desperate, but his desperation instead of quickening his wits seemed rather to cloud them; and Dr. Brewer, who frequently passed back and forth from Titusville to New York, and who was cognizant of most of their transactions and difficulties, relates how one day—it happened to be a day when the desperation of Mr. Crosby's prospects had sunk his mercurial temperament to the very lowest notch—he chanced to learn that Messrs. Eveleth and Bissell were about concluding a sale of several hundred shares of stock to a Connecticut gentleman at two dollars and-a-half per share; and regarding the knowledge in the light of a special providence—a plank that would save him from being engulfed in a sea of troubles—he reached out and grasped it; in other words he sought an interview with their client, and offered him the remnant of his own stock at fifty cents per share. The result was what any one might have foreseen—what Mr. Crosby himself could not have failed to foresee, only that he was blinded by desperation—the man knowing neither of the parties and suspecting a swindle, refused to take the stock from either, and peremptorily dropped the transaction.

The consternation of Messrs. Eveleth and Bissell, who had now expended seven or eight thousand dollars, without receiving a cent, and had calculated on this sale to help them out with their own obligations when they learned of the failure of the transaction, and the aggravating circumstances by which that failure was brought about, may possibly be imagined but cannot be described.

But as it was useless to offer stock for sale, while Mr. Crosby had any to sell, they found it expedient to buy for themselves, the little remnant of stock he found it impossible to sell to any one else, and he readily parted with it for such a meagre sum, as enabled him to reach the paternal roof at Hanover! And thus forever subsided, that luminary to whom it pleased the writer of the paper in the Atlantic, to gushingly ascribe the development of Petroleum!

The enterprise continued to hang fire. True it is, that neither of the partners was able to give his exclusive attention to its management. Their legal business claimed their attention, and so far there was nothing in prospect for the stock company, to encourage the thought of giving up a thrifty legal business, to assume the more active management thereof. But they engaged the services of

Mr. Sheldon, a superannuated minister from Connecticut, and kept him to fan the little flame of interest, manifested by a circle of gentlemen in New Haven, who eagerly watched the progress of Professor Silliman's analysis. To say that he was not earnest in his work, would be doing him injustice. He was enthusiastic. He bought several hundred shares himself, for which he gave a note that he had about as reasonable a hope of paying, as the immortal Micawber, when he negotiated his paper at the Canterbury Inn, and then, poor man, he became not only an enthusiast but a fanatic.

Some two thousand shares were transferred to him to sell, and the lowest price fixed. Dreaming the same fond dreams of sudden riches that have ever been the fatality of oil stocks, he fell frantically to work. The following letter from him will throw some light on the way they were obliged to dicker in the disposal of the stock:

NEW HAVEN, *April 11th*, 1855.

My Dear Sir,

Professor Silliman has not yet completed his photometrical examination of the Rock oil, in comparison with other burning fluids; but will probably wind up his analysis in all, this week.

The experiments last evening were favorable, and are to be renewed again this evening, and continued until the work is done.

The oil will not work well in the Carrol Mechanical Lamp, but burns finely in the camphene lamps, and will be tested in those now in general use. The value of the oil depends mostly on its properties as a burning fluid.

In this respect the analysis, in its results, has been highly satisfactory. Several gentlemen here have signified a desire to take some of the oil stocks, and pay for the same in town lots, but I have not as yet been able to satisfy my own mind as to their value and hence have not closed any negotiations.   *   *   *   *   *

This difficulty in disposing of stock was not occasioned more by the complete prostration of the money market than by the laws of the State of New York, which bore heavily on such enterprises by rendering the shareholder in a joint-stock company liable for its debts to the extent of the par value of the stock he owned.

During the preceding year such enormous frauds had been perpetrated by taking advantage of this law, that it was even difficult to give away stocks of just as good character as theirs. It was not considered the safest investment in the world for a man having no means of knowing the financial condition of a company, except by tedious investigations, which it was not possible for every one to make, to take shares at two dollars whose par value was twenty-five, when, for aught they knew, they might be called upon any day by decree of court to pay the whole twenty-five to the company's creditors.

With the opening of Spring, however, the partners took the matter more actively in hand. About the middle of April, Professor Silliman's report was handed in, and after being printed, was distributed wherever it was desired to obtain notice for the enterprise. On the 11th of May, writing from New Haven, their agent says:

"Silliman's report is now generally in the hands of the monied men of this place, and the impression it has created is decidedly favorable to the P. R. O. Company. But with the present state of feeling existing here in reference to joint-stock companies formed under the laws of the State of New York, and doing their business in the City of New York, I do not think that any great amount of stock will be taken by capitalists in this city.

The history of the New York and New Haven Railroad, and also the Western Empire Company, is still remembered with sorrow. Many had been ruined by the frauds committed by these companies, so that by them many others had sustained losses."

Under the circumstances, it is not at all strange that monied men should be cautious.   *   *   *

Some of the most prominent business men here have signified a desire to take stock in the Company, provided it be reorganized under the laws of Connecticut, and New Haven made the place of its business operations. In this state the property of the stockholder is not liable for the debts of the Company."   *   *   *   *

From the above we may obtain a view of the situation and the difficulties that trammeled them. To begin with, they were not rich; and the cost of the land, and the expense of the analysis—including the photometrical comparison about twelve hundred dollars—together with all the innumerable smaller expenses of organization, had absorbed the greater part of their available means. The previous Fall they had employed Mr. Angier to take charge of the spring, and run the rude machinery for pumping which he had himself invented and erected, and now when the spring opened, he was engaged to resume operations, while they hastened to do that which they saw must plainly be done before they could succeed—to organize a new company under the laws of Connecticut. This accomplished, a number of men promised to subscribe liberally for the stock.

Therefore a new company was formed in New Haven with a nominal capital of three hundred thousand dollars, and preparations made to take the property of the old company at twenty-four thousand dollars, and raise by immediate assessment a sufficient sum to undertake the development of the property by trenching on a large scale.

The deed to the first company had never been recorded, and it was thought the simplest course, to call in all the stock—on every matter, since they owned the most of it themselves, and the rest was held principally by their agent, who, poor man, rejoiced at the thought of cancelling his obligations by returning it—and thus after extinguishing the former corporation, make the deed directly to the new company.

When everything was ready for the transfer, Mr. Bissell had occasion to visit Titusville, where he was detained over Sunday.

A drizzling rain prevented his walking out. While lounging in the parlor of the once miserable little inn of the hamlet, he chanced to pick up a copy of the Pennsylvania Statutes, used by the Justice of the Peace, who held court in the room, and therein, to his amazement, he saw re-enacted the old English statutes of mortmain, devised and enacted three hundred years before, to check the absorption of the landed property of the realm, by ecclesiastical institutions too easily manipulated by the encroaching power at Rome.

The statutes there framed for a wise and beneficent purpose were here perverted so as to render forfeit to the state of Pennsylvania, the lands of any corporation organized beyond its borders!

He hastened at once to apprise the new company of this fortunate discovery, and on the 20th of September, 1855, executed a deed to Asahel Pierpont and William A. Ives, of New Haven, who gave a bond for the value of the property and promptly *leased* it for ninety-nine years, to the new company legally formed two days before, by the publication of the following articles of association :

ARTICLES OF ASSOCIATION

OF THE PENNSYLVANIA ROCK OIL COMPANY.

Be it known that we the subscribers, do hereby associate ourselves as a body politic. and corporate, pursuant to the provisions of Title 3d, Chapter 14th, of the Statute laws of the state of Connecticut, entitled " Of Joint-stock Corporations," and the act in addition thereto, and in alteration thereof, and the following are the articles of our agreement and association.

ART. 1. The name of the corporation shall be the Pennsylvania Rock Oil Company.

ART. 2. The capital stock of said corporation, shall be three hundred thousand dollars, and the said capital stock shall be divided into twelve thousand shares of twenty-five dollars each.

ART. 3. The purposes for which the said corporation is established, are the following, viz : to raise, procure, manufacture and sell *Rock Oil*, coal, paints, salt or any mineral or natural productions which may be found in any springs or mines, or on any lands that may come into the possession of said company by deed or lease, and generally to perform all acts and transact any business incidental to or that may be necessary in the prosecution of said business.

ART. 4. The statute aforesaid entitled "Of joint-stock companies," is hereby particularly referred to, and made part of these articles : and the corporation hereby established, and organized under and pursuant to the said statute shall have the powers, and shall proceed according to the regulations described, and specified in said statute.

ART. 5. Each subscriber to these articles, agrees to take the number of shares annexed to his name of the capital stock of said corporation, each share to be twenty-five dollars as aforesaid.

ART. 6. The said corporation is established and located in the city of New Haven, county of New Haven and state of Connecticut.

SUBSCRIBERS' NAMES.	No. of Shares.
George H. Bissell	1200
J. G. Eveleth	1200
Asahel Pierpont	1000
Prof. B. Silliman, Jr.	200
Henry L. Pierpont	200
James M. Townsend	500
John Hannah	150
Ebenezer Brewer	160
William A. Ives	1000
Brewer, Watson & Co.	1200
Edwin B. Bowditch	500
Eveleth & Bissell	4690
	12,000

By order of the Board of Directors.
New Haven, September 18, 1855.

It will thus be seen that Eveleth and Bissell retained a controlling interest in the affairs of the new company. There were, indeed, in the published articles of association, a number of other names on the list, among them Sheldon's; but they never took their stock, and it was retained by the partners. The consideration for the land was $24,000, and though retaining so much of the stock themselves they had now about received the amount of their expenditure, and felt inclined to hold it and wait the result of further development of the property.

A small fund was raised for the treasury, and Mr. Pierpont, an eminent mechanic, was sent out to examine the spring with a view to the improvement of Mr. Angier's machinery. It was, however, though rude, perfectly adapted to that mode of development, and no other had been yet thought of.

Mr. Pierpont would have resumed more extensive operations, but the inharmony that forever afterwards characterized the management of this company had already begun to manifest itself, and it was found impossible to raise more money for the treasury.

While Mr. Bissell and his partner held a majority of the

shares they were crippled by the by-laws to which they had subscribed, and which provided that a majority of the board of directors should be chosen from the New Haven stock-holders.   The one thousand shares that Mr. Ives had taken were paid for in local securities that were after-wards proven to be worthless at the time, and out of the vituperative charges that followed this discovery, sprang the spirit of dissension that always thereafter divided their counsels and circumscribed their usefulness.

In a letter to Mr. Bissell in October, Dr. Brewer, speak-ing of a new trench, says: "Mr. Angier took six gallons from it, though it had been gathered the day before," and in a postscript to the same letter he adds, " As I have no interest in the matter only the wish to see it go on to per-fection, of course I can have no object in magnifying its resources, but from fifty to one hundred gallons per day may be had by the judicious expenditure of five hundred dollars."   But the five hundred dollars were not forth-coming.   Even Mr. Angier's services were dispensed with.

CHAPTER V.

COMMENCEMENT OF DEVELOPMENTS ON OIL CREEK.

PRE-OCCUPIED by a course of specious reasoning, it is wonderful how completely the human mind may ignore the inductive logic of facts. It is a lamentable, and apparently, an incurable frailty which more than all other human infirmities, retards the progress of knowledge. It is the fault of an ancient system of speculative philosophy which accepted the plausible as conclusive—which, taking anything for granted, rejected that as exceptional to the law which could not be warped to the support of its theory—a philosophy which, while it encouraged reflection, forbade experiment, and thus left much uncertain, that might have been rendered positive by the simple turning over of a chip—a philosophy which received its death blow from Bacon, the experimentalist, and Franklin, his follower. But enough of its dreamy essence still lingers to tone the wild, progressive spirit of the age.

Without bringing into question here the plausibility or correctness of that theory which referred the origin of Petroleum to coal, can we help but express our wonder at the perversity of those minds, which, preoccupied with such a conclusion, steadfastly overlooked the fact, that in every important case to which they could refer, it had been found very far *beneath* the coal measures?

While clinging to and reiterating a theory that was perfectly indisputable, namely : that the oil was forced to the surface by the expansibility of the gas with which it is invariably accompanied, they overlooked the fact

that the same gas should have prevented its ever settling
to such a depth—should have forced it to the surface
when it was first expressed from the carboniferous stratum instead of forcing it downward through impervious
underlayers.

With infinitesimal modifications the " coal theory " is
the one that has mostly obtained—none are without some
insuperable objections; and, considering the organic
nature of the fluid, it is in many respects exceedingly
plausible.

But the day may come when Geology will discover or
invent a " period of Petroleum plants," which, after
revising her nomenclature a little, she can fix somewhere
anterior to the " period of coal plants," and simplify the
theories of the origin of Petroleum, which are now too
numerous and ponderous to be mentioned.

Although a perusal of the report of the State Survey
would have shown them that the last traces of the coal
fields of Northwestern Pennsylvania faded out in a thin
stratum at the tops of the highest hills a few miles
farther down the creek, and that, geologically speaking,
they were a great many feet below the coal measures,
still the inhabitants of an insulated hamlet like Titusville,
numbering less than three hundred souls, and offering
no facilities for extended investigations, were quite ex-
cusable for clinging to the supposition that the hills which
rose abruptly on either side of the little island on which
their famous oil spring was located, were filled with a
highly bituminous coal from which the Petroleum slowly
leaked into the valley of the Creek, and coming in contact
with water, was forced by specific gravity to the surface.
In the light of present events, this may seem sufficiently
absurd, but it was not without an appearance of great
plausibility to even reflecting minds in that day.

But a new day was dawning—a day which witnessed

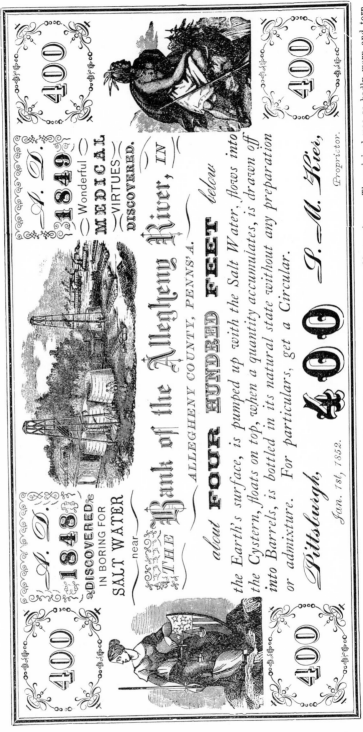

The above is as near a correct copy of the label referred to on the opposite page as we are able to give. The original was so badly worn and torn, as to render it impossible to be copied exactly.

the birth of an idea that gave a new direction to human thought, and developed an industry which will forever mark an era in the progress of the world.

It was the idea of obtaining Petroleum by means of artesian wells. It was a simple thought, but significant —a thought which, as Professor Silliman remarked, was the one of all others most naturally suggested by the various phenomena that had attended the discoveries of Petroleum in the salines of the Muskingum and Kanawha, described in a former chapter of this work—the first idea that *should* have been suggested to a mind cognizant of all these circumstances; and yet, though himself editor-in-chief of the periodical in which the circumstances were described, he very candidly confessed, that throughout the five months he was prosecuting the analysis, the thought of artesian boring, never once occurred to him. And yet of all in any way connected with these first transactions, he was the only one of whom we had a perfectly reasonable right to expect such an idea; but Professor Silliman's interest in the matter terminated with the conclusion of his elaborate analysis, for though he perfectly comprehended its value, he never expected to see it obtained in any great quantity, and the two hundred shares of stock he held were given him in order to make him president of the company, and thus secure the prestige of a name renowned in science.

The idea came from another quarter, and was suggested by an incident as trifling as that which disclosed the law of gravitation. While seeking shelter beneath the awning of a Broadway drug-store, one scorching day in the summer of 1856, Mr. Bissell's eye fell upon a remarkable show-bill lying beside a bottle of Kier's Petroleum in the window. (See copy of this label on opposite page).

His attention was arrested by the singularity of displaying a four-hundred-dollar bank-note in such a place; but a closer look disclosed to him the fact that it was only an advertisement of a substance in which he was deeply interested. He stepped in and requested permission to examine it. The druggist took it from the window, and having plenty of them, told him to keep it. For a moment he scanned it, scrutinizing the derricks and remarking the depth from which the oil was drawn, till instantly, like an inspiration it flashed upon him, that this was the way their lands must be developed—by artesian wells. It seems a very simple thought, but how astounding have been its results. It has added more than a thousand million of dollars to the material wealth of our country, and its history is only just begun. Already it makes, after wheat and cotton, our most valuable commodity of export, and throughout the world, must furnish the means of subsistence to more than half a million of people. Its influences upon civilization are incalculable. Yet all this by the birth of a new idea. Thus, step by step the world improves, moving on toward knowledge.

The idea was simple—at first it may almost seem to have been self-evident, but reflect that the mind which grasped it must also have taken in a better conception of the philosophy of the *existence* of Petroleum than had any other mind before.

It is not unlikely that the mind of this man may have been prepared for the reception of such an idea, by long reflection. It is quite likely that Newton had seen many an apple fall before the one that gave him an idea, and it is just as unlikely that he would have ever drawn the conclusion from the incident if the necessity of the law of gravitation had not previously occurred to his mind.

When Mr. Bissell disclosed his theory to his partner that gentleman embraced it with enthusiasm, and they promptly canvassed the practicability of putting it to the test.

Their first notion was to attempt the experiment them-selves, but even if they could induce the company to help them in their scheme they reflected that such a step would necessitate the loss of their legal business, and even if it should prove successful, which was all an uncertainty, they never dreamed of flowing wells that would make them millionaires in a day.

In this dilemma they imparted their convictions to Mr. Havens, of the firm of Lyman & Havens—real estate brokers on Wall street, N. Y.—a man who had been largely iden-tified with the construction of the first railroads in West-ern Pennsylvania, and he was so favorably impressed with the theory that he signified a desire to take part in the enterprise himself, and after a few days reflection, offered them five hundred dollars to secure him a lease of the property from The Pennsylvania Rock Oil Company. But that was a company very much inclined to thwart any plan proposed by the New York stock-holders, and though several of the New Haven parties, prominent among whom was Mr. Pierpont, readily accepted the new idea, they having faith—if the expression may be pardoned —and though after much delay the lease was finally granted, it was not till Mr. Havens had been overwhelmed in financial embarrassments which prevented his going on with the contract.

By the terms of the contract he was to pay the Com-pany twelve cents per gallon for all oil raised for fifteen years, and a year was given him to begin operations.

When Eveleth and Bissell conveyed their title to The Pennsylvania Rock Oil Company, they reserved to the lumbering firm, the use of the mill race and the right of way over the property, in consideration for which the Oil Company received a lease to take "oil, salt, or paint" from all other lands of the firm in Venango county for a term of ninety-nine years.

But the wives of the members of the firm had not joined in the power of attorney by which the agent conveyed the lease, and would be entitled to dower in the event of the death of their husbands.

The idea of artesian boring was too fascinating to be forgotten. It grew reasonable, upon reflection. It was sustained by all the phenomena of Petroleum. It was encouraged by every written account. It grew into such favor with the New Haven stockholders that they formed a scheme to monopolize its value.

Before the year allowed for Havens to begin operations had nearly expired, Mr. Townsend, then President of the Company, in lieu of Professor Silliman resigned, employed Mr. E. L. Drake, whom, in the darker days of its prospects he had cajoled into purchasing two hundred dollars' worth of his own stock for the ostensible purpose of going to Titusville, to rectify the oversight mentioned in the lease, though the real object was not less to have him inspect the locality with a view to what followed, while it might be done at the expense of the Company.

That this *was* their plan of operation, will be perfectly plain to any one who follows the progress of the cunning development of their scheme.

First, and foremost the legal hitch might just as easily have been fixed up, by sending the documents by mail; for it was merely an oversight, and the women had no objections to signing. Then Mr. Drake, though an intelligent gentleman, was the last one to choose for the performance of legal business, as no occupation of his life had prepared him for such duty; besides in order to give a pompous turn to the transaction in the eyes of the backwoodsmen, the legal documents, together with several letters were mailed to "*Colonel* E. L. Drake, care of Brewer, Watson & Co.," before ever the man left New Haven.

The title was the pure invention of Mr. Townsend,

who generously acknowledges his *pius fraudum*, and in the oil region and elsewhere, he has ever since been known as *Colonel* Drake. On his way to Titusville, he stopped to examine the salt borings at Syracuse, New York, and about the middle of December, 1857, was trundled into the little village of lumbermen, on the wagon that brought the mail from Erie, Pennsylvania. Prepared as they had been for his coming, he was received with ostentatious hospitality.

Finishing that part of the legal business, which could be accomplished in Titusville, he spent a few days examining the various indications of oil on the lands, and then proceeded to Pittsburgh, to add the signatures of Mrs. Brewer and Mrs. Rynd, to the instrument of conveyance, and after visiting the wells at Tarentum, the picture of which on Mr. Kier's advertisement had suggested the idea of boring for oil, he hurried back to New Haven, enthusiastic to conclude the scheme. On the 30th of December, the three New Haven directors, constituting a majority, executed a lease to Edwin E. Bowditch and E. L. Drake, by the terms of which, they were to pay the Pennsylvania Rock Oil Company, only *five and a half cents a gallon royalty* for the oil raised for fifteen years. At the annual meeting of the directors, eight days later, this lease was brought up, and notwithstanding the protest of the two other directors, George H. Bissell and Jonathan Watson, representing a trifle over two-thirds of the whole stock, it was ostentatiously ratified. The thing however was so palpably unjust that Mr. Bissell and Mr. Watson withdrew, protesting, from the council. Perceiving that they had overdone the matter, and might possibly lose all if they persisted, they at once changed the consideration to one-eighth, in kind, of all the "oil, salt or paint" produced, and determined to defy every protest against this. The deed was at once

sent to Franklin and recorded. But refusing to concur
in terms more favorable than those granted in Havens'
lease, Mr. Bissell threatened to restrain despoliation of
the property by injunction, if they attempted operations.
Their position was quite untenable. They knew it, and
rather than undergo the scrutiny of a legal investigation,
and being determined not to let the prize slip from their
fingers, they yielded. A supplemental lease was recorded,
making the terms the same as in Havens' lease, but extending the time to forty-five years. To this Bissell and
Watson cheerfully agreed.

On the 23d of March they formed themselves into an
association under the title of "The Seneca Oil Company."
They had the grace to shun publicity, and the publication of
the articles of association, required by law, was effected in
an obscure little weekly, published in one of the villages
of New Haven county.

The basis of their association was the lease. Mr.
Drake appeared as the principal stockholder; but no
stock was ever issued.

It was in effect only a partnership, the members of
which sought protection against each other under the
laws for joint-stock companies. From the little influence
he possessed in the management of their affairs, it is evident that Drake could have furnished but little of the
capital. He was not in a situation to do so. For eight
or ten years previous he had been a conductor on the New
York and New Haven railroad, at a salary of seventy-five
dollars per month, and the little he had been able to save
from such a pittance, had been swept away by an unlucky
investment the year before.

He was engaged at a thousand dollars a year, and
about May, 1, 1858, arrived in Titusville with his family.

He had been provided by the company with a fund of
a thousand dollars on which to begin operations.

His first step after getting settled was to start up the old works abandoned by The Pennsylvania Rock Oil Company, and then he cast about for a practical artesian driller to sink a well.

On the 2d of July, 1858, he writes:

"Here I am digging along yet in search of oil and other valuables. The month of May was a hard one, and the first eleven days of June, but since then we have had dry weather, so that I have got the start of the water, and am now gathering about ten gallons of oil per day— at the same time sinking a well for the purpose of taking what oil there is on the island.

I have found some difficulty in getting a borer. All were engaged on jobs that will last until fall. Yesterday Dr. Brewer wrote me that he could get one for me at Allegany, who will bore and tube for three dollars per foot, which is the best offer I have had. I wrote the Doctor to send him along at once. Yesterday I set some men to opening a new spring, so that things begin to look greasy."

In justice to his partners, it is due to say that Mr. Drake was well supplied with money. In the oil region there has been a general belief to the contrary, but this is entirely without foundation. In his private affairs possibly, he was embarrassed, but in his last quarterly statement to the company before striking the oil, he reported a fund of two hundred and eighty dollars on hand.

On the 16th of August, '58, he writes as follows:

"I received on Saturday at Erie, Aug. 14th, a package containing $472.67, from the Treasurer of the Seneca Oil Company, and gave the Express Company my receipt.

"I shipped two barrels of oil to Mr. Pierpont at New Haven, as he said he could make a market for it.

"In sinking our well last week, we struck a large vein of oil, but the same thrust of the spade opened a vein of water that drove the men out of the well, and I shall not try to dig by hand any more, as I am satisfied that boring is the cheapest. I should have had my borer here, but I wrote him on the 1st, I was not ready, as I did not know that you could raise the money, but money we must have if we make anything. I have abandoned the idea of boring and pumping by water, as I could not have the exclusive right to the power, but must be subject to the sawyer, the turner, and the blacksmith, so that after consulting the best salt and oil men at Tarentum, I have contracted for an engine to be ready for boring by the first of September.

"I have got out the timber for my pump-house, and am having it framed to-day. We shall get it up this week. I shall send in a statement of my stewardship on the first of September, but if in the mean time the Company should feel too poor to furnish a thousand dollars more by the 10th of September, please let me know at once. Money is very scarce here. The lumbermen could not sell their lumber for cash this summer, and the people all depend upon the lumber trade."

The Company did not send him the thousand dollars as soon as the 10th of September; his engine was not ready; and when he finally was prepared to start, the driller had taken another job and operations were suspended for the winter. In February '59, Drake went to Tarentum and engaged a driller to come up in April. But April came and no driller appeared. The man having been able to get a better job nearer home, affected to believe that Drake was crazy—a monomaniac on the subject of Petroleum. It afforded him the simplest exit from his contract.

When Drake went down to look after him, Mr. Kier recommended him to engage William Smith and his two sons who had done a great deal of work on his salt wells. About the middle of June, provided with a complete set of tools from Mr. Kier's shop, "Uncle Billy Smith" and his two sons arrived in Titusville.

Aggravating delays followed. In artesian boring it is necessary to begin on the rock to drill. This had been previously done by digging a common well-hole, and cribbing it up with timber. When the rock is within a few feet of the surface it is still the cheapest and easiest method, but in some localities to do so would be practically impossible. They started to dig a hole, but it so persistently caved in and filled with water when they got a few feet below the surface, that Drake determined to give it up, and try an experiment that had suggested itself to his mind. This was the driving of an iron tube through the quicksands and clay to the rock. If this is exclusively his own invention, which is probable, it is a pity he did not procure a patent on it. The royalty would have afforded him at least a competency, though the driving pipe is not so much in use now as formerly.

The operators in the oil region have had the benefit of his invention without any return, unless indeed we except the good feeling which prompted them to send him a present of $4,200, when they heard he was sick and in need.

The pipe was successfully driven to the rock—thirty-six feet—and about the middle of August the drill was started. The drillers averaged about three feet a day, making slight indications all the way down.

Saturday afternoon, August 28th, 1859, as Mr. Smith and his boys were about to quit for the day, the drill dropped into one of those crevices, common alike in oil and salt borings, a distance of about six inches, making the total depth of the whole well 69½ feet. They withdrew the tools, and all went home till Monday morning. On Sunday afternoon, however, "Uncle Billy" went down to the well to reconnoiter, and peering in could see a fluid within eight or ten feet of the surface. He plugged one end of a bit of a tin rain-water spout, and let it down with a string. He drew it up *filled with Petroleum.*

That night the news reached the village, and Drake, when he came down the next morning, bright and early, found the old man and his boys proudly guarding the spot, with several barrels of Petroleum standing about.

The pump was at once adjusted, and the well commenced producing at the rate of about twenty-five barrels a day. The news spread like lightning. The village was wild with excitement; the country people round about came pouring down to see the wonderful well.

Mr. Watson jumped on a horse and hurried straightway to secure a lease of the spring on the M'Clintock farm, near the mouth of the creek. Mr. Bissell, who had made arrangements to be informed of the result by telegraph, bought up all the Pennsylvania Rock Oil stock, it was possible to get hold of, even securing much of that owned in New Haven, and four days afterward was at the well. His views of the matter had ever been the broadest, as his transactions had been the boldest.

While others were seeking for surface indications before leasing, he rushed forward, and secured farm after farm down the creek and along the Allegany, where there were no surface indications whatever. The result has proven the wisdom of his conclusions. Drake unfortunately took a narrower view of the matter. He pumped his well in the complacent conviction, that he had tapped the mine! He was probably led into this supposition by what seemed to him the remarkable incident of having struck a crevice. No money was paid on most of the leases at first taken; a royalty of an eighth or a quarter, only being reserved by the easy old farmers who owned the land, and without a cent he might have secured any quantity of territory. He was repeatedly advised to do so, by shrewd men who were themselves laying the foundation of fabulous fortunes; but it was his fatal misfortune to disregard that advice.   When several other wells had been struck, and his eyes were opened to his mistake, it was too late—the golden opportunity had fled.

The well fell off slowly till toward the end of the year, it produced only about fifteen barrels per day.   It was never pumped at all on Sundays, and averaging the production at twenty barrels per day—an average probably much too high—and granting a hundred and five working days shows the production for the year 1859, to have been twenty-one hundred barrels.   But there were many days in succession when it was not pumped.   At one time Mr. Smith, approached one of the tanks with a light when the gas caught fire, and the derrick, pumphouse, oil vats and all were completely consumed, and it was nearly a week before operations could be resumed.

Probably two thousand barrels even would be twenty-five per cent. above the actual production of the four months of that year, and yet the production is generally quoted at 82,000 barrels.

The second well was promptly started by Barnsdal, Meade and Rouse, and at the depth of eighty feet, in November it was pumped for two or three days, but yielded in all less than five barrels of oil, till it was sunk to about one hundred and sixty feet, when in February 1860, it was again started, and produced from forty to fifty barrels per day.   The third well was sunk by Mr. Angier for Brewer, Watson & Co., in the spring of '60 on the M'Clintock farm, and oil was struck about the middle of December following, but both these last wells had been put down without the aid of an engine—"stamped down with a spring pole" they called it—and after pumping by hand one day at the third well, producing twelve barrels of oil, so much water came up, that operations were suspended till an engine could be got down from Erie and set up, which was not till the middle of January. Considering everything, the difficulty of disposing of the oil was much less than might have been expected. Kier contracted to take part of it at fifty-six cents per gallon, and the rest was disposed of through Scheifflin Brothers of New York. And here ends the history of Petroleum developments on Oil Creek prior to the year 1860.

NOTE.—It is claimed by Peter Wilson and R. D. Fletcher, of Titusville, that the Company refused aid to Drake during the drilling of the well, and that *they* assisted him by indorsing notes, enabling him to raise money, and thus supply the "sinews of war," when the work might have been abandoned.

CHAPTER VI.

DEVELOPMENTS ALONG OIL CREEK.

AFTER oil was struck on the island in Oil Creek, at the junction of Pine, the development spread rapidly thence to its mouth, where now is situated the thriving town of Oil City, containing about eight thousand inhabitants, and fast growing into a solid, business centre.

With one stride, indeed, the whole territory was virtually thrown open to development, for the very day that Drake's well began to pump, Brewer, Watson and Co. leased the Hamilton M'Clintock Farm, below Rouseville, where the " surface indications " were even better than on their own tract, which had passed entirely from their control to The Pennsylvania Rock Oil Company. Here they immediately began to sink a well—the third one ever sunk for oil—and all the intermediate territory for sixteen miles along the narrow valley of Oil Creek was soon taken up.

At first, it is true, speculators looked for "surface indications"—"pebble rock," oil springs," &c.—and leased only where they were found; but when a few older or better informed minds took the initiative, others followed, and a few months' experience proved that there was no reliance to be placed in " surface indications," and that good wells were as likely to be obtained half a mile away, as beneath the most productive natural spring.

The second well, on the Watson flats, below Titusville, was begun within a few days after the completion of the first, by Messrs. Barnsdall, Meade, Abbott and Rouse. It was situated a little above, and almost within stone throw of the first, and though it was tested in November, 1859, it had to be sunk deeper, and was not finally completed until February of the following year. Like the third, sunk by Mr. Angier, for Brewer, Watson and Co., on the M'Clintock farm, this was also put down by means of a spring pole; and indeed the same is true of several hundred of the first wells drilled along the Creek; nor was the system entirely abandoned before 1865, and to strong men, whose means were limited, it afforded a ready mode of development that answered a good purpose in opening up "shallow" territory; but it was a means totally inadequate when they began to sink wells below four or five hundred feet.

In February, Captain Funk and also Messrs. Phillips & Co., began operations on the Upper McElhenny Farm, about a mile-and-a-half above Petroleum Centre. On this farm in June, 1861, the first-flowing well was struck. In more than one respect the well may be considered remarkable—it being the FIRST drilled to the THIRD sand rock.

This was the "Fountain Well" on the upper McElhenny or Funk Farm. To the astonishment of all the drillers in the neighborhood it commenced flowing at the rate of 300 barrels a day. Such a prodigal supply of oil upset all calculations, and it was confidently predicted that the supply would soon stop. It was an "Oil Creek humbug," and those who had direct interest in the property in the well looked day after day to see the stream stop. But like the old woman who sat down by the river-side to let the water run itself out, that she might cross dry-shod, they waited in vain. The oil continued flowing with little variation for fifteen months, and then stopped (See chapter on early-flowing wells.)

Long before the Fountain Well had given out, the wonder in regard to it was overshadowed by a new sensation. On the Tarr Farm, the "Phillips Well" burst forth with a stream of three thousand barrels daily! Not to be out-done by the territory down the Creek, the Empire Well, in the immediate vicinity of the Fountain Well, suddenly burst forth with its three thousand barrels daily, figures which subsequent flowing wells vainly endeavored to equal.

Kier & Co., of Pittsburgh, began to develop the J. W. McClintock Farm, on which the village of Petroleum Centre now stands. Hibbard & Co. began on the John McClintock Farm, Henry R. Rouse, S. Q. Brown, John Mitchell, and others, on the Buchanan Farms; Crossley & Fletcher on the Stoeppel Farm—all before the first of February, 1860. "The Crossley" was the third well completed; and upon the flats below Titusville and up the valley of Pine Creek as far as Enterprise, there were not less than twenty compa-nies and individuals preparing to put down wells as soon as the spring opened.

The number of farms between the island on which oil was first struck and the mouth of Oil Creek, a distance of between sixteen and eighteen miles, was forty-three, and though most of the statistics of the early operations are now hopelessly lost, from the few preserved we are able to glean the following:

The Island tract, embracing one hundred and five acres, originally owned by Brewer, Watson & Company, was transferred to George H. Bissell and Jonathan G. Eveleth, of New York, in 1854, and by them put into a joint stock company called The Pennsylvania Rock Oil Company. In 1858 the property was leased to The Seneca Oil Company, which put down the first well.

The lease ran for forty-five years, and as the decline in the price of oil made it ruinous to pay the royalty of twelve and a half cents per gallon, The Pennsylvania Rock Oil Company came to an agreement by which The Seneca Oil Company took a small portion of the land in fee, and gave up the lease in the summer of 1860.

After this was accomplished George H. Bissell purchased the interest of The Pennsylvania Rock Oil Company, of which he was himself principal shareholder, for fifty thousand dollars, and that portion of the tract was thereafter known as the Bissell Farm.

Mr. Bissell at once began actively to develop the farm, putting down ten or twelve wells, six of which were productive, and yielded for some time eighty barrels per day, which was much better than any other territory in the immediate vicinity. The farm a few years later was sold to the Original Petroleum Company.

Next below this is the Griffin Farm, located on both sides of Oil Creek, and containing the ruins of several derricks. The property was pretty well developed after passing into the hands of the New York and Pennsylvania Petroleum Company, and below this is the Crossley Farm, on the east side of the creek, which, though well developed, was never very productive territory; and this again is followed by the second Bissell Farm, owned by George H. Bissell & Co., which, like all the rest of the territory which came under the management of this energetic man, was thoroughly and successfully developed.

Below the second Bissell Farm are the two Stackpole Farms, partly covered by one of the large dams used in the production of pond-freshets.

The upper Stackpole Farm passed into the possession of the Northern Light Oil Company, and Brewer, Watson & Co.; the lower one contains ten wells and two abandoned refineries. Then follows the Pott Farm, on which there have been no producing wells, and next below is the Shreve Farm, owned by the Great Western Consolidated Oil Company, but like the farm above, it has not been productive territory.

The Shreve farm is followed by the J. Stackpole Farm, which came into the hands of Brewer, Watson and Co., and was well developed by sub-lessees, but without remarkable success.

The Flemming Farm, next below, owned by Mrs. Flemming, was found to be, if not entirely unproductive, at least unremunerative territory, and is now without any evidences of development.

The same may be said of Henderson Farm, which is just below it.

The Jones Farm, which is next in order, though thoroughly tested by sinking nearly twenty wells, was never proved productive territory, yet from surface indications it was as promising as any farm along the Creek.

The second Flemming Farm, a little more than four miles below Titusville, is next after the Jones Farm, and the beginning of better territory. The flats on this farm were thoroughly tested, and several good wells obtained; one, a flowing well, was successfully operated for some time, when the owner, hoping the more completely to shut off the surface water and increase the flow of oil, drew up the tubing to change the locality of the seed-bag; but after re-arranging it, from some unaccountable cause, the well not only ceased to flow, but never again produced oil.

The Miller Farm, now a station on the Oil Creek R. R., and formerly the scene of great enterprise on the part of the Pit Hole and Miller Farm Transportation Company, is the first below the Second Flemming Farm on the Creek.

Though formerly excellent territory, having a great number of good flowing and pumping wells, it now produces little or no oil, and owing to the improvements in the manner of transporting oil by rail, the labor of three or four men sufficing to fill a whole train of the modern tank cars in a few hours, and the fact that many of the

refineries once operated here, are now abandoned, the enterprise that once marked the place has almost entirely
subsided.

Its capacity of iron tankage is nearly two hundred and
fifty thousand barrels, being, except Oil City, the largest
pipe line station in the region.

From the Miller Farm to the mouth of the creek at Oil
City, the territory has been incomparably the best ever
discovered, producing up to 1868, probably two thirds of
all the oil ever brought to the surface.

The Shaffer Farm, which is next, though containing
but little more than fifty acres, was formerly one of the
moderate producing farms on the Creek. This farm, though
in the year 1864, containing less than a half dozen buildings, was for a time the terminus of the Oil Creek Railroad, and immediately became a shipping station of great
importance, for the oil was then shipped in barrels, and
not less than fifteen hundred teams were employed in
hauling it to the cars from the well, and, together with their
drivers, and other auxiliaries, these supported the innumerable stables, hotels and eating-houses, that sprang up
in a night like mushrooms. Though at one time the
village on Shaffer Farm, numbered over two thousand inhabitants, there is hardly a house now remaining to
mark the scene of former activity.

When the road was extended, the buildings were taken
down and removed to the next station.

The Sanney Farm, which is the first after Shaffer,
though not unproductive, has not been remunerative territory; and though once the seat of a number of small
refineries, is now completely abandoned; and the same
may be said of the Gregg farm which follows.

The Beaty Farm just below, lying at the mouth of
Hemlock Run, has done better, and once contained a
number of good wells, several of which were up the run.

The Farrel Farm, though containing but thirty-six acres, has been probably the most remunerative bit of country property of its size in the whole oil regions. Its original owner was James Farrel, but a part of his interest was sold to the Commonwealth Oil Company. It is situated on Bull Run, and the East side of Oil Creek. The renowned Noble and Delamater well, which flowed three thousand barrels of oil a day when first struck, commenced in 1863, and continued to flow till 1865, and is estimated to have produced upwards of three million dollars worth of oil; and another well the Craft, on the same farm, produced over one hundred thousand barrels.

Besides these there were several other good wells on the tract, but most of them are now producing little or nothing. Opposite this, on the other side of the creek, is the Foster farm, rendered scarcely less famous by the great Sherman well, which began flowing in 1862, at the rate of two thousand barrels per day, and for nearly two years is said to have averaged nine hundred barrels per day. It commenced to flow in May 1862, and ceased in February 1864; but for a long time afterward, it was successfully pumped. On the other side of the creek again, bounded by both the Foster and Farrel farm, lies the Caldwell farm.

The famous well of the same name, struck on this farm in the spring of 1863, being found to have a subterranean connection, with the Noble and Delamater well, the owners of the later offered to buy it, and it was sold with an acre and a half of land for the sum of $145.000. (See sketch of Orange Noble.)

The upper McElhenny Farm, which is next below these, was one of the first to be thoroughly developed, and has always been one of the best producers. A great number of wells were sunk on this farm, in the years 1860, 1861 and 1862, and though none of them were large enough to be remarkable, nearly all were remunerative.

The Espy Farm adjoining, is remarkable for the old Buckeye well, and has proved valuable territory. The flats on this farm, were developed among the earliest, but the up-lands tested several years later, were very productive.

The Benninghoff Farm, which lies between the two McElhenny Farms on the creek, at the mouth of Pioneer Run, was at one time remarkable for the great number of its flowing wells.

Most of the wells on this tract flowed, when first struck, and though none of them were large, all were lasting, and the territory very sure.

A number of joint-stock companies, were chartered to work leases on this farm, for it was brought into market at the period of the great excitement.

The lower McElhenny Farm, situated on both sides of Oil Creek below the Benninghoff Farm, was one of the earliest developed, and for many years continued to be one of the most productive.

It was purchased, like the Upper McElhenny Farm, of the original owner, by Hasson and M'Bride, and L. Haldeman & Co.

The most remarkable wells on this farm were the Empire and Crocker; the former of which started off at two thousand five hundred barrels a day, and after yielding an average of two thousand barrels a day for nearly four months, finally dropped off to three hundred, and then ceased altogether. Among the other best wells on the farm were the Burtis and the Davis.

The Boyd Farm, next below this, has singularly enough proved very poor territory, though it is entirely surrounded by that which is exceptionally good. It was at one time the seat of several small refineries, all of which have been abandoned.

The Stevenson Farm in this vicinity, but not reaching

to the Creek, was developed in 1865, and proved to be valuable territory, though possessing no remarkable features.

The James S. McCray Farm has been so exhaustively described in another part of this work, that we pass it here to notice the J. W. McClintock Farm, on which the once flourishing city of Petroleum Centre is situated.

The farm contains two hundred and seven acres, and was leased in November, 1836, by George H. Bissell and Co., and in the following February put into a joint stock Company, called the Central Petroleum Company of New York.

It embraces, besides the land on which the town is situated, the circular ravine to the left, called Wild-Cat Hollow; nearly every square rod of which has been perforated with a well hole. Not less than a hundred and fifty wells have been drilled on this territory, and nearly eighty per cent. of the whole number have been remunerative; and this was undoubtedly—until the opening up of the Parker's Landing field, which has been for the most part very sure territory—the best showing made by any farm in the region.

The remarkable results shown by this farm are undoubtedly due quite as much to the excellency of its management, as to the superiority of the territory, and stands a bright example of the result of the judicious and economical management of an oil farm. Instead of giving out leases at random, to parties willing to pay a large bonus for the sake of getting leases on which to found speculative joint stock companies, and then in all probability—having enriched themselves by the sale of stock, neglect the development of the land—instead of pursuing this course, so common in the oil region, they only gave leases to actual operators, and at a uniform royalty of one half the oil.

In 1865 a bonus of one hundred thousand dollars and "one half the oil" was offered for ten leases on the farm, but promptly refused, and the result proved the wisdom of the decision, for while the owners may not have realized more than this, they have retained the unrestricted control of their property.

Three million dollars worth of oil, or more, must have been taken from this territory, and the revenue from ground-rent, for building, must at one time have been very great. The village retains of its former population only about fifteen hundred, and as the farm is now exhausted, beyond the hope of another excitement, must continue to decay.

Scarcely less decided was the success that attended the development of the Hyde and Egbert Farm, on the other side of the creek along the foot of the McCray hill. It was purchased in 1859, of the original owner, Davidson, and soon after tested with fair results, but it was not until 1863 that the flowing wells, which rendered this territory the subject of such wild speculation, were struck.

The Maple Shade, and the renowned and singular Coquette well are the most remarkable in the history of this farm.

The first operations here were the least successful, and thus it came to be looked upon as doubtful territory for some time.

The Hayes farm below the Hyde, and Egbert is situated on both sides of the creek, and though pretty well developed, and always with fair success, it has never been so well managed as many of the farms by which it is surrounded.

On this farm was attempted the experiment of *digging* a shaft five feet by seventeen, to the oil-bearing rock known as the "third sand." The enterprise was undertaken by the Petroleum Shaft and Mining Company, and the shaft was commenced in the rear of the Maple Shade, near the line of the Hyde and Egbert farm. Though unfortunately for the cause of science, the gas would have rendered the enterprise impracticable, it was abandoned for want of means, when down less than a third of the distance.

The Story Farm, better known as the Columbia Farm, lies directly below the Hayes. It was bought in 1859, by Pittsburgh parties, for thirty thousand dollars, and some other contingent benefit, and shortly afterward, put into the Columbia Oil Company, which has ever since owned and managed it. It has been, all in all, the best managed piece of oil territory in the region, and is still paying large dividends. The sale has been the subject of much litigation, and the Story family lately obtained a judgment for about $20,000 against the property. George H. Bissell, had determined to buy the farm, and was willing to pay all the owner asked; but the old lady refused to sign the conveyance.

Seeing that it would be for his interest to buy the wife's good-will, he returned, decided to offer her a fair sum for her contingent interest—the bargain as far as the old gentleman was concerned, was considered by both parties as made—but the agent of the others had been a few hours ahead, and having the documents all ready for signing, induced the woman to concur in the sale of a property, that has been worth at least five millions of dollars, by the promise of a new silk dress! To any one wishing a thorough knowledge of the profits, and risks of the oil business, when judiciously and economically managed, no better insight is afforded than by the study of the exhaustive annual reports of this most excellent company.

Next below, and on the opposite side of the creek, is

the Tar farm, one of the very best on the creek, and renowned for the Phillips and the Woodford wells, the first
of which, when struck, is said to have produced 3,000 barrels per day, and the latter 1,000 barrels per day, though
in both cases it is probable the figures are too high.

They were situated within two rods of each other, and
the subterranean connection between their sources of supply, was so manifest that when the "Woodford" pumped,
the only remedy left the "Phillips" was to draw the tubing
and let the surface water down to shut off the oil from
both.

Enormous sums were offered by both owners, but as
either had it completely in his power to render the
property of the other worthless, neither was willing to
settle until both wells were nearly ruined by the surface
water, and the consequence has been that many smaller
wells than either, have yielded as much oil as both.
These famous wells are located on the bank of the creek
to the left (going down) of the railroad bridge, and still
pumping, they are a subject of interest to travelers, who
view them from the car windows.

THE BLOOD FARM was one of the earliest to be developed,
and therein may be said to have been its misfortune, for
in 1861, and 1862, when oil was almost valueless, often
selling, in bulk, as low as ten cents per barrel, and not unfrequently suffered to run to waste as utterly worthless,
this farm produced more than all of the oil region together, and before the time of speculation and high prices
was exhausted, or nearly so, there were a great number
of good wells, many of them flowing, and one that flowed
twenty-five hundred barrels per day. Below this, at the
mouth of Cherry Tree Run, is the RYND FARM, which,
though now producing comparatively little, was once good
territory. Not less than a dozen different companies had
interests in this farm, and the whole flat has been perfo-

rated with holes, though it was rather uncertain territory. The Widow McClintock's or Steele Farm, next below, has also been good territory. The farm was the property of the widow McClintock, who was herself burned to death in 1863, while lighting the fire with oil, and left the farm, together with all the accumulated money of two years' production to her adopted son, John Steele. It has since passed from his possession, and is now in the hands of a company. Its production at present is very small.

The John McClintock Farm, below this, at the mouth of Cherry Run, commenced producing in 1860, and like those immediately above it, produced at a time when oil was worth least.

The number of wells sunk on this tract cannot now be ascertained, but must have been very great. Though but few of them were large enough to be singular, nearly all were remunerative.

There were also several refineries on the farm at one time, which are now abandoned.

The Buchanan Farm, situated on either side of Cherry Run, being mostly upland, neither of them were thoroughly developed until the speculative fever of 1864 and '65 brought them into the market as the basis of the formation of joint-stock companies, but the narrow flats along Oil Creek had previously been tested with considerable success.

The village of Rouseville, still a flourishing town, with a population of nearly three thousand, is situated partly on both of these farms.

Though both farms have been quite productive, and have had several large wells, the percentage of dry holes has been very great.

There were several smaller refineries on both those farms, and the number of stock companies was beyond all belief.

The Ham. McClintock Farm, containing three hundred

and fifty acres is situated at McClintock's Station on the
Oil Creek Railroad, and lies on both sides of the stream.
It is one of the first farms that came into the market as
oil territory—being in fact the second—for here were
found the surface indications which for the first year or
two were thought by most operators necessary to warrant
the sinking of a well.

For some years before the idea of adopting artesian
boring in the development of Petroleum had dawned upon
the world, the owner, Hamilton M'Clintock, had collected
oil from a spring that bubbled up in the middle of the
creek, and around which he built a crib in order to prevent the oil from being borne away on the current of the
water. By occasionally skimming the pool inside this
crib, and sometimes agitating the ground with a pick or
crowbar to the depth of a few inches, he collected several
barrels in the course of a year without giving himself
much trouble, and it was all disposed of with some profit
to the surrounding farmers, and was sent to Pittsburgh by
the lumbermen in the spring of the year. But though
fifty barrels *might* have been annually collected, it is
doubtful—reports to the contrary notwithstanding—if
ever more than five actually *were*. The third well to
produce oil on the creek was sunk in this crib for Brewer,
Watson and Co., by J. D. Angier. The town, which took
its name from the farm, and which was once a brisk, little
place, has been nearly obliterated.

The Clapp Farm, purchased of the original owner in
1859, by George H. Bissell and Arnold Plummer, was
thereafter at once thoroughly developed, and though there
was a large percentage of dry holes, the number of paying wells was also very great, and the fact that Mr. Bissel was at the same time conducting the largest barrel
factory in the oil region, enabled him often to ship
his oil to market when others either sold it to speculators

for ten cents a barrel, or let it run to waste because they could not afford to pay two, and two and a half dollars apiece for barrels to ship it in.

The total daily product of all the wells in June, 1860, was estimated at 200 barrels. By September, 1861, the daily production had reached 700 barrels, and then commenced the flowing well period, with an addition to the production of six or seven thousand barrels a day. The thing was monstrous, and could not be endured! The price fell to twenty cents a barrel, then to fifteen, and then to ten! Coopers would sell barrels for cash only, and refused to take their pay in oil, or in drafts on oil shipments. Soon it was impossible to obtain barrels on any terms, for all the coopers in the surrounding country could not make them as fast as the Empire well could fill them. Small-producing wells were forced to cease operations, and scores became disheartened and abandoned their wells. The production during the early part of 1863 was scarcely half that of the beginning of 1862, and that of 1864 was still less. In May, 1865, the production had declined to less than 4,000 barrels per day.

In the winter of 1864, and 1865, the "United States well," at Pit Hole, was struck, and flowed, as estimated, January 7, 1865, 650 barrels per day. By this well came the famous Pit Hole excitement, which must ever stand prominent in the history of the Oil Region of Pennsylvania. Pit Hole City was commenced May 24, 1865, and soon the town contained a population estimated at 8,000. (See the history of flowing wells.)

At one time Pit Hole City had fifty hotels; several of them were palatial in size, and truly gorgeous in their equipment. The cost of the Chase House, was over $80,000; the cost of the Morey and Bonta Houses, equally as large. It had miles of streets, lined with buildings, including banking-offices, school-houses, churches, an opera house, and other appendages of a first-class modern city.

Soon the production of oil at this point decreased to a mere nominal figure; fire swept away whole streets of the town; the signs of decay were evident. A general collapse took place, and piece by piece Pit Hole City was carted away and it is now but a mere " settlement " of a few hundred inhabitants.

Commencing at Titusville in 1859, the tide of development swept over the valley of Oil Creek, and along the Allegany river, above and below Oil City, for a considerable distance, then Cherry Run in 1864; then came Pit Hole Creek, Benninghoff and Pioneer Run—the Woods and Stevenson farms, on Oil Creek, in like succession, in 1865 and 1866. Tidioute and Triumph Hill, in 1867, and in the latter part of the same year came Shamburgh. In 1868 the Pleasantville oil field furnished the chief centre of excitement.

A lull now took place, to be followed by the developments on the McCray Hill, at Petroleum Centre, and on the Shaw farm near Rouseville, Foster farm and East Hickory Centre, and then the great " down the river," or lower oil field loomed up to become the principal centre of production at the present time. (See chapter on lower oil field.)

"THE SOUTHERN IMPROVEMENT COMPANY."

During 1872, unparalleled excitement prevailed in the petroleum business. This year witnessed the formation and collapse of the most formidable speculative combination ever formed for the purpose of controlling the markets and production of petroleum.

Early in the year " a ring " of railway officials and refiners, incorporated by the Legislature of Pennsylvania, and termed "The Southern Improvement Company," entered the field. This company possessed extraordinary powers by virtue of their charter, and forthwith made contracts with the principal railway companies, under which they were to receive enormous rebates. An advance of *one hundred per cent.* on all freight charges on crude and refined, was the result of this combination.

The storm of opposition which "The Southern Improvement Company" met with from the producers, forced it to dissolve, and its charter was abrogated.

Subsequently, about four-fifths of the refiners of the United States organized themselves into an association for the purpose of forcing a margin between crude and refined. The producers, in order to protect themselves against the exactions of the refiners, formed an " Agency' and a " Union," which embraced, probably, three-fourths of all the producers. The control of the production of Pennsylvania, and of the markets of the world, was given to the " Union," but after one or two vain attempts to regulate them, the producers came to the conclusion that it could not be done without the aid of the refiners. So a coalition was effected with the refiners. The contract between the two interests was a singular instrument. The producers agreed to stop a certain portion of the drilling and pumping wells, and the refiners agreed to keep the price of refined in New York from falling below twenty-seven cents per gallon, and take immediately 200,000 barrels of crude from the " Union " at twenty-five cents above the then market price. Neither of the contracting parties carried out their engagements, and the coalition was not long-lived.

Thus we have briefly sketched the discovery and development of this great industry to the present day. The discovery of Petroleum must be placed in the front rank of the important events of the present century, and no one doubts but it will give light to coming generations

and to nations yet unborn.   By the accident of its discovery it was found that the Creator had placed beneath
the crust of the earth, a reservoir of oil, giving as brilliant
a light as any discovered substance.   We have not attempted to picture the scenes of excitement of the early
days of the business—the anxious drillers, the smiling,
wealthy fortunates, the downcast, ruined unfortunates,
the busy teams conveying the barreled liquid to the
water, the oil-begrimed and mud-besmeared boats, the
eager barterer and the earnest seller.   The scenes of to-
day but mirror those of the days of the commencement
of petroleum mining.

THE OIL FIELDS OF AMERICA.

WEST VIRGINIA AND OHIO.

It is almost impossible to describe or rather define the limits of what is called the oil region of West Virginia. Streams which empty into the Ohio river, as far as fifty miles above Marietta, afford the usual surface indications of oil. One hundred and fifty miles below Marietta, the Ohio river touches the northeast corner of Kentucky, and, on the streams which empty into the Ohio at this point, oil is said to abound, and to have been discovered in paying quantities. From Fishing Creek, which empties into the Ohio, fifty miles above Marietta, there is a country running back from ten to thirty miles from both banks of the Ohio, that affords surface indications. Of course the existence of oil throughout all this vast region is doubtful.

On the Ohio side of the river are the counties of Washington, Athens, Morgan and Noble, in which oil has been drilled for and found. On the Virginia side, there are eight counties,—Tyler, Calhoun, Roane, Jackson, Kanawha, Wood, Wirt, Richie, and Pleasants,—in which oil has been found. These twelve counties embrace a territory which extends away from Marietta in every direction, and whose extent is from thirty-five to sixty miles. Great excitement prevailed on the discovery of oil in this region in 1860, and was followed by speculations in land and the formation of oil companies. As a specimen of the producing character of this oil territory in the early days of the business, we may instance the Dutton well, on Duck Creek, Ohio, which was struck in 1860, at a depth of fifty feet, and averaged, it is said, from seventy to one hundred barrels per day for a considerable time. Another, called the Steel well, on Duck Creek, produced some five hundred barrels per day for some weeks, and in 1865 was producing five barrels per day. This well was drilled to a depth of one hundred feet. In the Summer of 1864, a well called the Dixon, was sunk eight hundred feet, when it commenced flowing at the rate of thirty barrels per day. The Bull Creek region, on the Virginia side, had in 1865, some forty or fifty wells, the largest of which yielded sixty barrels per day.

In 1860, when the excitement was at its height, a crisis occurred. The price of petroleum suddenly went down, until the barrels, as they came from the hands of the cooper, were of more value than the oil that filled them. Two causes led to this—the world had not learned the uses of petroleum, and the early surface-wells, threw forth so many barrels of oil that the supply was larger than the demand, and the market became overstocked. This disheartened capitalists, and lands fell. Then came the war. Virginia seceded, and the line of the Ohio became contested ground. McClellan crossed, but his forces were too busy with the Baltimore and Ohio railroad to think of protecting the oil-hunters, then swarming along the Kanawha. Although there was no organized army of Confederates in Western Virginia, there was nevertheless a body of guerillas who were constantly harassing the country. The result was a panic. In a week the whole party left. The derrick stood. in the field over the half-drilled well, the oil gushed up and over-spread the ground, the houses were torn down for camp-fires, and the whole enterprise perished. As soon as the turmoil of war ceased the drill was again set in motion, and operations have continued with singularly uniform success up to the present time. The present centre of the oil-producing region of West Virginia is Volcano, formerly called White Oak, which consists of two narrow belts of land, only a few rods wide, running parallel with each other at a distance of about a stone's throw apart. Their direction the longest way is north, 10° east and south, 10° west. Here, within a distance of six miles in length by one mile in width, is embraced the whole of the West Virginia heavy oil-producing territory. Within this narrow limit is produced all the heavy lubricating oils known as the Globe, Peninsular, Grant, Hillsdale, Mount Farm, White Oak, Gales Fork, Volcanic, &c., which have become so well known in this country and in Europe. In one respect the geological formation here is remarkable. Upon either edge of these oil belts the rock, upon the surface, stands at an angle of about 80° to 90°, and is precisely similar in character to the rock found in the surrounding territory at a distance of six hundred feet below the surface. While the lower stratum of rock entirely surrounds the oil belt, it is missing under it, or rather, instead of lying six hundred feet below the surface, it here appears at the surface. The conclusion is irresistible that this belt rock once formed part and parcel of the lower stratum, and that at some period by some convulsion of nature it has been forced to the surface. We suppose it is to this circumstance that Volcano is indebted for its name. It is somewhat limited in numbers and territory, yet its productions aid not a little in making up the aggregate of the material wealth of the nation. There are at present quite a number of new wells being put down, both for heavy and light oils. The heavy oil is found at a depth of about 600 feet, while for the lighter oils it is necessary to go down to a sand rock 1200 feet below the surface. The wells yield, on an average, about three barrels per day.

The territory known as the Glantz tract, and owned by the Oil Run Petroleum Company at Volcano, is considered one of the best producing tracts in West Virginia. From twenty-three wells, during 1872, this company had an average production of 3,750 barrels per month, of all gravities ranging from 29° to 35°. The oldest well on the tract, the Moore well No. 1, produced alone 7,735 barrels of 35° gravity. The second well struck, the Shafer and Stein Well, has produced 2,748 barrels of oil, 29° gravity.

A recent writer at the Ohio oil field, and particularly of the

Cow Run region, says :—For the encouragement of oil producers, and men interested in the production of oil in Ohio, and more especially to show that Pennsylvania is not the only oil-producing territory, I present you the following figures and data of a two acre lease, situated at Cow Run, Washington Co., Ohio. Grecian Bend Company's Well, No. 1, struck oil in April, 1869, at a depth of 300 feet, and produced 966 barrels until January 1, 1870. Well No. 2 struck oil December 25, 1869, at a depth of 700 feet, and produced 21,985 barrels until January 1, 1870. Well No. 3 struck oil June, 1870, at a depth of 400 feet, and produced 1134 barrels until January 1, 1871. Besides this, the Company have paid the Transportation Company over 500 barrels charged for evaporation. This is the production of only one company. The School House Company have produced and sold, the past year, over $60,000 worth of oil. And Perkin's, Harvey & Co.'s wells have produced and sold, to August 1, 1870, $212,566.33 worth of oil. This is the production of only three companies, out of a host of good companies located at Cow Run.

The West Virginia and Ohio oil fields are justly celebrated for the production of lubricating oil, which is held in great estimation in England and on the continent of Europe.

The combined production of Ohio and West Virginia for 1872, was estimated at 320,000 barrels.

In connection with our sketch of the Oil Region of West Virginia and Ohio, we would offer some facts in relation to the discovery of what has been termed *Crystallized Petroleum.* In noticing it we simply give the reports of two eminent scientific men—Prof. Lesley and Prof. W. F. Roberts.

Professor Lesley's Report.

The following report is taken from the printed proceedings of the American Philosophical Society:

Professor J. P. Lesley communicated a notice of a remarkable coal mine or asphalt vein, cutting the horizontal coal measures of Ritchie county, West Virginia.

Mr. Lesley said, that through the kindness of R. H. Gratz, Esq , of Philadelphia, a descriptive letter and a map had been submitted to him, which exhibited geological facts of more than ordinary interest to those who are studying the origin of the rock oil deposits of the West.

The curious points of the case require careful investigation, but there seems to be no good reason to doubt the essential correctness of the statement.

The coal-beds of West Virginia pass horizontally through the prong-like ridges from valley to valley. Some of these ridges run as narrow on top and as regular as railroad embankments, for three or four miles, and in nearly straight lines, between equally straight vales terminating bowl-shaped against some cross ridge.

It is across such vales and dividing ridges, that the asphaltum vein of Ritchie county makes a straight course, "two thousand three hundred and twenty-three feet long, as at first measured, but since then traced in both directions still further, so that now it is known to extend more than two-thirds of a mile." Explorations beyond this line have failed to find it. Its outcrop, four feet ten inches thick, was discovered crossing a ravine fifty feet wide at the bottom, and rising on each side with slopes of nearly forty-five degrees. On one of these hill-sides at a height of ninety feet, the outcrop showed the same thickness, but at a height of one hundred and eighty-five feet, it was found to be but two feet six inches thick. It is not certain that this diminution is in a vertical direction; it may be lateral; for the slope between the ninety and the hundred and eighty-five feet levels is more gradual, especially upon the western side.

In the bottom of the ravine, a vertical shaft was sunk to a depth of thirty-four feet upon the vein, which continued uniformly four feet ten inches thick, the asphaltum being filled in pure and clear, without the least admixture of earthy or foreign ingredients, between the smooth and almost perfectly vertical walls of yellowish-greenish sandstone, lying in horizontal layers, through which this gash or fault was once no doubt an open fissure, communicating with some reservoir of coal oil which still, it may be, lies beneath it undisturbed. The most interesting part of the phenomenon for structural geologists is this gash.

The substance which fills this gash-fault in the coal measures of Northwestern Virginia, resembles the glossiest, fattest caking coals, and has a decidedly prismatic structure; breaks up into pencils, with flat, lustrous faces and sharp edges, but the faces not set at any fixed angles to each other, so that the effect upon the eye is rather that of a fibrous than of a prismatic structure. At the same time there is not the slightest appearance of layers, but the aspect of complete uniformity or homogeneity. Pieces are taken out, it seems, a foot in diameter; and that portion of one of these pieces which I have, shows a plain face on one side, as if it had encountered one of the walls, and is covered with a delicate film of a dead black substance like charcoal dust, which is probably the dust of the vein substance itself.

Pieces lying at the surface of the ground are said to yield as much oil as specimens taken out six or eight feet down. By the ordinary dry distillation the substance is reported to yield as much oil as the Albert coal. By a different process, the first and only trial, at which six hundred pounds in one charge were used, forty-four and a half gallons of superior oil were obtained. Retorts are now upon the ground.

Geological and Mineralogical Report of Prof. W. F. Roberts.

McFarland's Run is a noted locality in the great oil formation of West Virginia. A vertical crevice filled with crystallized or solidified petroleum in a direct line, is found crossing the deep-cut gorges of small streams and rising to the summits of the ridges bounding them.

In the month of June last I made a special visit to this part of the country for the express purpose of making a full and particular examination of this phenomenon, if I may so term it, in geology.

I travelled from Cairo station on the Parkersburg branch of the Baltimore and Ohio Railroad, over a road then in progress of grading by the Ritchie Coal Oil Company for a branch railroad to connect their property containing this solidified petroleum deposit with the main road, and during this journey, I could not detect anything remarkable or different in the general geological structure of the country to that shown in some of the other oil-producing sections in the West Virginia "oil belts," with the exception of an opening made on the line of the road on the Ritchie Coal Oil Company's lands near McFarland's Run, where there is a vein of a peculiar substance, resembling somewhat some of the most glossy kinds of bituminous coal. Having secured specimens, I continued round the point of the hill, and entered a deep-cut gorge formed by a small run, a branch of McFarland's, and at about half the distance from the head of the run, I reached a shaft sunk upon the line of a fissure, or crevice in the strata, in this peculiar kind of substance, of the same quality and characteristics of the specimen taken from the place above referred to. This crevice is a vertical one, four feet four inches wide, and the strata adjoining it on both sides is horizontal, a common micaceous sandstone, in their plys of a yellowish-green color, of the carboniferous formation.

The shaft I was informed was sunk thirty-four feet, and the crevice continued of the same width downward. It was perfectly filled with solidified petroleum. The course of the dyke or opening in the horizontal coal strata run in a course S. 75½ W. and to N. 75½ E. which I traced in both directions. I traced the opening which had been made in the line of this crevice up the steep-sided ridges and over their summits, and I found from the specimens visible at the several shafts that the solidified or crystallized petroleum rose to the surface, or nearly so, in all places. The west hill bounding the ravine where the dyke crossed over, I judged to be about three hundred feet above the level of the ravine where the deep pit was sunk. The east hill-side is about two hundred feet above the ravine. Developments of shafting have been made proving the continuation of this petroleum-filled crevice in solidified form more than one mile in a direct line, and bounded by a flat or horizontal formation of shales and sandstones of the middle carboniferous series, similar in all respects to other ridges in oil-producing sections in West Virginia. The walls of the crevice are perfectly smooth and regular, and exceedingly well defined.

The crystallized petroleum has a fibrous structure. It is very glossy in appearance, of the color of the purest specimens of richest and fattest bituminous gas coal. It melts under heat readily and runs like pitch. This peculiar mineral has been wrongly called "asphaltum." Its fracture, lustre, and general appearance are altogether foreign to the Albert coal, or to any other mineral of that class. By experiments made upon this crystallized petroleum, it has yielded from one hundred and forty to one hundred and sixty-nine gallons of oil to the ton.

Developments will prove the continuation of the crevice filled with the same material—the crystallized petroleum—into and through the properties I am reporting upon, and in consequence of its embracing within their boundary lines two deep-cut valleys and high ridges intervening, an immense quantity can be mined above water level, and one cannot put an estimate too high upon this property, containing as it does, this valuable mineral substance.

How deep this solidified material may continue down beneath the level of the valleys is not determined. The crevice may get much wider, and still be filled with this solid petroleum. One thing is however certain, that it has its source from some immense subterranean lake or large opening in the strata of the lower measures of liquid petroleum. The numerous gas and oil springs closely contiguous and ranging with this dyke show that there are beneath the surface large cavities filled with oil.

At the junction of the streams which meet in the southern part of this tract is excellent boring territory, room enough for a large number of oil wells. The geological structure of the strata shows great disturbance underneath the surface, and here may be seen the pure oil oozing out from the joints of the rocks, and gas springs bubbling up on the surface of the water, throwing off oil in rainbow-colored tints. The nature of the formation, the geological structure of the strata and the contour of the surface, as well as other indications, show that this tract of land is located in an exceedingly rich petroleum section of country, where proper developments should be prosecuted without delay. One thing more may with propriety be mentioned, that this solidified petroleum in all places where it has been shafted upon is free from any deleterious foreign substance. It is as pure as oil generally is found in the best oil-producing localities of West Virginia.

A few years ago thirty-two barrels of this mineral were sent north, and all of it was retorted, a large portion being put through on a commercial scale in the city of Brooklyn, and resulted as follows :

YIELD OF ONE TON.

Illuminating Gas,	7000 feet,	@	$2.00	$14.00
140 Gallons Oil,		@	.60	84.00
17 Bushels Coke,		@	:12	2.0
				100.04

The analysis of Professor R. Ogden Doremus gives the following result :

100 PARTS MINERAL DRIED AT 212° FAH.

Ash,... 2.15
Hydrogen,..8.45
Carbon,..75.96
Oxygen,... 12.75
Nitrogen,... .69

KENTUCKY AND TENNESSEE.

Along Boyd's Creek, Barren County, Ky., three miles from Glasgow, are the best oil wells of this state. Glasgow is a town of three thousand inhabitants, situated on a branch of the Louisville and Nashville Railroad, and one hundred miles from Louisville, and eighty from Nashville, Tennessee. At the present time there are fifteen wells in operation, producing oil. The Armell wells in this district are drilled to the depth of 130 feet, three of which are flowing about three barrels each per day. Two of these wells have flowed for six years, and at one time produced 400 barrels per day. Chess, Corley & Co., oil refiners at Louisville, who own a large scope of territory in this section, have met with good success as oil operators, having thus realized a large fortune.

Their Well No. 1 is producing 35 barrels per day. At first it produced 70, and has now been pumping three years. A Pittsburgh Company own some land here, on which they have put down a number of wells, producing on the average $3\frac{1}{2}$ barrels per day.

The pioneer oil men of this region are Messrs. Chess, Corley & Co. and Messrs. Graham & Thomas. Both these firms commenced operations early in 1865.

During 1865 and 1866 considerable interest was manifest on Boyd's Creek, and many wells were put down, all of which pumped or flowed more or less oil. With the increased production, the price fell from $1.50 to 25 cents per barrel, at the wells. The oil was 40° gravity, and tainted with sulphur. With the limited knowledge of refining at that time, it was difficult to deodorize it. Operators became discouraged, and the territory fell at once. Some wells, within a few days of completion, were abandoned, and the oil adventurers went back to their various homes, many discouraged, others hopeful that when oil would become scarce, and science overcome the deodorizing difficulty, all would again be prosperous. In 1869 active operations again commenced, and with improvements in the refining process, the prices rose from 25 cents to $2.00 per barrel.

There was quite an excitement in 1867 near Burksville, Cumberland county, Kentucky, which was occasioned by the striking of the Crocus well at a depth of 300 feet, which flowed 300 barrels heavy oil per day, after discharging salt water for three months, but this promising field was abandoned by reason of the extreme cost and difficulty of shipment and consequent low price of oil. There is one well in this section, which is only 80 feet in depth, and has flowed six dollar lubricating oil, for the last six years. The oil from this well is teamed 40 miles to the river, and shipped down the Cumberland to Nashville, Tennessee. Navigation on the river is only open five months in the year. In Cumberland county, on one of the tributaries of the Cumberland river, a well exists which was put down to a depth of 400 feet, and has flowed about 60 barrels per day into the stream, for upwards of four years. The oil from this well is quite black, very heavy, and of a rank smell. Along Scrub, Indian and Greasy creeks, on the Cumberland river, there are many oil springs, which produce from one to five barrels per day of surface oil. These wells or pits are dug down to the blue clay to the bed rock. Cumberland county is the foot of the mountain range, and is hilly and rocky. Burksville, the county seat, is connected with a railroad by a forty mile stage route.

Near Bowling Green, Ky., a well was sunk to the depth of 80 feet, which produced oil in considerable quantities, for over twelve months. The owner built a small refinery for the manufacture of his own production. This was the only well drilled in this section. Within a few miles of the Mammoth Cave, near Green river, there are several oil springs. The surface rock here is so impregnated with oil and gas, that a match will ignite them. At Boston station there is a gas well, some 90 feet in depth, which produces a considerable quantity of gas.

There is little doubt but the north-western part of Kentucky is rich in coal and oil. We know that one well was drilled in Henderson county, near the Ohio river, which passed through a four-foot vein of coal, at twelve feet from the surface; and another six-foot vein, at 50 feet from the surface, and at 450 feet a small well of the best lubricating oil *ever* produced was discovered.

It is the opinion of practical oil men who have visited Kentucky, that the best oil lands have not yet been touched, and that all that is required to make this rich mineral State prosperous, is capital and energy.

Little is known of the producing capabilities of Tennessee, but it bears the evidence of being an important field of enterprise for the production of oil. At White Bluff, Dixon county, thirty miles west of Nashville, there is a small green oil well of the same character and quality as that of Butler county, Pennsylvania. This well is 500 feet in depth.

Mr. Eugene Scott, of Karns City, Butler Co., Penna., informs us that he visited the wells on Boyd's creek, late in the fall of 1872. He says, "I was astonished to learn that in this enlightened age of petroleum mining, the people here knew so little about the business. The derricks were only 30 to 45 feet in height; some with one bull wheel, and some with two. The whole rig is of white wood. * * * * * * They dressed their bits in the shape of a wedge, and reamers, why, they are beyond description. Drilling is only done in the day time, and it is usual to take six weeks to get down 150 feet. The seed-bag is put on the tubing, which is frequently drawn in testing. The oil-bearing rock is a kind of shell and sand-rock mixed, and of a white color. The stream (Boyd's creek) had been tested for three miles north and south, and finding the best wells in the centre of the line of tests, and that the hills had not been operated on, I concluded that the belt crossed the creek, and ran parallel with the Pennsylvania belt, therefore I took a large lease southwest of the best producing well on the creek, and at the depth of 250 feet struck 28 feet of loose sand, full of oil and salt water. Two miles northeast of this, and on a line with the best wells on the creek, my partner drilled a well with a Pennsylvania oil rig, and in six days reached the sand-rock, but there was little oil; the well produced only one barrel per day."

"It is usual when a well is dug in this country, to pump it for a few days, and then tube with three-quarter inch pipe, and let it flow—then remove the machinery, and drill another well. The bed rock of the Barren County oil field is white limestone, which in the valley is about eight feet from the surface. The surface rocks are usually flat, and so porous that water will run through them readily.

We are furnished by the gentleman above named with the following well record:

30 feet, white limestone;
40 " shale;
60 " limestone;
20 " oil-bearing rock, a kind of shell and sand rock, mixed, and of a white color.

OIL REGION OF INDIANA.

In the western part of Crawford County, Ind., there is an oil region that has never been properly tested by that only sure test—"the drill." The surface indications of oil extend five miles in width by over ten miles in length, and consist of a tar spring, oil springs, and oil rock, of several varieties and in great abundance. The tar spring has been known ever since the county was settled, and is nine miles from Leavenworth; it is about half way up a large hill which is probably one hundred and fifty feet high; it flows after each heavy rain, and in the course of a year throws out tons of tar or asphaltum. It is between the Otter fork and the West fork of Little Blue; the nearest well to it is the Dexter well, which is two miles distant. The oil springs are found on Otter fork, and West fork, and on hollows tributary to them, there are quite a number of them from which small quantities of oil can be collected. The oil rock is found in great abundance. At one place on Otter fork, the bed-rock of the creek is a soft, black sandstone, and contains over thirty per cent. of oil If a piece of it is put into a fire it blazes like a candle, and will continue to burn till it loses one-third of its weight. In the vicinity of the Dexter well on the

West fork, the oil rock is in great abundance, and in almost every hollow—and their name is legion—in all that locality, rock more or less impregnated with oil may be found.

The Wells.—In 1861 a well was sunk to the depth of ninety-seven feet, by a man named Custerman. He obtained no show of oil, and never pumped the well. The war caused him to stop work, and he has never been heard of in that locality since. In 1863–4 several wells were sunk—three on the west fork. The Dexter well is 580 feet deep, and has a fine show of oil; it rises constantly in the conductor; a quart can be taken off every morning. It is a lubricating oil of good quality. It now belongs to the Dexter Oil and Salt Company, and they intend to sink it deeper and pump it this year. The Clark well, located three miles below, is 640 feet deep; found no oil; a little gas and saltish water. The well has been left open and is filled with sediment. The Eaton Sulphur Well, three miles further down the creek, is only 275 feet deep; flows white sulphur water. It is becoming a great resort for invalids.

These are all the wells that have ever been sunk on the West fork. On the Otter fork, there have been five wells sunk. A small show of oil was found in two of them. The Golden Salt well is one of them, and is over 1,000 feet deep. It is a good salt well. Two of the other wells are 700 feet deep. All of these wells are below the indications. No well has ever been sunk on either creek above the indications except the one mentioned.

The Geological Position.—The locality is in the sandstone which underlies the great Indiana coal field; the first strata below is the carboniferous limestone, which is about 800 feet thick; the next strata is known as the knob sandstone, and is about 400 feet thick; the next is a black shale or slate, 100 feet thick, and is the last strata through which they passed at Terre Haute, in sinking their well, which is the only successful oil well yet sunk in the State. It is 1,625 feet deep.

"The Crawford County Petroleum and Mining Company" are now sinking a well at Leavenworth. It is to be sunk through the black slate penetrated at Terre Haute, which will be a depth of about 1,300 feet.

The marked difference in the geological formation in the Indiana region and those of Pennsylvania and West Virginia is in the limestone, and consequently wells will have to be deeper here to reach the main source of the oil, which lies below it. While the thick strata of limestone will make it expensive to sink wells, it is a good evidence that the oil exists in abundant quantities, or it would not rise so far through it and make so great and extensive surface show.

The Louisville, New Albany and St. Louis Railway passes through the center of Crawford County.

CALIFORNIA.

The discovery of petroleum in California adds another to the already widely varied products of that State. It has been known for a long time that deposits of petroleum existed near the coast, but as yet the production of this oil has been unimportant. The Leaming Petroleum Company, recently organized in San Francisco, has fairly commenced operations, with results which augur abundant success. The crude oil is found in the mountains of the San Fernando District, thirty-five miles north-west from Los Angeles, from which point there is railroad communication of thirty-one miles with the Port of San Pedro. The cost of transportation from the mines is $2\frac{1}{2}$ cents the gallon, and the cost of refining 2 cents per gallon for small quantities, and $1\frac{1}{2}$ cents for large lots. Up to the present time several shipments of crude oil have been received at San Francisco, aggregating some hundreds of barrels, which have found immediate purchasers at remunerative rates. The Gas Companies are prepared to take and use not less than 1,000 barrels per day, and the district owned by the company is sufficiently extensive and prolific to warrant the delivery of that or even a greater amount, so soon as the requisites can be obtained to furnish suitable works. These shipments have, so far, been the products of natural springs from which the oil was dipped with buckets; one spring is now producing two barrels per day; but when a system of scientific operations shall have been introduced, the yield will be greatly augmented. It is the opinion of capable gentlemen who have carefully examined the mines, that they are of sufficient capacity to furnish nearly all the oil required for consumption on the Pacific coast.

THE OIL FIELDS OF CANADA.

THE CANADA OIL REGIONS.

THE first discovery of Petroleum oil in Canada was made in the township of Enniskillen, in the County of Lamberton, in the extreme western section of the province of Ontario.

Among the settlers on the almost barren and unproductive soil of this section of the county, was a lean, swaggy Down-Easter named Shaw, who had emigrated thither from the State of Massachusetts about the year 1836. Shaw is represented as being luckless, thriftless, and poverty-stricken to the last degree. He had ever been considered a visionary, a schemer of impracticable projects, and many were the undertakings which he broached to his friends and neighbors, only to be laughed at. In the years 1855–6, he set the seal to his reputation as a demented visionary, by his vain endeavors to get up a stock company for the purpose of boring down through the swamps, with a view to extracting from the bowels of the earth a substance which he averred existed there in exhaustless quantities. He waxed eloquent on this theme, and declared his conviction that beneath the barren marshes, a source of untold wealth lay hidden. By these and similar wild representations, Mr. Shaw ere long came to be regarded as a lunatic. He was wondered at by some, abused by others, and laughed at for a fool by all.

It must be premised that ever since the original settlement of Enniskillen, a dark, oily substance had been observed by the settlers floating on the surface of the water in the creeks and swamps. Good water was not to be had. No matter how deep the wells were dug, the water was brackish and ill-smelling, and in some localities totally unfit for use: while a surface of black, oily slime, frequently arose an inch thick, as cream rises on new milk. Here and there, in the forest, the ground consisted of a gummy, odoriferous, tar-colored mud, of the consistence of putty. These places were known by the name of "gum-beds," and in two or three instances were of considerable extent. At the present day such "surface indications" as these would reveal the secret to the most casual observer; but to the primitive, untutored minds of the then residents of the township they failed to convey any other impression than that their lines had not fallen in very pleasant places.

Finding it impossible to arouse any enthusiasm, or get up any agitation on the subject, Shaw quietly subsided into his normal condition of mental torpidity; in which condition he remained until the summer of the year 1857, when he was seized by another of his intermittent attacks of enterprise, and this time he determined to commence operations on his own account. He inaugurated proceedings by digging an ordinary well, as if for water, in a hollow about a hundred yards from his dwelling. He worked away at intervals for some days without attracting any notice, and without assistance from any one. Finally, he was one day aroused while eating his dinner by a loud explosion, accompanied by what seemed to be the shock of an earthquake. Upon running out to ascertain the nature of the disturbance, he perceived a huge fountain of what seemed to be black mud bursting with great violence from the hole where he had been digging. The "mud" emitted a very offensive smell. The "jet," when he first cast eyes upon it, was, as nearly as he could judge, about a foot in diameter, and it every moment increased in volume, frequently shooting high up into the air. As a necessary consequence, the ground was soon flooded, and had not his dwelling been built on an elevation, considerably higher than the mouth of the well, it would have been partially submerged.

Upon examination, the substance thus ejected proved to be crude petroleum. The well continued to flow, with occasional brief cessations, for upwards of sixty-seven hours, and this in a large and swift stream, which poured into the adjoining creek, and the contents were thus carried away and lost. The neighbors for miles around came and lent their assistance; but of course, owing to the want of tanks and barrels, only a very inconsiderable quantity could be preserved.

The neighboring proprietors followed Mr. Shaw's example, and leased portions of their lands to the highest bidders. In an inconceivably short space of time, enterprising operators from all parts of Canada and the adjoining republic began to pour in. The needy denizens of the Enniskillen swamps began to realize the fact that their slimy morasses might be confidently counted on to yield a revenue such as could never be hoped for from the richest and best cultivated agricultural soil in America. Land changed hands rapidly, and from one thousand to two thousand five hundred dollars an acre was paid for territory which, previous to the great discovery, could hardly have been given away. The more common method resorted to, however, was not to transfer the fee-simple of the lands, but to grant "oil-leases" thereof, whereby the lessee acquired the exclusive right to operate for oil upon the land for a specified number of years, in consideration of which privilege, a certain sum of money—usually several hundreds of dollars per acre—was paid down to the lessor, who, by the terms of the instrument, was further to receive a percentage—generally one-third—of the oil produced; called a "royalty." Stores, taverns and dwelling-houses sprang up all around with marvellous celerity, forming a village, to which the name of Oil Springs was given. Postal and telegraphic communication was shortly afterwards granted, passable roads were constructed, connecting the region with civilization, and for some time all went merry as a marriage bell. The supply of oil was supposed to be almost inexhaustible; but it was soon found necessary to spend much more time, and to incur much greater expense, than Shaw and his immediate successors had been compelled to do. In other words, the wells would not flow, but had to be pumped, and it became necessary not only to sink a surface-well down to the rock, but to bore, by means of the drill, through the rock, until the vein was reached; whereupon oil was said to be "struck." Frequently a vein of water, instead of oil, would be struck, which had to be exhausted before operations could be proceeded with, and this sometimes occasioned great delay. Meanwhile, Mr. Shaw was regarded as a public benefactor, and received as great credit as if he had discovered the philosopher's stone. He found himself in the possession of enormous wealth, arising partly from the lease of his lands, and the royalties therefrom; but chiefly from the product of his wells, the first of which continued to yield from three hundred to six hundred barrels daily. The market fluctuated considerably; but oil seldom brought less than six dollars per barrel at the well, and was generally much higher. For two days it reached the enormous figure of sixteen dollars and twenty-five cents.

Mr. Shaw, through injudicious speculation, spent all his splendid fortune. He came to the Pennsylvania Oil Region in 1868, and was employed as a common day-laborer at Titusville, up to 1870, at which place he died broken-hearted and quite unknown.

About thirty miles to the south-east of Oil Springs, near the village of Bothwell, another large vein of oil was struck, and ere long, a busy community sprang up there, scores of wells were put down, and fortunes made and lost. Oil was soon after discovered five miles north of Oil Springs, to which place was given the suggestive name of Petrolia.

In the year 1864, Oil Springs contained a population of more than three thousand inhabitants, several spacious and well-conducted hotels, and at least a dozen or more places of entertainment, which could lay no claim to remarkable distinction. It moreover contained two private banking-houses, a Board of Trade, a printing-office and weekly newspaper, and two hundred and forty-seven wells, all in active operation, yielding a large number of barrels per diem, and more were being put down. Money was plentiful, and gambling was practised on a scale which, for Canada, might well be pronounced gigantic. Every night, Sunday night *not* excepted, was consecrated to the unholy rites of "poker" and "euchre." Dancing assemblies were formed, tri-weekly fandangoes were held, and the whole place presented the appearance of a California in miniature. But the end was not very far distant. Already the deposit had indicated tolerably palpable symptoms of exhaustion. The Shaw well, after having yielded a sum total of thirty thousand barrels—and this in addition to the immense quantity which was lost as already described—suddenly collapsed, and refused to yield another drop. The Twenty Friends well, which was second only in importance to Shaw's, followed the example of its predecessor; and a gradual falling off was perceptible throughout the entire district. Towards the end of the year, Hendrick's well revived the hopes of the operators somewhat by a daily yield of seven hundred barrels for about a week; but in the course of a fortnight these seven hundred barrels were reduced to from thirty to forty, and shortly afterwards to fifteen. Operators could no longer count with certainty upon striking a vein of oil wherever they chose to bore, and many wells had to be abandoned as "dry holes." Many even of the yielding wells did not produce oil in paying quantities. About this time, charlatans professing to be endowed with the mystery of the "Divining Rod" made their appearance on the scene; and, though laughed at and condemned by nine-tenths of the operators, they contrived to realize something more than a good livelihood out of the other tenth. Success generally followed their predictions, even when dry holes were becoming alarmingly numerous in the district; but, of course, the secret lay in their former extensive experience of surface indications in Pennsylvania. They could form something more than a plausible guess as to where oil was to be found, but their science gave them no information as to the quantity; and the wells sunk by their directions generally exhausted themselves in a day or two from the time the vein was struck.

Though Western Pennsylvania has produced numerous flowing wells of wonderful capacity, there is no quarter of the world where the production attained such prodigious dimensions, as in 1862, on Black Creek, in the township of Enniskillen. The first flowing well was struck there on January 11, 1862, and before October not less than thirty-five wells had commenced to drain a store-house, which provident nature had occupied untold thousands of years in filling for the uses—not for the amusement—of man. There was no use for the oil at that time. The price had fallen to ten cents per barrel. The unsophisticated settlers of that wild and wooded region seemed inspired by an infatuation. Without an object save the gratification of their curiosity at the unwonted sight of a combustible fluid pouring out of the bosom of the earth, they seemed to vie with each other in plying their hastily and rudely erected "spring poles" to work the drill, that was almost sure to burst at a depth of a hundred feet, into a prison of petroleum. Some of these wells flowed three hundred, and six hundred barrels per day. Others flowed a thousand, two thousand, and three thousand barrels per day. Three flowed, severally, six thousand barrels per day; and the "Black and Matthewson" well flowed seven thousand five hundred barrels per day. Three years later that oil would have brought ten dollars per barrel in gold. Now, its escape was the mere pastime of full-grown boys. It floated on the waters of Black Creek to the depth of six inches, and formed a film on the surface of Lake Erie. At length the stream of oil became ignited, and the column of flame raged down the winding of the creek in a style of such fearful grandeur as to admonish the Canadian squatter of the danger, no less than the inutility of his oleaginous pastimes. From detailed determinations, Professor Winchell says, " I have ascertained that, during the spring and summer of 1862, not less than five million barrels of oil floated off on the water of Black Creek—a national fortune totally wasted."

The extent of the field is very great if we include every place where oil is found. Reckoning in that way, a district of two hundred square miles would not more than cover it. Most of these places, however, present only oozings from limestone rocks, such as occur in various other places in the world unconnected with qualities of any practical value.

The region of value, as developed by actual borings to this time, and the existence of the proper rock, is confined to the western part of the Dominion, and extends from near Lake Erie to Lake Huron, and from the St. Clair river eastward seventy-five miles or more; thus being about fifty miles north and south by, say, one hundred east and west. Its outside lines are somewhat irregular, but such is the general range of it. The part of practical value is, thus far, limited to the "Carniferous Limestone" (so-called from containing nodules of flint resembling a harp), and which is mostly confined to the southern part of Canada.

Within this range Petrolia, Bothwell and Oil Springs have produced nearly all the oil. The latter had the largest wells, though the former now produces more than nine-tenths of present amount. Petrolia is about sixteen miles southeast of the outlet of Lake Huron; Oil Springs seven miles south of that, and Bothwell, about thirty-five miles from that.

The surface of the country above described is nearly level, except where the ground descends to the bed of streams, and being largely covered with a dense forest of hard-wood trees, is often wet and muddy, and, in a rainy season, would wonderfully help a traveller to appreciate the "slough of despond." Time, muscle and money will however cure this, and while the timber now keeps out the drying process of the sun, it is furnishing most valuable fuel for the oil wells.

Western Canada has no coal. The land descends gently to

the southwest, and the general dip of all the rock formations is westerly.

In drilling, the first thing found is a yellow clay, and sometimes sandy soil, five to fifteen feet deep. Next, a compact blue clay of even consistency and appearance, from 50 to 100 feet deep. This rests on a thin shell of limestone, resembling a stalactite formation, which seems to have been crystalized out of the water as it drained from the clay into the next bed below, which is composed of gravel from two to eight feet thick.

Next comes a slate rock (called Hamilton shale), usually fifteen feet thick, in the region of good wells, and thinning out to nothing eastward. The surface wells, formerly so productive, were found in this gravel, held down by the clay and thin limestone; and when, in digging, the pick broke through this thin shell, the oil and gas would rush up so rapidly the laborer would often be compelled literally to flee for his life.

Next below the shale lies the carniferous limestone already referred to, the upper layer of which is about forty feet thick. This alternates with thirty to fifty feet of slate; then comes about forty feet more of limestone, then a similar amount of slate again, and then (being now at the depth of about 250 feet from the surface), is again found the limestone, which continues 250 feet more, making a total of about 500 feet. All the oil is found within that range, being regularly in veins in the limestone, and the deeper veins usually the larger. All the rocks below the clay are more or less saturated with it.

Next below the oil-bearing limestone, is a stratum of hard blue limestone, averaging four feet thick, and immediately underneath that a vein of salt water apparently inexhaustible.

At this point commences, with few exceptions in the oil regions, what is called the Onondaga salt group, which is a formation of unknown thickness, and in which is found the salt of Syracuse, N. Y., and also of Goderich on Lake Huron, at a depth of 1,100 feet. It has been penetrated five hundred feet in several places near Petrolia without producing a barrel of oil.

To prevent mistake we ought to say that the salt of Western Virginia is taken from quite another formation, lying geologically much higher, and coming very near the coal, being the "conglomerate" of the coal measures.

In some places, as near Oil Springs, the upper strata seem thinner, and the lower thicker, thus bringing the oil-bearing part nearer the surface, giving rise to large surface wells, and also to the gumbeds there, which resemble the residuum after the oil has evaporated, or been drawn off, leaving a black deposit like asphalt covering, in one or two instances, a hundred acres.

The oil-bearing limestone varies from a close compact structure to quite open, and these characteristics are presented often in bands or belts, and the best wells are expected where the rock is most open. In all its parts it is largely composed of marine shells and other fossils peculiar to that geological horizon. This seems to have strengthened the theory of the animal origin of the oil.

The thought is quite romantic—perhaps poetic—that the little animals which occupied these shells ages before men appeared, unwilling to be of no use in the future, built up the rocks out of the ruins of their dwellings, and then, by some subtle chemistry, allowed their substance to be converted into oil to fill them, and thus, with true charity, even "gave their bodies to be burned." Other facts, however, point to other causes of the oil, at least in a majority of cases.

The rocks of Canada, it will be seen, differ widely from those of Venango county, Pa., these being almost exclusively limestone, and others sandstone, alternating with slate.

The odor of the oil is rather unpleasant, and this arises from the sulphur and other substances often found in limestone. Sulphuric acid occurs occasionally in the water, corroding tools and tubing, and sulphuret of iron is found in the rock. Its decompositions and recompositions are taken up by the oil. An English deodorizing process, called Allen's, is now effectually used. The gravity of the oil varies from 33° to 43° Beaume. Refining produces about 80

per cent. of illuminating fluid, with less benzine and more tar than
Pennsylvania oil. The color of the crude is dark green, shading into
black. Very little paraffine is deposited in pumping, and benzine
in wells is rarely used. Casing six inches in diameter is put down
to 280 feet, or say 30 feet into the hard limestone, which shuts off
fresh water and prevents the soft shale rock from caving. Tankage of the oil is accomplished in part by the use of wooden tanks,
but mostly by making use of the blue clay above described. It is
found to be almost impervious to water and quite so to oil. Excavations are made in it from ten to twenty feet in diameter, and
sometimes seventy-five feet in depth. Curbing is used to prevent
the possibility of caving. The top of the excavation is planked
and covered with earth. Here is stored the oil, both crude and refined, free from waste and safe from danger until wanted, when it is
pumped out, sometimes by engines and sometimes by spring-holes.
This blue clay, which the farmers in some parts of Northern Ohio
and elsewhere regard with exceeding dislike, is here one of the
best friends of the oil-producer.

Exhaustion of the oil is not to be anticipated for several generations. Enough is produced for the present wants of the Dominion,
and as Canada develops and her population increases upon the surface, the regions below will respond to their wants. Nature does
not display all her treasures at once, but opens one storehouse after
another as man's needs may require.

Glaciers, it is quite evident, once moved over that country, for
whenever excavations are made down to the solid rock, scratches
and grooves are found, varying from mere lines to the size of fifteen inches, and nearly all running in a uniform course of north-
east to south-west. They dip up under Lake Erie and appear
along its southern shore, where they have been seen often, even larger
than any found in Canada. Icebergs have also floated over the
same sections, and evidences of both are abundant from New England to the Mississippi and beyond, and from Lake Superior nearly
to the Southern states.

By these and similar means were many of the Pennsylvania hills torn down and valleys formed, and from these sources came most of the soil, gravel and boulders scattered along the Oil Creek country. Their home was in the North. It seems wonderful that in all the tumults, earthquakes and upheavals of the past, nothing has been permitted to disturb the oil, though it has been left comparatively near the surface and easily accessible to man.

FOREIGN OIL FIELDS.

SOUTH AMERICAN OIL FIELDS.

Much attention has of late been directed to this field. Peru, Ecuador, Bolivia and Chili, offer inducements to speculators in search for oil, but only in the two first-named countries are the prospects of such a character as to attract foreign capital for the present. The existence of oil in Peru and Ecuador has been known as far back as the tradition of these countries reached, but to a certainty it has been made use of for at least two hundred and fifty years, by the Spaniards (the first conquerors of the Peruvian empire, which included Ecuador,) for making pitch. The oil was collected in a manner similar to that employed by the Seneca Indians, to obtain the petroleum of Oil Creek ; viz: shallow pits were dug, and the oil which collected, was skimmed from the surface, and was then allowed to evaporate under the heat of a tropical sun till it became of a thick, glutinous consistence, when it was removed and boiled down to a hard pitch. This pitch was used for coating the inside of earthenware, and particularly *Aguardienti* or liquor jars. The Spanish government long held a monopoly of this trade, which yielded an annual profit of $35,000. That portion of the South American oil field lying in the neighborhood of the town of Payta, on the river Achira, was purchased by a Mr. Lama in the year 1830, who worked the mines or pits after the primitive mode. In 1868, a Mr. Blanchard C. Dean in prospecting along the coast, discovered the works of the Lamas, and proposed a partnership agreement with them, and a joint prosecution of the work. They refused. He then "denounced" * a mine according to the old Spanish mining laws, which resulted in a law-suit. Mr. Rollin Thorne, a resident of Lima, assisted Blanchard in the litigation and won the suit, and possession of the land, which in extent is thirty-one miles in length and six in width.

Within the past year the courts of Peru have decided that petroleum is not denounceable. Happily for the Messrs. Thorne, who obtained their best oil territory by this means, the Peruvian law will not allow a decision already made to be affected by any subsequent one.

The Peruvian Oil Field is a belt on the west coast of South America, running along the thirtieth parallel of longitude, between Point Aguja on the south, (or needle point) and the town of Tumbez, on the north, and about seven degrees south of the equator. The known distance is about two hundred and fifty-one miles, running north and south along the Pacific coast, and about one hundred and fifty miles inland to the Andes. It is a singular coincidence that the oil belt corresponds with the Oil Region of Pennsylvania, as both are intersected by the 80th degree of longitude. The topographical structure of the Peruvian territory is broken and mountainous, and has evidently been subjected to volcanic action. It is also worthy of note that the mountain range of the Andes to the east, con-

* The proceeding is this. Any person who may have discovered a mine or vein of any metals, or as the Peruvian mining laws say, *juices of the earth*, can present himself before the Mining Tribunal and demand possession of said vein. The discoverer or claimant is called the *denouncer*. The Tribunal then awards him the possession of a piece of ground containing the vein, two hundred yards long and of the same width. Within ninety days he is obliged to commence work. He must dig a pit at least ten feet deep, by the same in length and width. When this is completed his next duty is to publish in the nearest daily papers, for thirty days in succession, calling and asking if there be any person, or persons, who can show a better right to the discovery. Within these thirty days, he must also post handbills in the village nearest the mine, and within the jurisdiction of the mining tribunal making his award, bearing the same import.

If no opposition is made or sustained, then the Tribunal decrees the ownership and title in the "*denouncer*," and that the former owner of the ground shall receive from the denouncer payment therefor at a price reckoned *without* the added value of the mine.

tains large deposits of anthracite coal, and that this coal deposit is
about the same distance from the oil field of Peru as our Allegany
anthracite deposit is from the Pennsylvania Oil Region. A coating
of sand about eight inches thick covers the entire surface of the
Peruvian oil field. A fossiliferous deposit of marine remains is
found on the surrounding hills, from 250 to 300 feet above the
level of the sea. The same deposits are also found on the bottom
lands, which proves that this part of the continent has at some period
of time been covered by the sea. The oil belt appears to belong to
the tertiary formation. The outcroppings of sand-rock are to be
met with everywhere, as we find them in Pennsylvania. Shale
exists below the top coating of sand, and is completely saturated
with oil; which, to say the least, is a good surface indication. The
shale varies from thirty to forty feet in thickness. These strata of
shale are interlined with thin lamina of bluish grey sand-rock, of
a fine texture. The first sand-rock is to be found at a depth of
from 130 to 132 feet, and is from five to ten feet thick. The
second sand-rock is found at 300 feet, and the third has not yet
been reached, as no well has been sunk to a greater depth than 350
feet.

In 1871, Messrs. Rollin, Thorne & Co. commenced drilling with
a common *pod-auger;* three wells were drilled and oil was found in
very considerable quantities at very shallow depths, viz., 226, 56
and 38 feet. Other wells followed with good results. In July,
of last year, a new well was commenced at Point Pavinas and
drilled to a depth of 351 feet, or 51 feet in the second sand, through
which the tools have never yet penetrated. No crevice was perceptible to the driller, as a volume of gas and oil at this depth was
suddenly ejected from the well, compelling the abandonment of the
drilling. The well was finally tubed, and both valves and sucker-
rods put in, to diminish the flow, which was calculated at 1,000
barrels per day. The well is now flowing 250 barrels per day
with the lower valve in. From all the wells struck on this territory
great volumes of gas continually issue. The gravity of the oil is

from 40 to 50 degrees Beaume. In color it is a little darker than Pennsylvania oil, and it has the same odor as an oil of the same gravity. It yields from 70 to 75 per cent. of 110 fire-test kerosene when refined, and is a superior article.

A contract has been entered into with Messrs. R. Thorne & Co. by an Anglo-Peruvian firm for two hundred barrels crude oil per day at $5 gold, extending over a long period. This firm is now building a refinery of two hundred barrels capacity at the point of operation.

In the year 1864, Messrs, G. H. Bissell and James Bishop, of the city of New York, leased of Don Diego Lama, his estate of Prancora, consisting of 4,500,000 acres. A company called the Peruvian Oil Company was formed, with capital of $5,000,000. Operations were soon commenced by the company, on the northern portion of the tract at Zorritos, twenty miles south of the Tumbez river, immediately on the coast.

This company has put down a number of wells, with good success. In the early part of 1868, the company struck a well which produced 300 barrels per day for nearly a year, when it *caved in*, and the production ceased. The company refine their own oil, and find a ready market for this product on the Pacific Coast, Australia, and New Zealand. The President of the company is Mr. Geo. H. Bissell.

In one well on the company's lands, the following is the order in which the rocks were found:

> Soapstone and slate,
> Sandrock and slate,
> Conglomerate limestone,
> Hydrate of iron, or reddle,
> Cretaceous sandstone,
> Carboniferous slate,
> Gravel-pebble, in which the oil was found.

Oil was met at 18 feet.

ECUADOR.

The oil fields of Santa Elena, in the Republic of Ecuador, is spoken of in the following terms by Raymond De Peiger, Engineer and Geologist to the Government of Ecuador, in his report to the President of the Republic. He says: "Petroleum is to be found in the country in very large quantities. On a surface of about four square leagues from the sulphureous spring of San Vicente to the sea shore, wells have been sunk, and the bituminous matter obtained in a liquid state. Its consistence is not the same in the different wells. In some of them it is fluid, like whale oil; in others, it has the consistence of butter at ordinary temperature. At the surface, or upper part of many wells, it can be seen in hard compact masses, which probably have been formed by the evaporation of the liquid. This oil has a dark-brownish color, which gets darker with the greater consistence of the oil. In one place where it oozes from the bed of a dried stream, the bituminous matter has a greenish color.

"Its smell is not disagreeable, which is generally the case with many of the American, and especially the Canadian oils. As the inhabitants have neither the knowledge nor the implements required, the works are very rude. Pits from ten to twelve feet deep are dug into the sand till clay is reached, and when the oil, which oozes from all sides, has filled them, it is dipped out.

" Near the wells are primitive furnaces, built with sun-dried clay, on which are open iron boilers. The bituminous matter is thrown into these vases and cooked until all the volatile products disappear, and leave a thick pitch.

" In * * * Santa Elena it is not admissible to suppose that petroleum has been formed in the upper sandy deposits. Its presence there can only be explained by the escape of the bituminous matter from the fissures in which it was contained. * * * * * We may then safely admit that, although large quantities of oil are to be found in the sand, it is only the mere waste of the real springs.

Deeper sinking will, without any doubt, be very profitable, and yield immense proportions of petroleum.

"Their proximity to the sea is another advantage of these mines. While great difficulties have been encountered in the United States for the conveyance of the oil to the seaboard, here it may be conducted at very small expense, from the wells to the port of Santa Elena, by means of pipes, if the crude oil is exported.

"Although I should think that it would be more profitable to refine at the place of production, the advantage expressed remains the same.

"By this extent, by the enormous quantities of petroleum that they contain, and by the short distance which lies between them and the sea, these mines have a real value. Intelligent capitalists will promptly appreciate it, and works will soon be established.*

"By building refineries at Santa Elena, enough kerosene might be produced for the use of the country, and for the markets of the neighboring republics."

What has been said of the topography and geology of the Peruvian field applies with equal force to that of Santa Elena. The climate of this region, though situated almost directly under the equator, is mild and salubrious, owing to the elevation of the country and its proximity to the Pacific, the cool breezes of the ocean exercising a favorable influence on the temperature. In the coldest season it is never below 50°, and in the hottest never above 85°.

These discoveries and these developments are destined to exercise a potent influence on the future of the trade of Peru and Ecuador. A powerful competition will be offered to Pennsylvania in the markets of Europe, Australia, and New Zealand, while in the Republics of Central and South America it can result in nothing less than entire exclusion.

---

* Since the above report was published, the Government of Ecuador leased the entire oil region of Santa Elena to Mr. Richard Linn, of Titusville, for a long term of years, with most valuable concessions. We understand that an association of capitalists are now preparing to operate under Mr. Linn's lease. The amount of territory covered by this lease is 70 miles in length, and 70 in breadth.

WEST INDIA ISLANDS.

CUBA.

In the early history of Cuba it is recorded that Havana was originally named by the early visitors and settlers Çarine—"for there we careened our ships, and we pitched them with the natural tar which we found lying in abundance on the shores of the beautiful bay." Petroleum springs are in number near Havana, rising from fissures in the serpentine rocks at Guanabacoa, and have been known for two centuries. "Allan's Manual of Mineralogy" says the whole of Cuba is impregnated with bituminous matter to a surprising degree, in cells and cavities in the rocks. The *Essai Politique sur l'Isle de Cuba:* "Petroleum leaks out in some, indeed in numberless places in this delightful island, from amidst the fissures of the serpentine, and perhaps has deeply-seated sources. We are acquainted with abundant springs of petroleum between Holquin and Mayari, in the eastern end of the island, and also possess notices in the direction of Santiago de Cuba."

SANTO DOMINGO.

On a stream called "El Aguatediondo," or stinking water, three miles north of the town of Azua, this spring makes its appearance as a stagnant, torpid pool, exuding slowly through a heavy gravel deposit. A very small area in the vicinity is covered with deposits of pitch; for half a mile down the bed of a rain-water stream, the gravel or sand, as the case may be, is more or less cemented by an impure pitch, sometimes plastic, oftener hardened to asphaltum. The pools of the spring and neighboring excavations contain a dirty water rendered brown by contact with the oil, and on the surface is a thin scum of petroleum dark brownish-green to reflected light, and a reddish-brown by transmitted light. An attempt was made during the oil excitement of 1865 to bore here; the usual tools were taken to the spot, but the undertaking was eventually abandoned. In the driving pipe yet remaining at the mouth of the well, may be observed an accumulation of oil, through which gas bubbles up. At the distance of a few yards from this well are several jets of gas. Over the whole area there is not a single blade of grass or any other vegetable.

BARBADOES.

An American gentleman in business on this island in 1864 visited the oil regions of Pennsylvania, and from his observations became convinced that a like article had been noticed by him oozing out of the rocks and lying on the surface of the ground on some of the plantations of this island, but being engaged in business demanding his entire attention, he made no efforts to test the facts in regard to it. During the year 1871, a firm on the island quietly commenced to secure it by sinking shafts and curbing as they went down; it was soon found that they shipped considerable oil, and that it was very valuable for lubricating purposes, netting them thirty dollars per barrel on the island. These facts coming to the knowledge of other parties, a company was formed and a favorable spot secured on a plantation having abundant surface indications. This company determined to take advantage of the modern Pennsylvania mode of obtaining oil by drilling and pumping. An experienced driller was engaged in Pennsylvania, who was furnished with a full rig—boiler, engine tools and wood work, and was dispatched to the island. In a letter dated the 2d of March the driller says: " We have drilled 168 feet, but the rock is soft soapstone, and not hard enough to prevent 'caving,' so we had to abandon the well. We then moved the rig from the ravine to higher ground." After the abandonment of this well it was found that it had filled up seventy-five feet with oil. The second well, for some cause unknown to us, has also been abandoned.

TRINIDAD.

In the island of Trinidad, three-fourths of a mile back from the coast, is a lake called the Tar Lake, a mile and a-half in circumference, apparently filled with impure petroleum and asphaltum. The latter, more or less charged in its numerous cavities with liquid bitumen, forms a crust around the margin of the lake, and in the centre the materials appear to be in a liquid boiling condition. The varieties contain more or less oil, and methods have been devised for extracting this; but the chief useful application of the material seems to be for coating the timbers of ships to protect them from decay. By the patented process of Messrs. Atwood, of New York, the crude tar of that locality, having been twice subjected to distillation and treated with sulphuric acid and afterward with an alkali, is then further purified by the use of permanganate of soda or of potash. Being again distilled it yields an oil of specific gravity 0.900, which is fluid at 32° Fahr.

THE CARPATHIAN PETROLEUM BELT.

The existence of rock oil springs and wells in Galicia, Moldavia and Wallachia, outside, or along the north, north-east and east foot hills of the chain of Alps which surrounds Hungary on the side of Russia, has been known for some years. In 1859 the Austrian geologist, M. Fœtterle, wrote of them in the Year-book of the K. K. Geological Institute. In 1866 Hochstetter and Prosepny published further observations in the same Annual; and Iicinsky, in the *Berg-ung-Hütten-wesen Zeitung*, No. 36–37. In No. 39–41, 1866, Prosepny gave another account of them; as Cotta did also in the East Austrian *Review.* Ellenberger in 1867 added something in the Annual K. K. G. R., and M. Coquand inserted his Memoir in the xxiv. vol. Bulletin of the French Geological Society.

We have now, however, a completer *resume* of all that is known on the subject from the pen of M. Emile Heurteau, Engineer of Mines, in the recently issued 3d part of the xix. vol. of the Annals of the Paris School of Mines, with a map of the Krosno-Dukla districts and sections of the petroleum-bearing rocks. He says that in 1869 he visited most of the points where oil was actually sought or obtained, but that the work was conducted by the proprietors of the land, no records of borings were kept, and scarcely any traces of what had been done were left to view.

The mountain range, in this part of it, runs north-west and south-east, and falls off gently to the great Miocene Tertiary plains of Galicia and Moldavia, in a series of parallel anticlinal and synclinal undulations, which are visible in the sections made by all the descending valleys and ravines. The mountain mass consists of cretaceous rocks and outcropping on the south-west flank, covered by Eocene Tertiary sandstones and clay-slate formations, almost vertical or a little overturned, so as to plunge south-westwardly, and rarely fossiliferous.

On these Carpathian rocks lie the Miocene Tertiaries, the lowest of which, outcropping all along the foot-hills, are the two thin beds of saliferous clay-slates which furnish the salines of the region. The salt mines of Wielisk and Bochnia, the gypsum masses of Podgorze, and the sulphur deposit in the gypseous marls of Schoszowice, are all in the Miocene.

Everywhere along the range of the salt-bearing rocks is a blackish clay, marl bed, more or less bituminous, of muddy consistency, strongly impregnated with salt, either crystallized in large grains imbedded in the mass, or condensed into large lenticular beds of impure rock salt ; or irregularly distributed. The whole saliferous formation is traversed by contorted beds of anhydrite gypsum alternating with beds of salt-clay, more or less pure. All stand vertical or plunge steeply south, growing less deep the further down they are followed, puzzling the observer with the appearance of passing *underneath* the older steeply south-dipping rocks of the mountain range. Heavy coverings of loss help the deception. It is of course necessary to suppose a long fault, the north-east country having settled down and curled the edges of its rock formation completely over. This fault is the key to the subject of the memoir.

On the Galician side, the oil belt, though extending for 200 miles, is explored at three principal centres : New Saudac, on the west ; Dukla, Krosno and Sanock, in the middle ; and Borslau, in the east, where the "mineral wax," *ozokerit* occurs in great abundance.

From time immemorial the peasants of Bobrka, on the banks of the Jasolka, between Dukla and Krosno, have noticed oil oozing from joints of the sandstone rocks, and standing, especially in dry seasons, on the little pools of water ; they collect it to grease their wagons, and fire it off on festival occasions. In 1860 M. Lucka-silwitch, hearing of the American petroleum wells, experimented with his own in the laboratory, and then commenced work on M. Klobassa's lands, but with very poor success. In 1861, he transferred his search to a place farther east, and struck oil in a bore 50 feet deep, which yielded 16,000 pounds daily. His second well yielded 600 bbls. Wells multiplied, until in 1870 the yield amounted to $70,000 per annum, giving a profit of $50,000. Seventy-seven wells are ranged along the axis of a sharp anticlinal, one-third of a mile long, none being more than 80 feet off the straight line, and the oil from all flows through a pipe to a common reservoir. Some of the wells are 350 feet deep ; but no law of depth has been obtained. Shafts 7 feet square are sunk about 70 or 80 feet to the sandstone, and bore-holes are continued from this downward. Gunpowder is used in shafting, and strong ventilating fans blow out the gases. Lights are forbidden, and accidents are few. The boring is very rude, being done by four hands without machinery.

On reaching the oil stratum a great quantity of carbonic acid gas mixed with hydro-carbons escapes from the well, followed by the oil, which rises to the surface of the water, filling the shaft. A small " Jewish " hand-pump is used to draw off the water and oil into barrels, from which the water is allowed to escape by gravity. It is evident that the oil is kept down by the weight of water, and must be relieved of this load before it will rise in any quantities.

Some wells yield per day 3,000 kilogrammes, others 600, 302 down to 80, and some mere traces of oil. The author gives interesting details of the lawless behaviour of the various wells of the group, with diagrams showing their relative situations and relations to the anticlinal axis, and confesses that no trace of a method of explanation has been obtained. The oil is always mixed with water, sometimes fresh, ordinarily saline. Between the two petroleum horizons the water is always salt. The following table shows the authenticated and official statistics of all the oil which has been transported and conveyed by the Carl Ludwig & Kaiser Ferdinand (Nord-Bahn) Railroads for eight successive years from the oil regions.

YEAR.	Oil conveyed and transported by two roads.	Into Austrian provinces and Prussia.	Consumed in the towns and cities in Galicia.	American oil consumed in Galicia.
	cwt.	cwt.	cwt.	cwt.
1862	32,295	26,725	5,570	
1863	67,336	53,796	13,560	787
1864	113,099	91,672	21,427	238
1865	133,356	117,043	16,313	114
1866	166,349	146,802	19,547	1,552
1867	155,589	139,059	16,530	395
1868	147,251	134,535	12,716	297
1869	81,398	72,701	8,697	

The above table will give some idea of the resources of the Galician portion of the Carpathian oil field, as it exhibits how many years the product of this field has been in the market.

A recent traveller says of the Wallachian portion of the Carpathian oil belt that "there is no country in the old world which has been so plainly proved to be a land flowing with petroleum." Associated capital has been brought to bear on its extraction and export, but strange enough, the mechanical appliances by which success has been achieved in western Pennsylvania have been but to a very limited extent introduced into Wallachia. The Romanian petroleum companies, situated on the same end of the Carpathian belt have been contented to adopt the primitive mode of collection in use with the peasantry, by making excavations into the earth, into which the oil saturating the strata flows. This plan of operations is not conducive to profitable commercial working on a large scale, but should attention be given to deep boring on the plan adopted in the oil region of Pennsylvania, there is every prospect of success.

In the course of this chapter mention has been made of " Ozokerit," a mineral wax or solidified petroleum. We consider the matter of so much interest that we give a brief account of this singular product of the field now under notice :

" Ozokerit " is a " mineral wax," and in the raw or native state is of a yellowish color, of light specific gravity and somewhat fibrous in its structure. It will not burn of itself, but will readily melt on a light being applied to it. On being roughly wrapped around a central wick, even in its native state, it is easily and readily consumed. In fact, a rude candle can be made of the raw material and a cotton wick. It is found principally in Austria, Moldavia, the Caucasus, and near the Caspian Sea, where it is obtained in great quantities, being largely used in those countries for illuminating purposes. It was discovered about two years since by a Russian military officer, who communicated the fact to a Mr. Gustav Siemssen, who has introduced it into England. In the premises where the candles are made, the native ozokerit is found in two conditions—in the one as dug from the earth, and in the other as roughly melted down for convenience of storage in transit. In the latter condition it forms a dark-colored mass, and is packed in barrels, the native or unmelted ozokerit being sent over in canvas bags. From the store, the crude material is conveyed into the melting-tanks, holding from two to three tons each, where it is melted down by means of a steam coil. From these tanks, which are situated in a gallery some fifteen feet above the ground level, the ozokerit is run off by gravitation to a series of stills placed outside the main building, and holding from two to three tons each, into which it is distilled over, partly by steam, and partly by bottom heat. The dirt and bottoms from the crude ozokerit are run off from the melting-tanks into another set of tanks beneath them, where they are remelted, the finer products being afterward distilled over, The ozokerit comes from the stills in the form of an oily distillate, which is run from the condensers into molds and allowed to cool. This gives a deep yellowish wax-like substance of a spongy nature, the pores being filled with oil, which exudes under a slight pressure. These cakes are packed between oil-skins and canvas cloths, and are placed in hydraulic presses, of which there are three of large capacity. The pressed cake after removal is put into reheating tanks and again melted down, and is pumped from these tanks by a steam pump into the acidifier, where it is treated with sulphuric acid. These acidifiers are steam jacketed, and are fitted with revolving agitators, by which the ozokerit and acid are agitated for a certain time, after which the mixture is allowed to settle. After settling, the purified ozokerit is drawn off from the lower part of the acidifiers—the acid remaining on the top—and run into vessels which are heated by bottom heat. This is the final heating, and from these vessels the fine stuff is drawn off into molds, the result being a hard white wax, the melting point of which is 140°, that of paraffine wax being only 128°. These blocks are sent to Messrs. Field's works at Lambeth, London, England, and from them the well-known ozokerit candles are made. There are several by-products, the chief of which is a very clear, colorless oil, and of very high illuminating power.

BURMAH.

The petroleum business in Burmah has long been in operation, the oil being used by the natives for heating purposes, for preserving wood, and also as a medicine. Thousands of wells have been excavated, and after working them so long as profitable, they were left and new ones dug out. Dry holes are as frequent as in Western Pennsylvania, broken-down operators as numerous, and lucky ones, who have succeeded in making their first million, just as few. The possession of the royalties of Burmese oil lands are still so valuable as to be deemed the most desirable gifts the sovereign of that country can bestow upon chosen favorites. Not only this, but English capital is largely invested there, and large quantities of the oil find a ready market in Europe.

The following interesting account of the wells of that distant country is taken from the journal of John Crawford, Esq., F. R. S., F. L. G., and ambassador of the Governor-General of India, to the Court of Ava, 1826. Though this report is of old date, it applies with equal truth to the present state of the business.

" At three in the afternoon, our whole party proceeded to the celebrated petroleum wells. Those which we visited cannot be further than three miles from the village, for we walked to them in forty minutes. The wells altogether occupy a space of about sixteen square miles. The country here is a series of sand hills and ravines, the latter torrents after a fall of rain, as we now experienced, and the former covered with a very thin soil, or altogether bare. The trees, which were more numerous than we looked for, did not rise above twenty feet in height. The surface gave no indication, that we could detect, of the existence of petroleum. On the spot which we reached, were eight or ten wells, and we examined one of the best. The shaft was of a square form, and its dimensions about four feet to a side. It was formed by sinking a frame of wood composed of the mimosa catechu, which affords a double timber. Our conductor, a son of the Myosugi of the village, informed us that the wells were commonly from one hundred and forty to one hundred and sixty cubits deep, and their greatest depth in any case, two hundred. He informed us that the one we were examining was the private property of his father—that it was considered very productive, and that its exact depth was 140 cubits. We measured it with a good lead line, and ascertained its depth to be 210 feet; thus corresponding exactly with the reports of our conductor, a matter which we did not look for, considering the extraordinary carelessness of the Burmans in all matters of this description. A pot of oil being taken up, and a good thermometer being plunged into it, indicated a temperature of 99 degrees. That of the air when we left the ship, an hour before, was 82 degrees. We looked into one or two of the wells, and could discern the bottom. The liquid seemed as if boiling, but whether from the emission of gaseous fluids or simply from the escape of oil itself from the ground, we had no means of determining. The formation when the wells were sunk, consisted of good, loose sandstone and blue clay. When the well is dug to a considerable extent, the laborers informed us that brown coal was occasionally found. Unfortunately we could obtain no specimens of this mineral on the spot, but I afterward obtained some in the village. The petroleum itself, when taken out of the well, is of a thin watery consistence, but this, by keeping, and in the cold weather it coagulates. Its color at all times, is a dirty green, and not much unlike that of stagnant water. It has a pungent, aromatic odor, offensive to most people. The wells are worked by the simplest contrivance imaginable. There is over each well, a cross beam, supported by two rude stancheons. At the center of the cross beam, and embracing it, is a hollow revolving cylinder, with a channel to receive a drag rope, to which is suspended a common earthen pot, that is let down into the well, and brought up full by the assistance of two persons pulling the rope down an inclined plane by the side of the well. The contents of the pot are deposited for the time in a cistern. Two persons are employed in receiving the oil, making the whole number of persons engaged on each well only four. The oil is carried to the village on posts in carts, drawn by a pair of bullocks, each cart conveying from 10 to 14 pots of ten viss each, or from 265 to 371 pounds avoirdupois of the commodity. The proprietors store the oil in their houses, and then vend it to the exporters. The price varies, according to the demand, from four ticals of flowered silver to six ticals per 1,000 viss; which is from five pence to seven pence half penny per 100. The carriage of so bulky a commodity, and the breakage to which pots are so liable, enhances the price in the most distant parts to which the article is transported, to 50 ticals per 1,000 viss. Sesamun oil will cost at the same place not less than 300 ticals for an equal weight, but it lasts longer, gives a better light, and is more agreeable than the petroleum, which in burning, emits an immense quantity of black smoke, which soils every object near it. The cheapness, however, of this article is so great, that it must be considered as conducing much to the comfort and convenience of the Burmans. Petroleum is used by the Burmans for the purpose of burning in lamps and smearing timber to protect it against insects, especially the white ant, which will not approach it. It is said that about two-thirds of it is used for burning, and that its consumption is universal until its price reaches that of Sesamun oil, the only other oil which is used in the country for burning. Its consumption, therefore, is universal, wherever there is water carriage to convey it—that is, in all the country watered by the Krowaddy, its tributary streams and its branches. It includes Bassien, but excludes Martaban, Tavoy and Mergui, Aracan, Tongo and all the northern and southern tributary States. The quantity exported to foreign ports is a mere trifle, not worth noticing. It is considered that a consumption of thirty viss per annum for each family of five and a half persons is a moderate average. If it were practical, therefore, to ascertain the real quantity produced at the wells, we should be possessed of the means of making a tolerable estimate of the inhabitants who make use of this commodity, consisting of the largest part of the population of the Kingdom. Of the actual produce of the wells we received accounts not easily re-concilable to each other. The daily produce of the wells was stated according to quality to vary from 35 to 500, the average giving about 235 viss. The number of wells was sometimes as low as 50, and sometimes as high as 400. The average made about 200, and considering that they are spread over 16 square miles, as well as that the oil is well-known to be a very general article of consumption throughout the country, I do not think the number exaggerated. This estimate will make the consumers of petroleum for burning amount to 2,066,721. In the narrative of one of my predecessors, Captain Cox, the number of wells is given as high as 520, and the average daily produce of each well is reckoned at 300 viss, which makes the whole amount produced 56,940,000."

We here give extracts in reference to Petroleum from the Narrative of Major Michael Symes, of the English Army, who was sent by the Governor-General of India as Embassador to the Court of Ava, in 1765 (published by Bulmer and Co., in London, in 1800), who says at page 261: After passing various sands and villages, we got to Yaynangheoum or Earth Oil (Petroleum) Creek, about two hours past noon. We were informed, that the celebrated wells of Petroleum, which supply the whole empire and many parts of India with that useful product, were five miles to the east of this place. The mouth of the creek was crowded with large boats, waiting to receive a lading of oil, and immense pyramids of earthen jars were raised within and around the village; disposed in the same manner as shot and shells are piled in an arsenal. This is inhabited only by potters, who carry on an extensive manufactory, and find full employment. The smell of the oil is extremely offensive. We saw several thousand jars filled with it, ranged along the bank; some of these were continually breaking, and the contents, mingling with the sand, formed a very filthy consistence. Mr. Wood had the curiosity to walk to the wells; but, though I had felt the same desire, I thought it prudent to postpone visiting them until my return, when I was likely to have more leisure, and to be less the object of observation.

PAGE 441.—We rode until two o'clock, at which hour we reached Yaynangheoum, or Petroleum Creek, of Benangyun. The oil drawers stated to us, that in cleaning out old wells, accidents sometimes happened from the fire-damps; and they pointed out a particular well at which two men had lost their lives from this cause.

PAGE 178.—The celebrated Petroleum wells afford, as I ascertained at Ava, a revenue to the king, or his officers. The wells are private property, and belong hereditarily to about thirty-two individuals. A duty of five parts in one hundred is levied on the Petroleum as it comes from the wells, and the amount realized on it is said to be 25,000 ticals per annum. No less than 20,000 of this goes to contractors, collectors, or public officers, and the share of the State, or 5,000, was assigned during our visit as a pension of one of the Queens.

PAGE 206.—The Petroleum wells of Renangyorong have been already described in the Journal. From the more accurate information, which I obtained at Ava, it appears that the produce of these may be estimated at the highest, in round numbers, at 22,-000,000 of viss, each of three and sixty-five one-hundredth pounds avoirdupois. This estimate is formed from the report of the Myo. Thugyi, who rents the tax on the wells, which is five in a hundred. His annual collection is 25,000 ticals, and he estimated, or conjectured, that he lost by smuggling 8,000, making the total 33,000. The value of the whole produce, therefore, is 660,000 ticals. The value of the oil on the spot is reckoned at three ticals per 100 viss, and consequently its amount will be as above stated.

PAGE 238.—I should observe, that Petroleum is universally used, wherever the navigation of the Irrawaddy and Ryendwen, with their tributary streams, will allow of its being conveyed, and that it is also carried to a place already noticed in our journey up the river. Dr. Buchanan partook of an early dinner with me, and when the sun had descended so low as to be no longer inconvenient, we mounted our horses to visit the celebrated wells that produce the oil, an article of universal use throughout the Empire.

PAGE 442.—The evening being far advanced, we met but few carts; those we did observe were drawn by a pair of oxen, and of a length disproportionate to the breadth, to allow space for earthen pots that contained the oil. It was a matter of surprise to us, how they could convey such brittle ware with any degree of safety over so rugged a road. Each pot was packed in a separate basket and laid in straw, notwithstanding which precaution, the ground, all the way, was strewn with broken fragments of the vessels, and wet with oil, for no care can prevent the fracture of some in every journey. As we approached the pits, which were more distant than we had imagined, the country became less uneven, and the soil produced herbage. It was nearly dark when we reached them, and the laborers had retired from work. There seemed to be a great many pits within a small compass. Walking to the nearest, we found the aperture about four feet square, and the sides lined, as far as we could see down, with timber; the oil is drawn up in an iron pot, fastened to a rope passed over a wooden cylinder, which revolves on an axis, supported by two upright posts. When the pot is filled, two men take hold of the rope by the end, and run down a declivity, which is cut in the ground, to a distance, equivalent to the depth of the well. Thus, when they reach the end of the track, the pot is raised to its proper elevation ; the contents, water and oil, together, are then discharged into a cistern, and the water is afterwards drawn through a hole in the bottom. Our guide, an active, intelligent fellow, went to a neighboring house, and procured a well-rope, by means of which we were enabled to measure the depth, and ascertained it to be thirty-seven fathoms ; but of the quantity of the oil at the bottom we could not judge. The owner of the rope, who followed our guide, affirmed that when a pit yielded as much as came up to the waist of a man, it was deemed tolerably productive; if it reached his neck, it was abundant ; but that which rose no higher than the knee, was accounted indifferent. When a well is exhausted, they restore the spring by cutting deeper in the rock, which is extremely hard in those places where the oil is produced. The government farms out the ground which supplies this useful commodity, and it is again let to adventurers, who dig wells at their own hazard, by which they sometimes gain and often lose, as the labor and expense of digging are considerable. The oil is sold on the spot for a mere trifle—I think 200 or 300 pots for a tackal, or half a crown. The principal charge is incurred by the transportation and purchase of vessels. We had but half gratified our curiosity, when it grew dark, and our guide urged us not to remain any longer, as the road was said.to be infested with tigers, that prowled about at night among the rocky, uninhabited ways through which we had to pass. We followed his advice, and returned with greater risk, as I thought, of breaking our necks from the badness of the road, than of being devoured by wild beasts. At ten o'clock we reached our boats without any misadventure."

PUNJAB, INDIA.

The Public Works Department of the Government of India a few years since engaged a gentleman from Pennsylvania, Benjamin Smith Lyman, Esq., to report on the commercial value of the oil lands of the Punjab. Mr. Lyman reports as follows:

The Punjab oil region is in the corner between Cashmere and Cabul, and lies wholly between north latitude 32° 31′, and 33° 47′, and east longitude (from Greenwich) 71° 18′, and 73° 5′; a nearly square space about a hundred miles long east and west, by ninety miles wide, north and south.

Just inside the north-east corner of this square is Rawul Pindee, the largest town of the region, with about twenty thousand inhabitants; just inside the south-east corner is Pind Dadun Khan, a town of about twelve thousand inhabitants; and just inside the south-west corner is the ancient uninhabited ruin of a walled town, now called Kafir Kot. Just within the north-west edge of the region, and less than twenty miles from its eastern edge, stands the little village of Shah kee Dheree, on the site of the ancient capital Taxila, where the King Taxiles hospitably entertained Alexander the Great. The small town of Attok, where Alexander crossed the Indus into India, is only ten miles north of the middle of the northern edge of the square. The famous Muneekyala Tope, built by King Kanishka, about the Christian Era, to mark the spot where Booddha in compassion gave his own flesh to satisfy the hunger of a starving tiger, stands a little outside the square, fifteen miles south-east of Rawul Pindee.

The river Indus enters the square about the middle of the northern edge, and leaves it at the south-west corner. The Jhelum river (the "fabulosus Hydaspes" of the ancients), one of the five rivers that gives its name to the Punjab, flows across the south-east corner, past Pind Dadun Khan, south-westerly toward the Indus. The center of the region is drained by the Sohan, which rises near Rawul Pindee, and flows west, south-west to the Indus.

The region lies, then, mostly between the Indus and Jhelum, in what is called the Sind Sagur Doab (two rivers), and it is mainly in the mountainous or hilly part (Kohistan) of the Doab. The oil has been bored for at Gunda, and at first fifty gallons of it a day were pumped from the well; but the yield, of course, grew quickly less (like the ordinates of a parabola), and after the whole amount had reached two thousand gallons (about five months) the daily yield was less than ten gallons. In the region, oil flows also at five other places from natural springs, from a gill to three quarts a day, and there are traces of it at yet two other places, making eight in all. Asphalt, or dried oil, is found in small quantities at four of these places, and at four other places—at two in notable quantities. At most of the asphalt places there are traces of rock tar or asphalt melted in the heat of the sun; and at one of them (Aluggud) as much as one hundred gallons. Besides these dozen places where oil or asphalt is found there are half a dozen places where there are small traces of one or the other, enough to attract notice in the minute examination of the country by its inhabitants. About half of all the places are in the north-eastern corner of the region ; about half toward the south-western corner, and one or two in the north-western corner toward the middle.

The Aluggud oil (now dried to asphalt) seems to have come from rocks of carboniferous age, to judge by their fossils, though other things would rather show that they were of later age. If they are carboniferous, then the nummulitic rocks are wanting above them, and have thinned completely away from a thickness of 2,000 feet only thirty miles distant. This oil is also the only case of oil outside of the older tertiary rocks anywhere in the whole region.

All the other oil springs or shows of oil in the southern part of the region are on the northern side of the Salt Range and in the nummulitic lime rock, or close above it. The northern ones are either in the nummulitic lime rock of the Choor Hills, the same probably as that of the Salt Range ; or in the Gunda rocks (chiefly sand rocks) that lie south of them, also accompanied by nummulites.

In every case the oil seems to come from a deposit of very small horizontal extent, sometimes only a few feet, seldom as much as a few hundred yards ; only in one case, that of the Chhota Kutta and Burra Kutta oil springs, near Jaba, does the deposit seem to extend as much as half a mile. Here, too, the oil comes from a thickness of about a hundred feet, and the natural springs yield at one place as much as three quarts a day. At all the other places the oil comes from a much smaller thickness of rock, from forty feet at Aluggud and twenty at Gunda and Punnoba downward. Scarcely do any two oil springs come from the same bed of rock.

The oil is dark green in color, and so heavy as to mark 25° of Beaume's scale, or even less. The Gunda oil has been burned a little by the natives with a simple wick, resting on the side of an open dish ; but the Punnoba oil is more inflammable, and needs a special tube for the wick, though the main opening of the dish or lamp may stay uncovered. The oil, generally, however, has been little used for burning, except at Punnoba; but has been sought for as a cure for the sore backs of camels. The asphalt was highly prized forty years ago by the natives as medicine, especially for broken bones. It was carried far and wide, and was called "negro's fat," because it was believed to have dripped from the brain of a negro who had been hung up by the heels before a slow fire.

It is perhaps needless to say that there is nothing whatever in the mode of occurrence of the Punjab oil, to uphold the chimerical belief that rock-oil ever passes by distillation, emanation, or otherwise, from one set of rocks to another; that it originates in any different rocks from those in which it is found; and nothing to show that it has been formed by any other method than the very natural and sufficient one of the slow decomposition of organic matter, deposited along with the other materials of the rock. Neither is there anything to show that the oil has been driven up by the upward pressure of water from the lower parts of a bed of rock through its pores to a higher part of the same bed; on the contrary, as the rocks near most of the oil springs dip pretty steeply, if such an action of water were possible, all the oil would long ago have been altogether forced out of the rock at the outcrop. Indeed, such an idea is quite inconsistent with the fact that even a slight amount of oiliness in the pores of a body is a complete bar to the entrance of water; much less could water (without soap) scour the oil from one mass of rock and make it flow into another mass filled with moisture. If oil wells are more numerous in some regions along the tops of rock saddles, the reason is clear, that the oil-bearing bed lies too deep for boring conveniently elsewhere.

Wild hopes have sometimes been entertained that a large amount of oil might, by boring near the oil springs, be struck in some cavity below the oil-bearing bed; but it is safe to say that they are not justified by anything whatever, either in the Punjab or in any other part of the world either in the practical experience of oil boring or in the general laws of physics.

CHINA.

Late accounts from China report immense oil fields, some of which are worked to a limited extent. The Chinese may justly claim to be the first to *drill* for oil, as for hundreds of years they have regularly bored their wells, and that to a very great depth. The celebrated traveller, Abbe Hue, discovered the existence of petroleum in many parts of the Empire. In describing the wells, he states, that many are drilled to a depth of 1,500 to 2,000 feet, the drilling being done very laboriously by a tube six inches in diameter.

A Catholic missionary who was engaged in the province of Slo-Tchouch in 1833—a territory which is celebrated for its fine wells, gives some very interesting particulars about the petroleum business in the Celestial Empire. After describing a burning well, and the method of quenching it, by turning the waters of a small lake upon the flames, the missionary states, that when the mouths of these wells are closed, the gas is conveyed to any place where it is needed through hollowed bamboos, and used for lighting the towns and villages. He also describes how in a province about 200 leagues from Canton the gas is used in the great salt mines for fuel. The gas is conducted under the boilers by bamboos from the well. These are tipped with earthenware, which keeps the bamboo from burning when the gas is ignited. So great is the quantity of gas produced that all the flame cannot be utilized, but much of it is allowed to escape to the surface of the earth by means of chimneys.

JAPAN.

A very extensive and valuable oil field exists in this country. Oil is found at shallow depths by sinking pits. An English Company recently purchased the necessary tools and machinery to commence developments. This Company took out with them a driller and refiner from the Pennsylvania Oil Region. From accounts received of this field, we conclude it is of a very promising character.

ALSACE.

The value of Alsace to Germany, and the consequent extent of the loss to France, commercially considered, are alike enhanced by the probable development of a considerable petroleum industry in that celebrated province.

Oil works on a small scale, already exist in the valley of the Rhine, near the village of Schwatwiller, within and near the borders of the forest of Hagenau. A thick alluvial deposit has first to be penetrated, beneath which are alternating strata of indurated clay, and micaceous sandstone, with seams of compacted sand. These last named seams, contain the petroleum, and are found at a depth of two hundred or two hundred and fifty feet. Indications of the presence of petroleum are observable in various parts of the forest, and bitumen is found and worked in the adjacent country. Borings to test the presence of the petroliferous sand, have been multiplied to some extent, and in all cases with satisfactory results. The mode of working very much resembles that of a colliery. We believe that at present there are only two oil pits existing, and one of these is of a very recent date. The pits are sunk in the ordinary way, and the seams of sand are worked by galleries, in a manner similar to that of getting coal. As the workmen cut their way through the compacted sand, the oil oozes out of it, running down the wall, of the gallery on to the floor, where it accumulates in shallow wells dug for the purpose. From these wells the crude petroleum is conveyed to the surface. But the process of draining does not remove all the oil, and the sand itself is accordingly taken to the surface, to be distilled in retorts. The crude oil which oozes from the sides of the gallery, and that which is distilled from the sand, are subsequently rectified by a further distillatory process, and the product is understood to be in no degree inferior to Pennsylvania refined petroleum. In working the existing pits, it is a singular fact that no water is found. Of the extent to which the petroliferous sand prevails, it would be premature at present to judge, but there seems no reason to doubt its presence over a considerable range of territory.

HANOVER.

Experiments have been made in regard to the well-ascertained deposits in different localities in Hanover, and borings have been prosecuted in the neighborhood of Helde, with the object of determining the extent and thickness of a remarkable layer of chalk, occurring at the depth of about one hundred and twenty feet, and saturated with petroleum. Several years ago this chalk deposit was examined to a depth of four hundred feet, and the first one hundred and fifty feet were extremely rich in petroleum, and various amounts were yielded as the drillings descended. It would appear that in consequence of the inefficiency of the apparatus, the engineer was unable to penetrate any deeper than four hundred feet, at the point pure petroleum was found. At present the borings are to be conducted more vigorously, and are to be carried down to a depth of one thousand feet, with a bore of the diameter of seventeen inches.

ITALY.

From time immemorial the inhabitants of Rivanazzano, a small place a short distance from the town of Voghera in the former kingdom of Sardinia, have been in the habit of using mechanically a certain fluid which issues in small rills from the Madonna del Monte, as well as of burning it as a light in their dwellings. At the top of this mountain there are traces of an extinct volcano, and some short time since wells were sunk at its foot, and their contents subjected to chemical analysis, the result of which was that petroleum of an excellent quality was found to be present in considerable quantities. The explorers then came to the conclusion that abundant subterranean reservoirs of this mineral oil must necessarily exist at no great distance from the scene of their operations, and they determined to trace the above mentioned rills to their sources.

Excavations were accordingly commenced on the borders of the pleasant slopes of Nazzano, about twelve kilometres from Voghera. At a depth of about fifteen metres a considerable issue of gas took place, and when thirty metres had been reached, salt-water strongly impregnated with petroleum was met with, a circumstance which the explorers remembered as always occurring in the oil springs of Pennsylvania. Following up the excavations, loud explosions of gas took place at a depth of ninety metres, and large volumes of salt water mixed with petroleum issued from a stratum of sandstone rock which was there met with. Pumps, on the principle of those used in America, under similar circumstances, were then introduced, and an abundant supply of petroleum obtained. Ultimately, a concession of this valuable property was granted by the Italian government to the explorers, as a reward for their exertions.

The petroleum thus obtained has now been refined, and found to yield a valuable lubricating oil, and one well adapted for mixing up paints and varnishes, while the oil for burning gives a very brilliant white light, and has been found remarkably free from the offensive odors usually existing in mineral oils. Our contemporary adds that the result of these explorations has created quite a sensation, and that it is to be hoped that capital will not be wanting fully to develop discoveries which have been pronounced by eminent engineers, geologists and chemists, to be most promising in a commercial, as well as important in a national point of view.

" At Salso the Marchese della Rosa," says an American gentleman traveling in Italy, " took me to see the place where he is boring for oil. The country has very much the appearance of that around Oil City, Pennsylvania. The Marchese said, that one could not stick a cane into the ground, without finding traces of oil. The work has now been carried down about one thousand feet, but not in paying quantities.

NEW ZEALAND.

In the vicinity of Taranaki there is an exhalation of gas, and
bubbles of bituminous matter, have been observed since the earliest
days of the settlement, at about half a mile from high water mark,
between the main-land and Moturoa, the highest of the Sugar
Loaf Islands; and, according to Dieffenbach, " was whimsically
attributed by the Maoris to the decomposition of an atua, or spirit,
who was drowned there."

It was not, however, until November, 1869, that any attempt
appears to have been made to search for this oil, by boring or sinking wells on land, and as these experiments have to a certain degree proved successful, much attention has been recently attracted
to this natural production.

Two companies were formed to test the oil lands of this island.
The Taranaki Company drilled two wells, and the Alpha Oil
Company one well.   Dr. James Hector, in his abstract report on
the progress of the geological survey of New Zealand, says, close to
the main Sugar Loaf, and to the foot of the cliffs is the Taranaki
Company's bore, No. 1, which has been sunk with much trouble
to a depth of 300 feet.   The derrick stands at ten feet above high
water; and for some time the water level in the bore, maintained
by this level, but after a time it sunk suddenly to 32 feet, which
would appear to indicate the existence of subterranean channels,
communicating with chambers where there is less than the external
atmospheric pressure, owing, perhaps, to the condensation of oil vapors.   At 254 feet a patch of grey, ferruginous tufa was passed
through, charged with oil, which was the only result.   In this bore
some patches of hard basaltic rock were encountered, but in the
whole there was no decided change in the character of the agglomerate.

Taranaki Company's bore, No. 2, is on the island on the north
headland, and is commenced on a shelf above the water level.

The bore was, in October, sunk to a depth of 145 feet, being 10 feet in the sand-stone, 95 feet in the agglomerate breccia, 30 feet in the consolidated tufa, and a few feet more in the agglomerate again. A few oil patches have been passed through, but no appreciable quantity has been obtained.

The third bore is that of the Alpha Company, which is situated a short distance from the north headland.

At 10 feet above high water, and close to the boulder-covered shore, into a high sandy cliff, a shaft was sunk for 60 feet into the agglomerate, from the sides of which, at 44 feet from the surface, oil was found to ooze.

This shaft was continued by a bore hole to a depth of 180 feet, oil being got at 80 feet, and again at the extreme depth.

When allowed to stand at rest, a considerable quantity of oil collected on the surface of the water in the well, * * * accompanied by the escape of gas. The oil was pumped into a tub along with the water. Recently, the well has been pumped more regularly, and yields, I am informed by the directors, about two barrels per week."

The general results of the chemical examination of the oil obtained from these wells are given as follows: One hundred parts of crude oil, as obtained from the wells, having a specific gravity of .963, give—

Distilled oil of specific gravity .... .874	.02
" " " " .... .893	.10
" " " " .... .917	.08
" " " " .... .941	.60
Solid bitumen	.06.1
Fixed carbon	.12.4
Ash	.01.5

100.00

The presence of petroleum has been reported in other parts of New Zealand.

We understand that recent developments promise success. During the summer of the present year a complete set of the most approved Pennsylvania drilling tools were sent out to be used at this oil field.

NOVA SCOTIA.

Attention has been attracted to the existence of surface-oil oozing from the sand rock exposed on the shore of Lake Anslie. A Company has been formed at Halifax, and are now operating under the superintendence of Mr. William Harrington. Two wells have been drilled to a depth of 800 feet, in which the tools were lost; a third was commenced of which we have had no report. The drill has revealed the existence of three sand rocks. The second well struck oil at 758 feet. Oil was produced from this well in such abundance as to give reasonable evidence of the existence of oil in paying quantities. The oil taken from the well was of unusually high gravity, and almost destitute of odor.

The oil field is distant from Halifax some 200 miles, and is near a fine harbor on the coast, which we understand is now connected by rail with Halifax.

—

CAUCASIAN OIL REGION.

The petroleum deposits of the region of the Caucasus are very remarkable. For many centuries the springs have been known, and the oil has been collected by skimming. On the eastern shores of the Caspian Sea, twenty thousand such wells, all of them quite shallow, existed in 1868. The wells are described as being often close to each other, and the opening of a new one, it is ascertained, does not affect the productiveness of another near it. One sunk in 1863, by the side of another, which for centuries had produced three thousand five hundred pounds per day, yielded forty thousand pounds per day, without affecting in the least the first. The American method has lately been introduced, and flowing wells have burst forth from a depth of two hundred and fifty feet, which have, until controlled, sent up a jet from 40 to 60 feet high. It is calculated that nineteen million pounds are annually produced in the Caucasus region. The present oil-producing region is 25 miles in length, and about half a mile in width. The oil is found in a porous argillaceous sand-stone belonging to the tertiary period. In the vicinity are hills of volcanic rocks, through which heavier sorts of petroleum flow out. It has been observed that from the central portion of the tract the oil is as pure as if refined, and by its faint yellow tint resembles Sauterne wine. That obtained near the sides of the tract is darker, changing to a yellowish green, then reddish brown, and finally to Asphaltum.

The oil is largely introduced into Persia, and over large districts no other material is used for producing artificial light. The following article by M. Sainte-Claire-Deville upon the properties of Caucasian petroleums will be found of much interest:

On the Physical Properties and the Calorific Power of some petroleums of the Russian Empire. By M. Sainte-Claire-Deville, Corresponding Member of the Academy of Sciences of St. Petersburg. Read April 21, 1871.

Rear Admiral Likhatchof desiring to know, in the interest of the transport trade on the Caspian sea, the value and composition of the petroleum oils of Bakou, sent to me samples of these materials. I have made a very attentive examination of them, persuaded that one day the employment of mineral oils as a combustible will be general in all countries where nature furnishes them abundantly to a regular and well organized exploitation.

Petroleums receive divers applications which necessitate a knowledge of certain of their properties and composition. These special properties and the result of their analysis will be the object of this memoir.

1st. The investigation of volatility.—Petroleums are employed in considerable quantities for lighting purposes. Lamp oils, to use an expression established in France (*les huiles lampantes*), ought at the same time to be very fluid in order to mount easily into the wick, and little volatile in order not to be too dangerous in their management. The more fluid these oils are, the more volatile are

they, and the lower their density.   When they distil, as they do
in America on a large scale, petroleum of low density, they only
devote to the manufacture of lamp oils the intermediate products
whose points of ebullition are above 150° and below 250° or 280°.
Those portions which go above 300° in the retorting possess a
viscosity which make them useless- for lamps, and puts them in
demand for the fabrication of lubricating matters, or for fuel.   To
handle a petroleum oil, one must know exactly the number which
represents the quantity of these volatile matters between 150° and
300°.   This number indicates the proportion of lamp oils which
can be extracted from the natural product.

All that which does not pass in the distillation below 280°
should be considered as properly furnishing the lubricating oils,
or to be employed without danger as a combustible of perfect
quality.   As to the volatile portions below 150°, they are composed of gaseous substances, such as hydrure of butyline, or of
matters possessing at ordinary temperatures strong volatility.   These
are the substances which cause such frequent accidents since the
development of a commerce in petroleums on so grand a scale.

The table which I am about to give, and which contains numerous figures relative to the volatility of the Caucasian oils, enables
them to distinguish at once those which are dangerous, those which
furnish lamp oils, and finally the parts of these oils which may be
employed for heating purposes.

It will suffice for this to state for each of them the quantities of
materials volatilized below 150°, between 150° and 300°, and
those which have resisted this temperature.

M. Likhatchof has sent me three specimens of oils, or products
of the works at Bakou.   On submitting them to distillation, the
following results are obtained :

No. 1. Raw Naphtha, from the Balchany Wells.

Volatile matters at 100°... 1.0 per cent.
  "   "   160 .... 5.0   "
  "   "   180 .... 9.3   "
  "   "   200 .... 14.0   "

Volatile matters at 220 .... 15.3 per cent.
   "   "   260 .... 29.0   "
   "   "   280 .... 37.0   "
   "   "   300 .... 41.3   "

No. 2. *Residue from distillation of the Bakou Works.*

   "   "   240° .... 1.0   "
   "   "   260 .... 2.3   "
   "   "   280 .... 4.3   "
   "   "   300 .... 7.7   "

No. 3. *Black oil from the Weyser Works, Bakou.*

   "   "   200 .... 2.3   "
   "   "   240 .... 8.0   "
   "   "   260 .... 14.0   "
   "   "   280 .... 22.3   "
   "   "   300 .... 33.7   "

Numbers 1 and 3 give a certain quantity of lamp oil, and number 2 can only serve as a combustible or lubricating substance.

The specimens from another source, but which have been collected at Bakou itself, and of which M. Likhatchof has sent me great quantities, have given the following results :

NO. 4.—LIGHT OIL.

Volatile parts at 140° .... 2.7 per cent.
   "   "   160 .... 7.0   "
   "   "   180 .... 13.3   "
   "   "   220 .... 19.0   "
   "   "   240 .... 23.3   "
   "   "   260 .... 29.3   "
   "   "   280 .... 36.7   "
   "   "   300 .... 75.3   "

NO. 5.—VISCOUS OIL.

Volatile parts at 200° .... 1.0 per cent.
   "   "   220 .... 1.3   "
   "   "   240 .... 3.7   "
   "   "   260 .... 1.0   "
   "   "   280 .... 6.0   "
   "   "   300 .... 9.7   "

These materials are those which have served to determine the calorific power which will be given further on.

2d. Density and co-efficients of dilatation.—I have had many times occasion to note the dangers which arise during the transportation of petroleums from their considerable dilatability. When a building is filled with barrels containing petroleum, a large empty space must be left in order to avoid their explosion ; the volume of which space can be calculated from tables now to be cited, constructed with reference to the changes of temperature to which the material may be exposed on its voyage or during its stay in the ports and warehouses of commerce.

I have taken the density at 0° and at 50° of the petroleum oils of Bakou, and have calculated with these numbers their co-efficients of dilatation. Supposing the oil to be exposed during its voyage to a change of temperature of 50°, which is prudent to admit, the value of the space which must be left empty in the vessel is found by means of the following formula : $v+k+50$; v being the volume of the vessel, and k being the co-efficient of dilatation given below :

No. 1. Density at 0°...................0.882, raw naphtha from the Balchany wells.
          at 50°:...................0 8473.
Co-efficient of dilatation...... ....0.000781.
No. 2. Density at 0°...................0 928, residue of distillation from Bakou works.
          at 50°........... ....0.888..
Co-efficient of dilatation............................ ....0.00091.
No. 3. Density at 0°...................0 897, black oil from Weyser works of Bakou.
          at 50°...................7.865.
Co-efficient of dilatation.. ......0.000737.
No. 4. Density at 0°...................0.884, light oil of Bakou.
          at 50°...................0.854.
Co-efficient of dilatation..0.000724.
No. 5. Density at 0°...................0.938, heavy oil of Bakou.
          at 50°...................0.907.
Co-efficient of dilatation..0.000681.

3d. Elementary composition.—The elementary analysis of petroleum serves principally to calculate the theoretical calorific power of these minerals. In default of direct determination, you can admit that the quantity of heat given by the combustion of the compound is the sum of the quantities of heat of the combustion

of the elements, and calculate thus the calorific power of these hydro-carbons. The number thus found for petroleums is always a
maximum that experience never permits us to reach, doubtless
because carbon and hydrogen in combining disengage heat, and
naturally this disengaged heat is no longer present in the compound. But, as M. Macquorn Rankine has very judiciously remarked, you obtain by this calculation an approximate number,
which, wholly inexact as it is, may be a guide in the comparison
of values, as combustible of divers mineral oils. Here are the
results which I have obtained by analyzing the petroleums of
Bakou. I designate them by the numbers which have already
served me to specify them in the preceding chapters.

No. 1. Hydrogen.. .... 12.5
        Carbon.. 87.4
        Oxygen.. 0.1
        ___
        100.

No. 2. Hydrogen......................................,................................ 11.7
        Carbon.. .... 87.1
        Oxygen... .... 1.2
        ___
        100.

No. 3. Hydrogen... 12-0
        Carbon.. .... 86.5
        Oxygen 1.5
        ___
        100.

No. 4. Hydrogen 13.6
        Carbon.. 86.3
        Oxygen... .... 0.1
        ___
        100.

No. 5. Hydrogen... 12.3
        Carbon.. 86.6
        Oxygen 1.1
        ___
        100.

To compute with these results the heat of combustion, deduct
from the number of hydrogen one-eighth of the oxygen found, multiply this difference by 344.62, multiply the number of the carbon
by 80.8, and get the sum of the two products thus obtained.

Thus calculated, the following table gives the theoretical heat of combustion of the Bakou oils:

No. 1.. 11,370 units of caloric.
  2.. 11,000   "   "
  3.. 11,060   "   "
  4.. 11,660   "   "
  5.. 11,200   "   "

4th. Calorific power or heat of combustion.—I have determined already the calorific power of petroleums by proceedings which have been described in the reports rendered by the Academy of Sciences of France (see volume lxviii., page 349). No longer having at my disposal the apparatus which has served for these experiments, I have had recourse to a method which seems to me to give also very good results, and which has the advantage of possible application whenever you have a steam-engine whose boiler is heated by mineral oils.

I have shown that the heavy oil of gas works has a very nearly constant composition, and furnishes with truly remarkable regularity the same quantity of heat when it is burned in a calorimeter rightly arranged. Under these conditions heavy oil at 0°, a density of 1.044, furnishes by kilogram 12k.77 of vapor, and produces in burning 8,916 units of caloric. These numbers being definitely fixed (see reports rendered, vol. lxvi., page 450), it is evident that a sufficiently exact relation could be obtained by burning successively representative heavy oil under the boiler of a steam-engine producing a known work, and then the oil to be experimented on doing also the same work, and burning the same quantity of matter. The quantities of water vaporized by the combustibles will be very nearly in proportion to their calorific powers. As you know the number for the heavy oil, a simple proportion enables you to determine the heat of the combustion of the mineral oil taken experimentally.

I operated upon an engine with a Belleville boiler of eight-horse power. I maintained constant, for less than a tenth of an atmosphere nearly, the pressure in the boiler, while the engine was condensing in a large iron reservoir of forty cubic meters, air at a constant pressure of two atmospheres three-quarters.

The air brought in by pumps escaped by a cock whose opening was conveniently arranged so that—the engine doing a constant work—the pressure in the reservoir remained itself absolutely invariable. Under the conditions which I have just mentioned, you can measure exactly the quantity of water volatilized in the boiler, the quantities of oil consumed to produce the constant work of the engine, and when you have made the two determinations successively for the heavy gas, oil serving as representative and the oil taken as experiment, you have given all that is necessary to calculate the calorific power of this last.

1st. Bakou oil, specimen sent by M. Likhatchof, and arrived in a sheet-iron box carefully closed, fluid oil and already studied above under No. 4.

Here are the results of its comparison with heavy oil:

Heavy oil has given—

Pressure of the engine... 3a.8.
Pressure of the air in the reservoir.............................. 2a.75
Temperature of the feed water..................................... 26°
Volatilized water... 161k.
Oil consumed... 18k.23

Oil No. 4 has given—

Pressure of the steam.. 3a.8
Pressure of the air in the reservoir.............................. 2a.75
Temperature of the feed water..................................... 26°.
Volatilized water .... 175k.
Weight of the oil burned... 15k.45

From this is deduced:

1st. Calorific power of oil No. 4, 11,460 cal. Quantity of vapor produced at an ordinary pressure, and without work, by 1 kilogram of oil, 16k.4.

2d. Bakou oil, specimen sent by M. Likhatchof, and arrived in a sheet-iron box carefully closed, oil very viscous, and already examined above under No. 5.

This oil, to flow easily in the pipes of conduit, requires that it should be put under a pressure of about 4 decimeters of mercury. To burn well an oil so little volatile, you must give to the vertical grating of my preparations (see their descriptions in the reports rendered, vol. lxviii., page 349), a little more height than for the fluid oils, and so dispose it (or such a disposition) that the air arrive a little more easily at the bottom than at the top of the grate, where the access of the air should be a little narrower.

Oil No. 5 has given—

Vaporized water.. 126k.6
Oil consumed.. 11k.852

From this is deduced:

Calorific power of oil No. 5.. 10,800 cal.
Quantity of vapor produced without work and at ordinary pres-
    sure by 1 kilogram of oil.. 15k.55

If you compare these calorific powers with those which were theoretically deduced from their composition, you find:

	Observed power.	Calculated power.	Diff.
Oil No. 4	11,460 cal.	11,660 cal.	200
Oil No. 5	10,800 cal.	11,200 cal.	400

If you admit that this difference which is in mean some 300 units of heat, between the real calorific power and the calculated calorific power, is the same for all the Bakou oils, you find for the specimens of the materials sent me by M. Likhatchof, which bears the numbers 1, 2, 3, and of which too small a quantity were sent me to make the experiment possible, the following results:

	Real power.	Calculated power.
No. 1	11,070 cal.	11,370 cal.
2	10,700 cal.	11,000 cal.
3	10,760 cal.	11,060 cal.

All these determinations are infected by a very slight cause of error, proceeding from the manner in which the calculations are established. But they demonstrate nevertheless that the oils of Bakou, compared to the American and European oils which I have examined previously, hold the first rank from the considerable value of their calorific power.

THE SHALE OIL BUSINESS OF EUROPE, ETC.

Scotland holds the first place in the manufacture of shale oil. It is estimated that 800,000 tons of shale are annually put into the retorts of the various Scotch oil works. The probable yield of crude oil from this source is reckoned at 25,000,000 gallons. To obtain this result, and also for the distillation of the crude, about 500,000 tons of fuel must be used. The principal product from the crude is burning oil, of which 300,000 to 350,000 barrels may be taken as the annual yield. Of lubricating oil, the demand for which appears to be increasing, there is produced about 9,800 tons. Also paraffine wax—of which the bulk is made into beautiful, semi-transparent candles, and the commonest of it is used in the manufacture of lucifer matches—say 5,800 tons. To these figures may be added some 2,300 tons of sulphate of ammonia, and several thousand barrels of coal oil spirit. The probable commercial value of these products is estimated at $1,260,000. All told, there are about fifty shale oil works in Scotland. Russia is advancing in this department of industry; probably the largest works of the kind in the world are to be found at Riazan. Works are now being erected at Taganroy of a very extensive character for the manufacture of shale oil. From experiments made with this coal-shale the result is said to have been satisfactory in the very highest degree. As regards the mineral, recent explorations have led to the discovery that there are apparently inexhaustible coal mines in Kharloff and Taganroy, and from these oil can be produced in extremely large quantities. This mineral is pronounced to be, some of it, anthracite, and some "half anthracite," while another quality of it is called "smolisteongle," or steam coal.

In France and Germany quite a number of these works exist, and are in working order. Of their number and capacity, we are unable to get reliable statistics.

Shale oil works are to be found in England, Spain, at Constantinople, in Italy, Bohemia and Australia, in which latter the works are very extensive. Under the present mode of retorting the shale gives a yield of 150 gallons of crude to the ton. This gives a net return of 50 per cent., or 75 gallons of clear illuminating oil. This company (New South Wales Shale and Oil Company) manufactures about 20,000 barrels per annum.

We understand that there is one company in Ireland who extract oil from the peat deposits, so plentiful in that country. Previous to the discovery of petroleum a large number of coal-oil works existed in the United States; at present there is but one, that of Henry R. Foote, at New Galilee, forty miles from Pittsburgh.

Drilling an Oil Well.

INSIDE VIEW OF A DERRICK.

GEOLOGICAL.

THE OIL FIELDS OF PENNSYLVANIA, Etc.

THE geology of the oil country is a subject upon which many theories have been wrecked. In dealing with it we purpose to present in this chapter a few quotations from the best authorities we can find upon the subject, and we would here acknowledge our indebtedness to that excellent little work of Henry E. Wrigley, Esq., C. E.*

It is well known that the Allegany mountains divide the United States geologically as well as geographically; that east of them lie the transition, the primitive and alluvial formations, and west of them the great secondary formation, or formation by deposition from water. This secondary formation extends across the continent, from the Alleganies to points far west of the Mississippi.

Whether the great valley, drained by the Mississippi, was once swept over by an ocean, of which the great lakes are but the remaining puddles, is not an object of immediate interest. That the Alleganies formed the shore or beach of some such body of water, and that along its edge were strewed animal and vegetable remains, it is undoubtedly safe to assume.

The presence of carbon, as the base of oil, shows that these deposits were either animal or vegetable, it being the base of the

* Practical Memoranda for the Use of Refiners, Producers and Shippers of Petroleum. Cleveland, 1872.

animal and vegetable world, as silica is likewise of the mineral world.

Of course, the drill does not reach these deposits. They lie perhaps almost uniformly under the edge of this formation, at a depth of from 30 to 40,000 feet. The heat at this depth, although only a matter of estimate, is doubtless very great, as we know that between 150 and 2000 feet in depth, there is an increase of 30°. It would seem, then, that these deposits of animal or vegetable matter are volatilized and thrown off into the upper rocks, and condensed there, by the lower temperature, into liquid oil. Into what rock the gas will enter, will depend upon the character of the rock. A close slate or sandstone will resist it; but wherever it finds a crevice or an open porous rock, it will force its way into it, and will condense there.

Consequently the rock itself is the guide of the driller in searching for oil, and the location of the oil producing spots, resolves itself into the existence of this porous sand rock.

All the oil-producing spots that have been found in this section of the United States are included in a belt of twenty miles in width, stretching from Western New York to Tennessee, in a line parallel with the Alleganies, and lying about fifty miles to the west of them. The producing spots themselves are in area but the smallest specks upon this belt, and are scattered over it in such an indiscriminate manner, that it is impossible to trace any connection between them, or, rather, to deduce the position of one producing spot from others, with any degree of satisfaction. It is equally impossible to trace any connection between these spots and the water-shed or river-drainage of the country.

A matter which will somewhat affect the question of production at the south end of the belt is the dip of the sand rock deeper into the earth as it goes south. Although this is, in a great measure, counteracted by the general slope of the water-shed of the country in that direction, it will still average, as near as can be ascertained by leveling and drilling, about thirty inches to the mile. Professor

Silliman says, that "Petroleum is uniformly regarded as a product of vegetable decomposition."

Professor Dana says: "Petroleum is a bituminous liquid resulting from the decomposition of marine or land plants (mainly the latter), and perhaps also of some non-nitrogenous animal tissues."

Professor Denton says: "It is a coral oil, not formed from the bodies of the coral polyps, as some have supposed, but secreted by them from the impure waters, principally, though not exclusively, of the Devonian times."

Professor Winchell says: "Crude petroleum is not a product of definite composition. It seems to be a varying mixture of several hydro-carbons, some of which, as naphtha, volatilize with rapidity when exposed to the atmosphere; others, as kerosene, slowly; while others, as bitumen, are nearly fixed. It contains also varying quantities of aluminous matter and other impurities.

Petroleum occurs in stratified rocks of all ages, from the Laurentian to the recent. It has even been observed in some rocks of a granitic structure. The mere presence of petroleum in a formation is far from being evidence that it exists in large quantities. Observation has shown that it does not exist in large quantities in any formation except under certain intelligible conditions. Its presence in small quantities is to be expected.

It is an opinion almost universal among geologists that petroleum has been produced from organic remains. Hence long before the discovery of the eozoön in Laurentian rocks, it had been inferred that organic life existed upon our planet during the accumulation of these rocks, because, among other reasons, they afford conspicuous quantities of petroleum. Geologists are somewhat divided in opinion as to whether animal or vegetable organisms have afforded most of the native oil. Little dissent exists, however, from the doctrine that most of the oil occupying the pores and pockets of fossiliferous limestone has been derived from animal bodies, while that saturating shales, and arising from shales, has had a vegetable origin. As the oil of commerce is probably derived from the latter course, it appears that we are to regard our commercial oil as a vegetable product."

Professor Winchell closes his article on the geological phenomena of petroleum * by presenting a synopsis of oil regions, and the formation tributary to their supplies.

I. The black shales of the Cincinnati group afford oil which accumulates in the fissured shaly limestones of the same group, and supplies the Burkesville region of Southern Kentucky, and Manitoulin Island, in Lake Huron.

II. The Marcellus shale affords most of the petroleum which accumulates in the fissured shaly limestones of the Hamilton group, and thus supplies the Ontario oil region, locally divided into the Bothwell district, the Oil Springs district, and the Petrolea district. The Marcellus shale affords also a large portion of the oil which accumulates in the drift gravel of the Ontario region.

III. The Genesee shale, with perhaps some contributions from the Marcellus shale, affords oil which accumulates in cavities and fissures within itself in some of the Glasgow region of Southern Kentucky. It affords also the oil which accumulates in the sandstones of the Portage and Chemung group, in North-western Pennsylvania and contiguous parts of Ohio. It affords also the oil which accumulates in the sandstones of the Waverly (Marshall) group, in Central Ohio. It affords also that which accumulates in the mountain limestone of the Glasgow region of Kentucky and contiguous parts of Tennessee, as also some of that which is found in the drift gravel of the Ontario region.

IV. The shaly coals of the false coal measures, aided, perhaps, by the Genesee and Marcellus shales, seem to afford the oil which assembles in the coal conglomerate, as worked in South-western Pennsylvania, West Virginia, Southern Ohio, and the contiguous, but comparatively barren regions, of Paint Creek, in Kentucky.

V. The coal measures may perhaps be regarded as affording a

* Sketches of Creation. Alex. Winchell, LL. D.

questionable amount of oil, which may have been found within the limits of the coal measures in the West Virginia and neighboring regions.

From this exhibit, it appears that the principal supplies of petroleum, east of the Rocky Mountains, have been generated in four different formations, accumulated in nine different formations, and worked in nine different districts. The sandstone beds in which the Pennsylvania oil is found, belong to the Chemung group of the Devonian formation. It is so called from the Chemung river, in the State of New York, where it is well exhibited.

HISTORICAL DATA.

MANUFACTURE OF PETROLEUM PRODUCTS.

Communicated to the Society of Arts, Massachusetts Institute of Technology, March 14th, 1872, by S. DANA HAYES, *State Assayer and Chemist for Massachusetts, etc., etc.*

BY referring to any authentic shipping-list, the number of thousand gallons of crude and refined petroleum sent away from the United States every day and week may be ascertained; and very little search in this direction develops statistics that are surprising to persons previously unfamiliar with them. As, for example, the total value of the crude and refined petroleum exported last year (1871,) estimated at a low average value of twenty-five cents per gallon, amounts to nearly *thirty-five millions of dollars*, in one year. And it is especially notable that a considerable proportion of this material is classed as "refined," and consists of products manufactured from the crude petroleum of the wells before shipment. There is certainly no other article of commerce exhibiting similar statistics of production and manufacture among the industries of this country.

The object of this memoir is to briefly sketch the history and present condition of the manufacture of petroleum—a manufacture which is of great importance, and which, after the diligent study given to it, and under skillful management, yields products superior to those obtained in Europe, and elsewhere, from the same crude material.

The literature on this subject is at present exceedingly meager, and generally in short articles, not always trustworthy, distributed through many journals and publications of different kinds; and very little correct history can be compiled from any records, excepting those of the Patent Office. But as the industry itself is not, at most, more than eighteen years old, we find, in the experience of practical chemists and manufacturers, a fund of very valuable and interesting information relating directly to the subject.

Having had uncommon opportunities for making myself familiar with the manufacture of petroleum products, after careful investigation, and in the correspondence of others, I find it generally acknowledged that to Mr. Joshua Merrill, manufacturing chemist of the Downer Kerosene Oil Company, of Boston, more than to any one else, belongs the honor of bringing this manufacture to its present advanced state; and, as an account of his labors and discoveries in this connection would provide a nearly complete history of the art, I take pleasure in recording some of them in this form.

COUP OIL.—The first coal-oil made for sale in this country, was produced at the works of the United States Chemical Manufacturing Company, in Waltham, Mass., by Messrs. Philbrick and Atwood, early in the year 1852. It was made, in connection with picric acid, benzole, and other products, from coal-tar; and was named by Luther Atwood, the inventor, "Coup Oil," after the *coup d'etat* of Louis Napoleon, which had taken place a few months before.

This was a lubricating oil for machinery, of which a hundred and seventy-five thousand gallons were made. It was used by many of the largest factories and railroads, and at that time was so highly esteemed, that Messrs. Atwood and Merrill were employed to make and sell it in Glasgow, Scotland, for Messrs. George Miller & Company, in 1855 and 1856. But if compared with a neutral hydro-carbon lubricating oil of the present day, it would be considered entirely unmerchantable, on account of its very offensive odor and other comparatively poor qualities.

EARLY EXPERIMENTS.—In 1856 Mr. Samuel Downer who

had previously been a successful sperm and whale-oil merchant, erected buildings in South Boston, and employed Mr. Merrill to manufacture hydro-carbon oils for lubricating purposes especially; and a great many experiments were tried there, on a manufacturing scale, with different forms of apparatus; and to determine the most suitable crude material from which to make these oils.

About four hundred tons of Trinidad bitumen, and one hundred tons of Cuban "chapapote," were consumed, and converted into lubricating and burning oils, during these early experiments. The experience gained in this way, and the many difficulties then overcome, proved of great service in the latter operations at these works.

THE FIRST ILLUMINATING OIL.—Light coal-oil products appear to have been used by individuals in this country, for illuminating purposes previous to this time; but upon the introduction of the Knapp and Dietz lamps, which were originally designed for burning resin and other oils, it was found that some of the light hydro-carbons obtained from these West Indian bitumens burned freely in them, yielding a bright and beautiful light as compared with that from the animal oil lamps and candles previously in common use. This was an important advance, and inaugurated the general burning of these hydro-carbons in lamps in this part of the country; the first illuminating oil having been made by Mr. Merrill, from Trinidad bitumen, in 1856.

ALBERTITE PRODUCTS.—In the spring of 1857, the first attempts to use the Albert coal, from Hillsboro', New Brunswick, as a source of lubricating and illuminating hydro-carbon oils, were made at South Boston. But the condensing apparatus, which had been used when distilling other coals and bitumens, was found to be unsuitable for this new material, as a black asphaltum-like substance passed bodily over out of the retorts, and often closed the cool pipes of these condensers. This difficulty was only overcome after six months spent in experimenting by Mr. Merrill, who then invented an atmospheric condenser, which, being constructed of large hollow disks, allowed this tarry distillate to pass through, with the more liquid hydro-carbons, from which it was separated afterward. With this improved apparatus, the Albertite proved such a valuable material, that, in the fall of 1857, six retorts, each having a capacity for twelve hundred pounds of coal, were erected in the open air, with the new condensers attached, and together yielded about three hundred and sixty gallons of crude coal-oil in twenty-four hours. Twelve more retorts were soon added to these six, out of doors. After Mr. Downer had made contracts with the Albert Mining Company for a regular supply of their mineral, thirty retorts were erected in a substantial brick building ; and these were followed by twenty more, the first eighteen having been worn out and removed. These fifty retorts were used for more than four years, and produced at the rate of nine hundred thousand gallons of crude, or six hundred and fifty thousand gallons of refined oils each year—quantities very much larger than were anticipated when the first experiments were made, in 1856.

A very large part of the products from this Albertite mineral were hydro-carbons used for lubricating purpose ; and their acknowledged excellence was entirely due to the untiring labor and skill of the manufacturing chemist. It is almost impossible for persons unfamiliar with the distillations of this mineral to realize at the present time, when similar processes are in common operation, how many obstacles there were to success in these early days. Many of the best forms of steam-distilling, and other apparatus now in use, were employed and perfected then.

"CRACKING."—One observation made by Mr. Merrill during the manufacture of Albertite products became of such practical and scientific importance in after years, that it is worthy of special description here.

The light or "thin" products, afterwards used as illuminators, were for a long time unmerchantable, and the production of them was undesirable ; but it was observed that every time the crude coal-oils, or the heavy lubricating oils made from them, were dis

tilled, these thin, light-colored hydro-carbons were invariably produced at first by the distillation, although the oils in the apparatus
had been previously fractioned with great care.

This caused so much loss of material, that every endeavor was
made to prevent it; such as surrounding the upper part of the stills
with heated flues, and covering their tops with sand, or other poor
conductors of heat. But these efforts to prevent the decomposition
only served to demonstrate that any distillation of these hydro-carbons is necessarily *destructive*, and that the light distillates were
produced by condensation of vapors in the upper part of the distilling apparatus, which falling back into the body of the heated fluid,
to be again raised in vapor, were thus decomposed, with the deposition of carbon, into lighter and thinner hydro-carbons. Further
experiments showed that these hydro-carbons could be so easily
decomposed, that the *continuous* production of light distillates,
having a specific gravity of about .818 (42 degrees Beaume) was
effected from hydro-carbon oils having a specific gravity of .880
(30 degrees Beaume) in an apparatus holding a thousand gallons,
by properly regulating the heat applied; the other products being
only uncondensed gases, and deposited carbon left in the apparatus
at the end of the distillation. These light distillates became valuable for use in lamps some time afterward, and the manufacturer's
difficulty was thus removed; but the unstable nature of these
hydro-carbons, and the ease with which they may be "cracked,"
was practically demonstrated when endeavoring to overcome an
annoyance.

KEROSELENE.—Any account of the Albertite products would
be incomplete without mention of the lightest naphtha obtained
from it, which was known as "keroselene." This was made by
Mr. Merrill in 1857, just after the first successful distillation of the
New Brunswick mineral; and it was used in considerable quantities for supplying carburetted air or automatic gas machines, which
would have been useless at that time without this material. It
was obtained in a crude condition by carefully refrigerating the

waste gases as they passed from the outlets of condensers attached to the stills and retorts; this crude material, after agitation with sulphuric acid, was redistilled by steam heat; and the exceedingly volatile keroselene, having a specific gravity of only .634, and which boiled at 85 degrees Fahrenheit, was manufactured in this way.

Mr. Merrill first noticed the anæsthetic effects of keroselene upon a laborer engaged in cleaning a tank or cistern which had contained it, at the works; and afterwards experimented further with it upon rats and mice. This discovery of its anæsthetic properties was recorded in the medical and surgical journals of that time, and many interesting results were obtained with it by different members of the medical profession.

PETROLEUM.—I have been unable to find any record of the distillation of American petroleum, taken from the wells, in a large way; although it was investigated, and the distillates obtained in the laboratory were burned experimentally, quite early. But Pennsylvania petroleum was not, probably, made into illuminating oil, for sale, before the fall of 1858; although there were fifteen establishments using petroleum exclusively, in the United States, by the fall of 1860.

During the years 1858 and 1859, several hundred barrels of petroleum were brought to Mr. Merrill from surface wells; and, on one occasion, a considerable quantity was sent to him for distillation from the famous Tarentum well in Pennsylvania. And afterward the heavy paraffine oils and residuary products from Western establishments, which were considered valueless there, were manufactured into lubricating and illuminating hydro-carbon oils at the works in South Boston. As the supply from flowing wells increased, the use of Albertite, as a source of fluid hydro-carbons, gradually diminished; although it was not abandoned until 1865.

After the trying experiences of former years, no difficulty was encountered in converting crude petroleum into naphthas, burning

oil (called kerosene or "wax-oil"—a patent trade-mark name),
lubricating oil and paraffine, similar to those made from coals and
bitumens.   Petroleum breaks up into thin hydro-carbons, by distillation, even more readily than the Albertite products; and when
large demands are made for burning oil, the distilling apparatus is
operated slowly, or modified in form, so that the condensed vapors
of the petroleum, or heavy oils, obtained from it are repeatedly
heated by being returned into the body of the still; and in this
way the yield of the lighter hydro-carbons may be increased at
will, the whole contents of the still being converted into burning
oil when desirable.   This principle is applied to the immensely
large wrought-iron stills, holding two thousand barrels, or eighty
thousand gallons each, that are now frequently used; they are
placed over a number of small fire-places, with the top and upper
part of the stills exposed to the outside atmosphere, for the purpose
of condensing and returning heavy vapors in the stills.

Petroleum yields, by distillation, nine distinct commercial products.

Name.	Specific Gravity.	Beaume Scale.	Boiling Point.
Righolene	.025		65° F
Gasolene	.665	85	120° "
C. Naphtha	.706	70	190° "
B. Naphtha	.724	67	228° "
A. Naphtha	.742	65	300° "
Kerosene Oil	.804	45	350° "
Mineral Sperm Oil	.847	36	425° "
Neutral Lubricating Oil	.883.	29	575° "
Paraffine	.848(?)		

Four of these products are especially interesting and valuable:
the first is Righolene.   It nearly corresponds to the keroselene of
the Albertite products, and is an extremely volatile hydro-carbon,
used for producing local anæsthesia, by its rapid evaporation,
during short surgical operations.   In 1866, Mr. Merrill was requested to make the most volatile fluid he could produce from
petroleum, by Dr. Henry J. Bigelow, the eminent surgeon of
Boston.   This was done by redistilling gasolene, which was the

very lightest petroleum naphtha, by steam heat, and condensing the first distillate by the aid of ice and salt. In this way, ten per cent. of the gasolene was converted into the lightest of all known fluids,* which was named righolene by Dr. Bigelow. Its specific gravity is only .625, and it boils at 65 degrees Fahrenheit. The evaporation of this fluid is so rapid at common temperatures, that it will depress the mercury in a Fahrenheit thermometer to nineteen degrees below zero, in twenty seconds. Several hundred gallons of righolene have been consumed for surgical purposes. We pass next to the neutral heavy lubricating oil. At the works in South Boston, the production of lubricating oils has always been a specialty, but with petroleum there came a scarcity of heavy and dense crude materials from which to make these oils; fortunately, however, other manufacturers of petroleum, finding the illuminating oil and light products more profitable, have been glad to dispose of their heavy residuum at these works; and the deficiency of material has been supplied from this source.

The lubricating or paraffine oils were always characterized by offensive odors and tastes, so that a person brought in contact with them became at once aware of their origin; and the desideratum with manufacturing chemists, from the earliest days of this industry, has been the production of dense *neutral* oils, or oils free from these offensive objections. Much time and study have been devoted to experiments having this object in view; and shallow stills, stills with double heads, repeated distillations, different processes of purification by chemical agents, and many other means, have been tried without success. But partly as the result of an accident, Mr. Joshua Merrill succeeded in making neutral oils in November, 1867.

Distillation had fairly commenced, from a still heated in the usual manner, by a direct fire underneath, and charged with nine hundred gallons of mixed heavy and light oils that had been pre-

---

* Cymogen is a still lighter and more volatile product of petroleum. Its gravity is 0.590=110° B. Its boiling point 32° F. [EDS.]

viously distilled, and which were too heavy for illuminating, and too light for lubricating purposes, when it was found that the condenser had partially closed from some accidental cause; and this, by the consequent pressure, soon caused leakage at joints about the bottom of the still, over the fire. Continued distillation increased the leakage, so that it was necessary to withdraw the fire gradually from under the still; although the distillation was continued for some time in an attempt to empty the apparatus by operating very slowly. When the fire was removed, and after distillation ceased, two hundred and fifty gallons of light hydro-carbons, had passed over through the condenser. The next day, the oil left in the still, having cooled sufficiently, was removed; and Mr. Merrill was surprised to find it different from anything that he had ever seen before. It had a bright yellow color; was clear, very nearly odorless, *neutral* and dense. Further experiments showed this result to have been obtained by the *removal of all light odorous hydro-carbons without decomposing* either the distillate or the oils remaining in the still, and that this had been accomplished by the moderate fire employed, and its gradual withdrawal.

This mode of operating was immediately applied to other distillations; and, after two months spent in determining the best mode of procedure, Mr. Merrill obtained letters patent for his valuable discovery, from which extracts will be made in a description of the present improved processes of manufacturing petroleum products, at the end of this memoir. This discovery consisted, first, in determining that the odor and taste of the heavy distilled oils arise from the presence in them of light and odorous hydrocarbons, formed during the previous and necessarily destructive distillations; and that when these are removed by distilling in a suitable apparatus, with the application of only sufficient heat to remove them, without decomposing or *cracking* the oils in the still, the latter are left nearly odorless. And, secondly, in perfecting the means for effecting this removal of the odorous bodies. The introduction of steam from an open pipe to the body of the apparatus during this distillation, aids greatly in effecting the separation, as it lifts the light vapors out mechanically into the condenser ; and it also serves to regulate the heat employed for distillation.

Thus, by extraordinary means, true fractional distillation of the heavy mixed oils is effected ; and Mr. Merrill says : " I believe it is impossible to prevent the cracking or decomposition of these hydro-carbon oils, except by keeping them in the still, at temperatures below their boiling points."

Many hundred thousand gallons of this neutral heavy hydrocarbon oil, which has frequently perplexed the most expert judges and dealers in oils, have been made by Mr. Merrill. It is almost odorless and tasteless, and cannot be easily distinguished when mixed with one-fifth part of its volume of the best bleached animal, sperm or other fat oil, as an examination of this specimen will assure you. No better estimate of its valuable qualities can be given than the statement that, in the year 1871, fifty thousand gallons of this oil were sent to England alone, where it was used for lubricating spindles, oiling wood and other purposes. Another important substance obtained from petroleum is

PARAFFINE.—This was one of the very first products made at the works in South Boston, in the early days: and it is interesting to know, that while Cuban chapapote bitumen yielded paraffine by distillation, and Albertite gave large quantities of it, the Trinidad bitumen never afforded any of this crystalline mineral wax. Mr. Merrill began to make paraffine from Pennsylvania petroleum in 1859, and since then has, at times, made the enormous quantity of fifty thousand pounds (or twenty-five tons), in one month. It is a product of *destructive* distillation of this petroleum, and does not exist already formed in the crude oil.

This substance is used principally in candles, also for rendering textile fabrics water-proof, and for many other purposes ; as one maker of friction-matches in New York has used one hundred thousand pounds, and a manufacturer of chewing-gum in Maine seventy thousand pounds, of paraffine in one year.

MINERAL SPERM-OIL.—The fourth product to which I wish to draw your attention is mineral sperm-oil. This is a burning heavy oil made from petroleum; and its valuable properties as a safe illuminating agent are such as to render this product one of very great importance. The following statement of its discovery and characters is given in Mr. Joshua Merrill's own words:

" In the summer of 1869, in connection with Mr. Rufus S. Merrill, I made an important discovery relating to burning the heavy or paraffine oils in lamps, for illuminating purposes. Mr. R. S. Merrill is a skillful mechanic, who has devoted himself for several. years to perfecting the construction of lamps and burners for hydrocarbon oils. While experimenting upon an apparatus for burning paraffine wax, with a view to increase the light from this beautiful substance over that obtained from common candles—the only form in which paraffine is burned—he one day put some lubricating oil into the lamp, instead of the paraffine wax, and we were both much surprised at the good qualities of the light yielded by it. But, after examining some days, we found this heavy oil to be impracticable as an illuminating material in its present form, and that some modification would be necessary. It occurred to me, that if this heavy paraffine oil was passed through a partially destructive distillation, cracking it enough to lessen its viscidity, but not enough to render it volatile, its increased mobility would cause it to ascend the wicks freely, and yet preserve its character as a fixed oil.

" After many trials, I obtained the product now called ' mineral sperm-oil,' which is sufficiently thin to fill the wicks perfectly; but it is so far from being a volatile oil that it is comparatively inodorous, and will not take fire at any temperature below 300 degrees Fahrenheit, or nearly a hundred degrees hotter than boiling water. Flames of considerable size, such as a large ball of wicking-yarn, saturated with oil, and ignited, when plunged beneath the surface of this oil, previously heated to the temperature of boiling water, are extinguished at once. It burns freely in the German student lamps, and with great brilliancy from the 'Dual' burner."

The manufacture of this oil is patented in this country and in Great Britain; and Mr. Merrill estimates the quantity that may be made as at least one-quarter of the whole production of petroleum, or about one hundred and sixty thousand gallons of mineral sperm-oil every day—a quantity more than twice that of the whale and sperm oils obtained in the best days of the whale fishery of this country.

The present time, when government authorities and scientific men are so generally cautioning against the "dangers of kerosene," and just as French *savans* have discovered that certain heavy petroleum oils may be burned in lamps,* seems peculiarly opportune for the introduction of this product of American skill and invention—namely, a hydro-carbon, or mixture of hydro-carbons, which seems to fulfil all the requirements of an oil to be burned in lamps, yielding a steady, brilliant and safe light. And practical indications of its appreciation may be found in the manufacturer's announcement, that the demands for this mineral sperm-oil is steadily increasing. It is used on ocean steamers plying between the United States and Europe, and also on several railroads.

MANUFACTURE OF PETROLEUM PRODUCTS.—It only remains for me to give you a brief outline of the manufacture of petroleum products, as conducted at a well-managed establishment in this vicinity.

The crude petroleum is received here from the West in round wooden tanks, one or two of which occupy a railway carriage, as you have undoubtedly observed them on the neighboring railroad; and these are sometimes emptied into bulk-boats, when the works are more easily approached by water. The petroleum is pumped out and underground into the iron reservoir tanks that closely resemble gasometers, where any sand and water that may be present are deposited and removed. The crude oil, drawn from these

* Comptes Rendus de l'Acad. des Sciences, July, 1871

reservoirs, is first pumped into large wrought-iron stills or upright cylinders, incased in wood to prevent loss of heat, which hold about twelve thousand gallons each. These are the naphtha stills, in which the petroleum is heated by steam alone; the distillates being collected by condensers, consisting of iron pipes surrounded with cold water. Only the naphthas, or about fifteen per cent. of the crude oil, are distilled from these large stills; but the four kinds (gasolene, A, B and C naphtha) are collected in different receptacles. Righolene, as stated before, is made by a second distillation of gasolene.

The steamed crude oil remaining in the naphtha stills is pumped from them into smaller stills heated by direct fires underneath, and holding about a thousand gallons each; the whole contents of these are distilled over and condensed, excepting that which passes into the air as uncondensable gas, and the separated carbon. This is the first complete distillation, and it is eminently *destructive;* the products being separated, by densities, into three grades—No. 1, crude burning oil; No. 2, intermediate oils; and No. 3, crude lubricating oil. Each of these is redistilled by itself in apparatus of the same size and construction as that used for the first distillation, and they all break up again into lighter, intermediate and heavy oils. No. 1 is thoroughly agitated with sulphuric acid and caustic soda successively, by revolving stirrers in large tanks, before its second distillation; and it then yields from the still, eighty per cent. of its volume of finished kerosene and mineral sperm, and nearly twenty per cent. of denser oil. No. 2 is redistilled before treatment with acid and alkali; it yields crude lubricating oil principally. No. 3, the crude lubricating oil, requires more careful manipulation than either of the others. It is first agitated with sulphuric acid, and then distilled with caustic soda present in the still, the product being mostly dense paraffine oil. This is placed in wooden barrels, in ice-houses, where it remains for from seven to ten days; and during this time the paraffine wax crystallizes, so that the masses retain the form of the barrels when they are removed. It is now put into bags made of strong cloth, which are arranged one above another, with sheets of iron between them; and, when submitted to heavy pressure, it yields crude scale-paraffine wax remaining in the bags, and heavy oil is pressed out. The crude paraffine is refined by repeated solution in naphtha, recrystallizing and pressing until it is perfectly white and pure, ready for sale. The heavy oil is treated by the patent deodorizing process. It is placed in stills heated by fires underneath; and the temperature is slowly and gradually raised, until from twenty to thirty per cent. of the contents of the apparatus is distilled over; it is then allowed to cool in the stills, and when removed, is ready for sale. The hydro-carbons that pass over to the condensers during this process have very offensive odors; but the oil remaining in the stills, if the operation has been properly conducted, is free from the characteristic odor of paraffine oil, and has only a slight odor, similar to that of fat oil. Live steam is generally used in the body of the oil during this operation, and the distillation is effected at as low a temperature as possible.

The very last distillates that are obtained from all the destructive distillation made at the works are highly colored, and known technically as "cokings;" these are accumulated and distilled by themselves, yielding crude lubricating oil principally. After every distillation of petroleum, or the products obtained from it, considerable masses of separated carbon are obtained as residuum; and, as caustic soda is frequently used in the apparatus, it remains in this coke; this is saved, however, by burning the carbon in a properly-constructed fire-place; and the ashes lixiviated yield the soda as carbonate. The sulphuric acid that has been agitated with the oils, known technically as "sludge," is carefully saved, and generally sold to makers of biphosphate of lime fertilizers, although some of it has been successfully reconverted into commercial oil of vitriol by an ingenious process that I need not detain you to describe.

I have thus hastily reviewed the history and *modus operandi* of an important manufacture, in the belief that some parts of this memoir are new and of general interest.

GAS WELLS.

PENNSYLVANIA, OHIO, NEW YORK, KENTUCKY.

CARBURETTED hydrogen is the chief component of the gas which escapes from the earth in wells and springs in many localities. It is evolved in the working of coal mines, and constitutes " fire damp." It is also a constant associate of petroleum, and always issues in greater or less quantity from oil wells. It is given off, too, in the decomposition of recent vegetable matter, and may be seen bubbling up through the water of all pools in which plants are decaying. When it escapes from the earth it may be generally traced to beds of bituminous matter, such as coal, lignite, carbonaceous shale, asphalt, oil, etc. From these substances it may be obtained by artificial distillation, and is evolved by the spontaneous decomposition which all organic substances suffer on exposure.

As carburetted hydrogen produces a brilliant light in combustion, it is largely manufactured and used for the illumination of cities and residences. So extensively is it employed for this purpose that it may he regarded as an indispensable element in our modern civilization. It is not strange, then, that efforts have been made to utilize the immense quantities of gas which flow from wells and springs in so many different countries. The Chinese have for hundreds of years used for lighting and heating, the gas which emanates from the earth in several provinces of their country. In the United States the gas which issues from the salt wells of the Kanawha Valley has been for many years employed as a fuel in the evaporation of the brine.

Of course the oil wells of Pennsylvania produce gas, and often in very great abundance, and it occasionally occurs, that wells drilled for the purpose of obtaining oil, produced only gas. On newly developed territory the sight to be witnessed at night in the many illuminations from this natural gas, is truly grand, causing the heavens to be lighted up, and the earth to be spread abroad with a brightness equal to the best artificial illumination of any modern city. To the reflecting observer the sight will prompt him to look from " nature up to nature's God."

The village of Fredonia, in western New York, has for more than forty years been fully or partially lighted by gas which issues from springs at that place. In the borings made for oil in the various oil districts of the Western States, the gas which has been produced so abundantly has been regarded as a useless, frequently inconvenient and dangerous product. Within a year or two past, however, this gas has been utilized in numerous localities, and already a large number of wells have been bored for the express purpose of obtaining it. In some cases these gas wells have been highly productive, furnishing an abundance of material for heating and lighting in its most convenient and manageable form, so that this deserves to be reckoned as one of the important elements in the mineral resources of our country. As this method of procuring carburetted hydrogen gas forms a new industry, and one which will probably assume considerable importance, a few words in reference to its present condition and prospects may not be without interest to the public. We therefore extract from our notes a few facts in regard to some of the most interesting of our gas-producing districts. On the Upper Cumberland, in Kentucky, gas accumulates in such quantities beneath the sheets of Lower Silurian limestone, that many hundred tons of rock and earth are sometimes blown out with great violence. These explosions have received the local name of " gas volcanoes." In Ohio, gas escapes from nearly all the wells bored for oil in the oil-producing districts. Of these, two bored by Peter Neff, Esq., near Kenyon College, in Knox county, present some remarkable features. These wells were bored in 1866, at the same geological horizon as that which furnishes the oil on Oil Creek, Pa. At the depth of about 600 feet, in each well, a fissure was struck from which gas issued in such volume as to throw out the boring tools, and form a jet of water more than 100 feet in height. One of these wells has been tubed so as to exclude the water, and gas has continued for five years to escape from it in such quantity as to produce, as it rushes through a two and-a-half-inch pipe, a sound that may be heard a considerable distance. When ignited, the gas forms a jet of flame three feet in diameter and fifteen feet long. The other well, which has never been tubed, constantly ejects, at intervals of one minute the water that fills it. It thus forms an intermittent fountain, one hundred and twenty feet in height. The derrick set over this well has a height of sixty feet. In winter it becomes encased in ice, and forms a huge translucent chimney, through which at regular intervals of one minute, a mingled current of gas and water rushes to twice its height. By cutting through this chimney at its base and igniting the gas in a paroxysm, it affords a magnificent spectacle—a fountain of water and fire which brilliantly illuminates the ice chimney. No accurate measurement has been made of the gas escaping from these wells, but it is estimated to be sufficient to light a large city.

At West Bloomfield, N. Y., is another gas well, not unlike those described. This is bored to the depth of five hundred feet, reaching down to the vicinity of the Marcellus bituminous shales. From some measurements made by Prof. Wurtz, it appears that about fifteen cubic feet of gas escaped from this well every second. It is proposed to utilize this large amount of valuable combustible by conducting it through pipes to Rochester, a distance of twenty miles.

At Erie, Pa., there are now twenty-five wells in successful operation, most of which have been bored for the special purpose of obtaining gas.

FIRST. H. Jarecki & Co. (Petroleum Brass Works) have two

wells; the first bored for oil in 1854, 1,200 feet deep. No oil was obtained, but brackish water and an abundant supply of gas.

At Conneaut and Painesville, Ohio, wells have been bored for gas with entire success, and others are being bored, in these localities, and at many points farther west.

Of two of the Painesville wells, a few notes may be of interest to the residents of the lake shore. First, is the well of Gen. Casement on the east side of the town. This well is 700 feet deep, and passed through the following materials :

One—Drift, clay and gravel, 40 feet.

Two—Erie shale, "soapstone rock," 648 feet.

Three—Huron shale, very black and bituminous, with strong smell of oil, 12 feet.

Gas was found in the Erie shale. The supply is abundant, and is used for all domestic purposes in Gen. Casement's house. The pressure of the gas was tried; but at 27 pounds per square inch part of the apparatus failed.

SECOND.—Well at the Erie Seminary, 725 feet deep at the time of observation. Strata passed through—

One—Clay and sand, 23 feet.

Two—Erie shales, alternations of sandy and argillaceous gray and green shales, 687 feet.

Three—Huron shale, black and bituminous, 15 feet.

The gas was found in the fissures or crevices of the "soapstone rock." In the seminary well four such fissures were found. The first gas was obtained at 300 feet.

Another well at the seminary is of similar character. The supply of gas from both wells is insufficient to light and heat the building. If collected in a gasometer, it would at least supply all the light required.

In the vicinity of Cleveland, as in many other localities in the eastern half of the state, gas and oil springs are frequently met with, and many wells have been bored for one or the other of these useful articles. Here, as elsewhere, there are two marked lines of gas and oil springs, connected with the outcrops of the two sheets of bituminous shale which underlie the surface. First, the Cleveland shale of Lower Carboniferous age, which crops out along the base of the hills that bound the Cuyahoga Valley. This bituminous shale is from thirty to sixty feet in thickness, and is the source of the oil of the East Cleveland and Kingsbury quarries, the noted gas spring at the brick-yard beyond East Cleveland, etc. The oil of Mecca and Liverpool is derived from the Cleveland shale. Second, the Huron or Great Black Shale, which passes beneath Cleveland, and rising westward comes to the surface in Huron and Erie counties, and forms a broad belt of outcrop, thence to the Ohio river. The Huron shale has a thickness of 300 to 400 feet, mostly black and highly bituminous, and is the source from which the gas of the Neff wells and the well on the lake shore, and the oil of Oil Creek are derived. In the valley of the Cuyahoga, in both Cuyahoga and Summit counties, a large number of wells were bored for oil some years since. Most of these yielded both gas and oil, but neither in large quantity. From a similar well in the valley of Rocky river a copious flow of gas has continued to escape for several years. Of the wells recently bored in this vicinity, a brief notice may perhaps be of interest to the people of Cleveland. First, well at mouth of Kingsbury run, bored by the Standard Oil Company for water. The depth of this well is 1,005 feet. The well head is about ten feet above the river. Before reaching the rock, 238 feet of clay, with partings of sand and gravel, were passed through. The rock penetrated was gray and black shale. Some water and gas flow from this well, but neither in large quantity.

Second, well bored by the gas company, at the gas works near the mouth of the river, well head about fifteen feet above the lake. This well is 835 feet deep. The rock was reached at 116 feet, the overlaying material being mainly blue clay, with sheets of quicksand and gravel. The rock passed through consisted of alternations of gray and black shale. Gas was obtained at several points, but not in remunerative quantity.

Third, well bored by Captain Spaulding between Cleveland and Rocky river. This well began about 100 feet above the lake. At the date of my visitation, it had been sunk to the depth of 715 feet. It passed through,

1. Sand and clay, 12 feet.

2. Erie shale, gray argillaceous shale, with bands of sandstone, 400 feet.

3. Black shale, with thinner bands of gray Huron shale, 303 feet. In this well gas was obtained at several horizons, and the quantity is sufficient to light a number of houses.

In the city of Erie there are some thirty gas-wells. These wells are for the most part drilled to a depth of from 500 to 700 feet. The shale is here reached at the depth of from 30 to 40 feet, and extends below any depth yet reached by the drill, and is composed of alternate gray, and black layers or veins—the gas being found only in the latter. The gas from a portion of the wells here furnishes fuel to three steam flour mills, the city water works, an oil refinery, two machine shops, a car manufactory, and four or five other steam works. The City Gas Company also have a well, and use a mixture of natural and manufactured gas for the use of the city.

At Buffalo, N. Y., a well was sunk to a depth of 640 feet, when a large vein of gas was struck. The volume of uprising gas showed a presence of 130 pounds to the inch. This gas is of remarkable purity.

At Cumberland, Maryland, a company started what they intended to be an oil well. Gas was struck, and soon afterwards was accidentally set on fire, and continued to burn for a period of two years. A Mr. Haworth, having heard of the burning well, went to Cumberland, tested the quality of the gas, and was satisfied that he could put in operation a scheme or plan of his own, for the manufacture of carbon from the gas. The well was leased, and a patent obtained for the manufacture of carbon from the gas. Mr. Haworth has now in operation 660 burners, each burner consuming eight cubic feet of gas per hour. The gas is allowed to burn against soapstone plates, on which the carbon is deposited in the shape of soot.

The carbon is used for the manufacture of ink, and these works, we believe, are the only ones of the kind in the country.

Some six miles east of Crab Orchard, in Lincoln county, Ky., there is a spring known as the "Burning Well," situated at the very base of the Cumberland mountains, on the banks of a small stream called Dix river. The water in this well is in a constant state of ebullition, and regularly, every day, between four and five o'clock in the afternoon, overflows. A large quantity of gas is liberated, said to be carburetted hydrogen gas, to which a light being applied, a flame, sometimes ten or fifteen feet in height, results. The only peculiar feature of this well is the diurnal and infallibly regular overflow.

REMARKABLE GAS WELL AT FAIRVIEW, PA.

In June, 1872, a well was drilled about two miles from Fairview, Butler County, Penna., to a depth of 1335 feet, for oil, and was abandoned on account of the strong flow of gas and salt water; so great was the flow of gas that the boiler had to be removed to a distance of twenty-five rods. After the well was abandoned some two months, the pressure of gas became so strong that it forced the water entirely out of the hole, and in the autumn of the same year a company was formed to utilize the gas, which was done by bringing it through a 3¼-inch casing to Fairview, and thence to Petrolia, three miles from Fairview. The gas is used to light the streets and heat residences and offices in both places. The pressure, as indicated on a steam gauge, is 80 pounds. This well has an escape through a 6-inch pipe, and the noise of the escaping gas, can be heard readily for a distance of two miles.

A correspondent of the Titusville *Herald* under date of September 3, 1873, gives the following graphic account of this remarkable gas well:—

"The roar of the escaping fluid was equal to the sound of Niagara, and the iron tools that had penetrated 1335 feet of solid rock were raised, and tossed in the hole with as much ease as a skiff is rocked upon the surface of an angry ocean; so strong was the gasgiant that one man might have helped the tools out of the well without the aid of an engine. A man might throw a one hundred pound rock into the escaping column, and it would be thrown with ease to the height of forty or fifty feet in the air; an ordinary club might be launched into the upward stream, and it would be toyed with as a fountain jet toys with a marble. It would raise a club seventy-five or eighty feet in the air, and when it would begin to descend it would be elevated again until it would escape the centre of the current, and then return to the ground. The voice of this giant can be heard for five miles distinctly, and it sounds like the approach of a train of cars on the railroad, or like the sound of a brake when letting down tools into a 1500 foot well. In the hole is a little salt water, just enough to make the gas appear like blue smoke. The water, under the force of the gas, is formed into a mist, and on approaching the well, appears like a column of smoke rising out of the valley, but 'woe be unto him' who touches a match in this giant's face, for his breath is explosive, and would, when lit, make heat enough to melt iron.

"For a few weeks this well blew, and howled, and whistled, making night hideous and day tedious with its ceaseless 'yells,' until the arms of science opened to receive the wasting fuel. A twenty-horse power boiler was stationed near the well and connected to receive it; to the boiler was connected a three-and-a-half inch tubing, which was laid for seven miles to Fairview, Petrolia, Karns City, and Argyle; to this seven miles of pipe are attached forty pumping and drilling wells, eight pump stations and different pipe lines, two hundred gas burners, and forty cook stoves, all of which burn the gas from this well. But they do not use it all. The well is only a five-and-a-half inch hole, but the waste pipe is a five and five-eighth inch casing, which fairly rings with the pressure of the escaping waste gas."

NEWTON GAS WELL, NEAR TITUSVILLE, PA.

Nearly all wells producing oil, yield small amounts of gas, which is often found in quantities large enough to make it available as fuel for boiler fires; but wells, producing large quantities of gas unaccompanied by oil, are comparatively rare. We have instanced a few in different parts of the country, and would here make mention of one quite famous well of this sort, at the mouth of East Sandy Creek, hereafter described, and another at Stewart's Run, both in the Pennsylvania Oil Region. But the most remarkable gas well yet discovered is the Newton well on the A. H. Nelson farm, five and one-fourth miles northeast of Titusville. This well is the second one drilled in this vicinity,—the first sunk in the fall of 1871, proved a dry hole. The usual strata of rock found in this region, were passed through in drilling, and there was no indication of oil, and but slight signs of gas, during the process of boring.

This well is drilled to the depth of seven hundred and eighty-six feet, and was finished on the 11th of May, 1872. A few minutes after the pump was set in motion the flow of gas commenced throwing up the fluid as fast as a two and a-half inch outlet would allow it. Soon the water was exhausted and the gas rushed out with a deafening noise, and with terrible force. The well was at this time tubed only to the second sand-rock, a depth of seven hundred and five feet. The casing was now lowered below the second, and the tubing to the third sand-rock, and pumping resumed, with about the same pressure from the third sand-rock, as from the second, but showing a difference in the quality of the gas, it being much purer and of higher illuminating power. On the 24th of June, the casing was removed and placed above the first sand-rock, leaving all the gas veins open below the casing. A sand-pump was then run down a few times for the purpose of exhausting the water, and agitating the well, so as to permit the gas to flow the more freely. On putting down the sand pump for the fourth time, the gas again rushed up, carrying the sand pump and line with it, faster than steam power could be made to draw it out; and for several minutes the well discharged a column of water to a height of at least one hundred feet, making a splendid sight, and a noise which is said to have been heard for a distance of ten miles.

As soon as possible the gas was divided into seven two-inch jets, one of which was sufficient to run the engine, the gauge showing a pressure of 75 lbs. to the square inch. Calculations, as accurate as it is possible to make, showed a total pressure of not far from 350 lbs. to the square inch, and a flow of more than 500,000 cubic feet of gas per day. Each day the volume of gas seemed to increase and then occurred to some capitalists, the feasibility of carrying the gas to Titusville to supply the many manufacturing firms and private families with it as fuel, and to this end the well was purchased by Henry Hinckley, Esq., of Titusville, who had the product of the well measured, which revealed the fact that it was producing over four million cubic feet per day! On the first day of August, 1872, the gas was conveyed through a two-inch pipe to the city of Titusville. This two-inch pipe was found, after a short time, inadequate for the demand, and a line of $3\frac{1}{4}$-inch pipe was laid down, and now supplies two hundred and fifty firms and private families with gas, for heating and lighting purposes.

Of the many striking features of the Pennsylvania oil region, Gas City, Cranberry township, Venango County, is one of the most remarkable. It is called Gas City because of the large amount of gas flowing from its wells, of which there are some thirty producing oil. The town has about forty houses, composed principally of hardware stores, groceries, restaurants, &c., and these are all heated and lighted by gas from one well. Each well has gas enough to make steam for its engine, and to light the engine-house —and each engine-house uses for light as much gas as would supply a large hotel, and wastes more than would supply a town of five thousand inhabitants. At each well there is pipe run from the boring to some distance, through which the waste gas is burned at an elevation of fifteen to twenty feet.

GAS WELLS AT EAST SANDY.

There is a remarkable gas well at East Sandy, in the Pennsylvania oil region, which was struck in the spring of 1869. It caught fire, and resisted all efforts to extinguish it, and it burned for a little more than a year, lighting up the surrounding country for a great distance. The rush of gas and flame, roaring like a cataract, could be heard for miles. After partial exhaustion, the gas was conveyed in pipes in some instances upwards of half a mile, for use on both drilling and pumping wells. The amount of gas produced daily by the well is not known, having never been tested, but some idea of it may be gathered from the fact that it has supplied gas to 20 pumping and drilling wells at one time. In some instances this gas was utilized directly into the engine, like steam as a motive power, the steam-gauge indicating a pressure of 80 to 90 lbs. to the inch.

The presence of large quantities of gas in the Pennsylvania oil region usually indicates the presence of an abundance of oil in the neighborhood. East Sandy offering such inducements, oil men, prominent among whom we may mention F. W. Andrews, of Titusville, commenced active operations. Success was not so great as expected, yet quite a number of good paying wells were found.

THE PHENOMENA OF OIL WELLS.

As every human being has his own set of features, tone of voice, and the like, so each individual well has its characteristics, whether it be a flowing or pumping well. All differ in regard to the flow of water, gas and petroleum. In one the flow of oil will be continuous and uniform, day and night, not ranging more from week to week than a spring-brook. In others the flow will be intermittent, but with precise regularity as to time; others again flow at irregular intervals.

It is recorded of "The Coquet Well"—Hyde and Egbert farm—that she emitted a succession of sounds as loud and as sharp as the exhaust of a small steam-engine, occurring in tolerable order every ten seconds, in such a manner as 1, 2, 3, 4; 1, 2, 3; 1, 2, 3, 4; and sometimes two or three coming off together by an extraordinary effort. "The Wild Cat" and "Yankee Wells" remained silent for forty and twenty minutes respectively, and then began to foam and flow, the oil coming off at first only in drops, but increasing by degrees, until it belched forth with terrifying force and power. These discharges then decreased in violence, and finally fell off entirely, after the lapse of from five to eight minutes. Each escape of liquid was accompanied by a sharp report, heard at the distance of a hundred yards or more. Some wells have remained quiescent for twenty-one or twenty-two hours in the day, and then have broken forth in one continuous flow, or a succession of belchings, for two or three hours. A few wells have run for six hours, and then subsided or distributed their favors over twelve hours in the twenty-four. "The Dunn Well," on the Watson flats, produced freely from morning till midday; then the supply diminished or stopped altogether, for the rest of the day, the pump bringing little but salt water. In the case of pumping-wells, with each revolution of the band-wheel, (all things being in working order,) a discharge of oil, or salt water, or both, takes place.

It is not difficult to account for the phenomena of flowing wells.

Gas seems to be the life-blood of these remarkable wells. Professor Winchell, in an article on gas-wells, intelligently disposes of the question, which we here append:

"The escape of oil at the surface of a well is caused sometimes by mere hydrostatic pressure, as water rises in a common artesian well. More frequently, perhaps, the oil is forced up by the elastic reaction of confined gases. An open cavity, or a porous portion of rock, bounded on all sides by impervious walls—which constitutes a virtual cavity—may be partly filled with oil, while gases occupy the higher portions of the cavity. Such a cavity, whether actual or virtual, may possess any form or extent, or may consist of a number of cavities connected by narrow passages or mere fissures. In nearly all cases, more or less gas accompanies the oil,

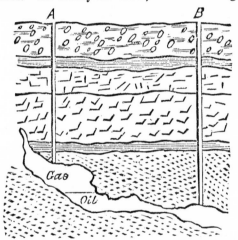

FIG. 1.   FIG. 2.

and subsists under a very high degree of pressure. The pressure in such cases is not the hydrostatic pressure of water, but a consequence of the continual generation of gas and oil long after the cavity had been filled. If the boring happens to penetrate the higher portion of such a cavity, (Fig. 1), the gas at once rushes forth with greater or less violence and persistence. As soon, however, as the tension is relieved, the escape ceases. No oil will be obtained in such a case without applying suction, since there is no hydrostatic pressure exerted from behind, and the reaction of the gas tends rather to confine the oil in the lowest ramifications of the cavity.

"Suppose, however, on boring a hole for oil, we happen to penetrate some of the lower portions of the cavity occupied by the oil. The elastic pressure of the confined gas above will at once force the oil up, and produce a spouting or blowing well. The flow must necessarily subside by degrees as the confined gas, by the escape of the oil, acquires more space for its accommodation. It may continue, however, until the cavity is exhausted of its oil, after which pumping will be of no avail. If the confined gas attains its equilibrium before the oil has been completely forced from the cavity, it is evident that the remainder must be obtained by pumping.

"Intermittent wells appear to act in some cases precisely after the manner of intermittent springs. More frequently, however, it is manifest that the combined action of gas and oil produces the phenomenon. In boring a well, suppose a stream of gas is struck over one hundred feet from the surface of the rock, and a small stream of oil twenty feet below the gas. The entrance of oil fills twenty feet of the hole, and begins to submerge the fissure at which the gas is escaping. The gas forces its way through the oil with a sputtering sound, bubble after bubble rising to the surface. As the oil ascends, the gas makes louder and louder complaints, till, finally, summoning all its accumulated energies, it hoists the super-incumbent column of oil to the surface, and pours it out in a few seconds' duration. The flow then ceases, and the same operation begins to be repeated. After a minute or more, renewed grumbling and sputtering, the pent up-gas again relieves itself, and thus the work continues. The same result would ensue if oil and gas found entrance at the same fissure, or even if the gas were admitted at any distance beneath the entrance of a small supply of oil."

In evidence of the truth of Professor Winchell's conclusions, we may instance a case in point. On the main street of Rouseville, Venango Co., Pa., is the site of an old well, drilled some ten years ago, which was pumped, and long since abandoned; it is now covered with earth, and hidden from view. This well, with the regularity of time, flows oil and gas once a month. The cause is obvious: evidently the gas accumulates, carrying with it oil, forcing itself through the resisting earth.

SALT WATER IN WELLS.

In the course of this work we have had occasion to mention the existence of salt water in oil wells. It is safe to say that in all oil wells salt water is to be found. But whence this salt, which, in the form of brine, gushes upward from depths of one hundred feet to the greatest depths to which wells have been drilled? Above the first sand-rock, the water is invariably fresh; between the first and second, it is expected to be such; below that stratum, it is certainly expected to be brackish or briny. As a rule, the surface water does not go down through the second sand rock, or the brine force itself above it, until an artificial opening exists, the exceptions being so few in either case as to establish the general principle here laid down. Besides, it is well known that water, in passing through sand or clay, is apt rather to part with impurities than take others up. Wright, in his work, says:

" The only satisfactory explanation of the existence of these salt springs is that the strata in which they abound, at one period in the world's history, formed part of the ocean-bed. This may have consisted of lime-stones, sand-stones, or conglomerates, all saturated with brine, and reposing on what had been beds of clay which contained carbonaceous ingredients; while the process of baking this clay into shales filled it with cracks and seams, that have since become so many veins, filled with salt water or petroleum expressed from the rocks subsequently formed; the whole being upheaved to their present elevation. But who shall fill up the picture of those ages, of which this is scarcely an outline?

The relationship between brine and petroleum is intimate; yet they are not invariably found together. Where oil is obtained in the second sand-rock, it has sometimes happened that the *show* of salt was scarcely perceptible in or above the oil vein. By sinking a few feet deeper, however, the brine would flow up so profusely as to completely monopolize the tube and choke off the more desirable product. At Tidioute, the alliance is so intimate that it has come to be considered a maxim: "No salt, no oil." We have stated, that brine is found in all wells; but it is not always found in large producing wells, when at their best, the appearance of salt water in good producing wells is generally a precursor of a decreased production. As to the why or wherefore of this connection, our superficial philosophy must place its finger on the lip, and be modestly silent—for the present.

LOCATION OF WELLS BY SPIRIT INFLUENCE.

The story related of the locating of the first well (by spiritualistic agency) called the Harmonical, No. 1, on the Porter farm, at Pleasantville, which led to the extensive developments in the Pleasantville district in 1868, is stoutly affirmed by the party who claims to have been the instrument selected for locating it. As a matter of history we give an account of the *location* as recorded at the time. Mr. James, the spiritualist, in company with a number of gentlemen, was on his way to examine some property a few miles south of Pleasantville. Seated in a buggy with a companion, they had proceeded but a little distance when Mr. James became violently influenced by what is termed his attending spirit-guide. This invisible power increased till, Paul-like, he hardly knew whether he was "in the body or out." The control soon becoming absolute, he was taken over the fence into a lot on the east side of the road, moving rapidly, and his companions following. Nearly unconscious, the locomotion seemed to him like being hurriedly forced over a fence. Proceeding towards the south side, then back and near the north end of the field, he moved more cautiously, as though tracing some lode or vein. Reaching a certain locality, he was thrown heavily upon the ground, and making a mark with his finger, thrust a penny some inches into the earth. He then fell upon his bosom stiff, and apparently lifeless. His eyes were closed, his face pale, the pulse feeble, and the limbs rigid as in death.

In this condition, he was given to understand that they were then upon a superior oil-producing territory, extending many miles in a certain direction,—that directly under their feet, were floating streams of oil that if opened would yield rich supplies. This was the spot—the precise location of " Harmonical Well, No. 1," which was struck in February, 1868, and produced upwards of 100 barrels per day. The striking of this well created great activity in oil developments in the district, and thus commenced the famous Pleasant-ville excitement of 1868. Mr. James has located many wells in the oil region by " spiritual guidance," many of which proved good producing wells. He became prominent as an enterprising and successful operator. Recently, however, he has located a well on the Clarion River, claiming the same manifestations as related in regard to the Pleasantville well. This well is now sunk to the depth of 1600 feet, and no sand and no oil, but drilling still continues. There are many operators in whose minds are yet fresh the implicit confidence placed in " Oil Wizards," and their power to successfully *locate* wells; indeed, they are to be found at the present time, and still a few of our oil men employ " Hazel-twig " manipulators to mark the spot to drill upon. That they were skillful, at least so far as manipulating the divining rod—or *" dowsing rod,"* as it was sometimes called—to the satisfaction of their employers and their own emolument, there can be no doubt, and that they have now almost become extinct is equally true. Still the system or practice has yet its defenders, and it seems not without some shadow of consistency. In 1826, a book was published by Count de Tristam, which gives a general history of its use and many details, which are curious as well as interesting. The French call it *" Bagnette Divinatoire ;"* and M. Chevreuil, in 1854, published a book combating the objections raised to it as a deception, and ascribing its action to philosophic causes.

The divining rod is a forked stick of either hazel or peach, held by the extremity of each prong of the fork in a peculiar way—the palms of the hands being upward, and the prong in either hand crossing the palm and being held by the thumb and tips of the fingers. The wizard then walks over the country he is to try, and as he approaches the greatest body of oil in the immediate neighborhood, the thick end or handle of the fork turns down in spite of all the efforts of the holder to the contrary. There is no doubt that, owing to the way in which it is held, it has, when once it begins to move, a mechanical tendency to turn, and this increased at the will of the holder, and in such a way as to remain undiscovered by even a close observer.

Wells have been located and by some diviners, with wonderful success, but whether through good luck or actual philosophy is a question not easily disposed of. It has been used with marked success in Europe, in discovering mineral lodes or deposits ; and even as late as 1863, we read of a French ecclesiastic making a handsome income by its use in discovering veins of water.

A case still later—1869—is quoted, in which what is called the Chiverton lode was discovered by its aid,—this in Cornwall, England. In the central counties of Pennsylvania, a well is seldom dug without first calling in the " water wizard," and making him " smell "—as they term the process—over the ground where water is wanted. This same water wizard is generally shrewd, and allows the rod to turn down where water will be most convenient, and then announces the depth at which it should be found, giving as nearly as possible the same depth as of other wells in the vicinity, —with due allowance for surface irregularity,—and the prophecy seldom fails.

Some of the best oil wells on Pit Hole were located by the use of this instrument, and this fact is often quoted by its defenders, but of course finds little sympathy with the incredulous.

A PHENOMENON WITH AN EXPLANATION.

At Pit Hole, in the early part of 1866, a singular phenomenon manifested itself. A fire occurred at the United States Hotel; the water used to extinguish it was procured from an adjacent well, but after a time it was observed that the supposed water was only adding fuel to the flame, and on examination it was found that the well from which the water was obtained, was covered with oil some inches in thickness; in fact, the oil was running into it. On further examination, other wells were found in like condition. Great excitement followed; many thousands visited the locality, and large prices were offered for them. It was not confined to the wells alone. Two or three springs in the vicinity of the water wells mentioned, were found to be covered with oil. Nature, it would seem, had become weary of the drilling and pumping process of obtaining oil, and poured out her treasures of her own accord

The first well in which oil was discovered, was only sixteen feet in depth, and from it over fifty barrels of oil were taken with a common pump. The second well was twenty-three feet in depth; from this well were obtained over one hundred barrels of oil. These wells produced at the rate of five to twenty barrels per day, when operated. From the spring near them, the owners dipped several barrels per day. Some parties sank wells to a moderate depth, and in cases obtained oil. One of these, only a few feet from the first-named, struck a crevice in the surface rock, at a depth of twelve feet, from which poured a fine stream of oil. The yield from these wells was of brief duration, and many who had wildly speculated in leases lost heavily. These wells were located about 150 feet above the level of the creek, on the second bench or table land, half a mile from the creek, where producing wells were located. The wild theories of the wise and learned, and the speculative, too, were soon after wiped away,—a solution of the mystery, and its explanation being conclusive. The Pit Hole and Miller farm pipe line had burst during the cold weather of the winter of 1866, at a point in the neighborhood of these wells, and some thousands of barrels of oil were lost. The oil taken from the wells described, was the product of this misfortune to the Pipe Line Company!

COST OF WELLS.

LIFE OF WELLS AND COST OF PRODUCING OIL PER BARREL.

The cost of putting down oil wells has varied through all the years of the history of the business, and also varies with the locality and facilities for the work. In the early years of this industry many and frequent were the mishaps, the losing of tools, and other vexatious accidents, the results of inexperience, which often terminated in the abandonment of the work of drilling. Experience and the improved facilities offered by many valuable inventions have almost entirely obviated the difficulties with which the early operators had to contend. Most wells at the present time are cased, with pipe six to six and a half inches in diameter, to a depth sufficient to shut off the fresh water. Previous to entering the oil-bearing rock, a six to eight inch hole is drilled down to this point, and then "seed-bagged." This device prevents the fresh water from falling below the lower end of the casing. The drilling now proceeds with a reduced bore of about five and a half to six inches in diameter, to the required depth, after which two-inch tubing is let down to or below the oil rock, and pumping is commenced at once. The power of engines used varies—from ten to twelve horse-power being the maximum. The cost of these is in accordance with the contingencies above mentioned. Wells range from five hundred to sixteen hundred feet in depth, and the entire expense for rigs, machinery, and drilling, averages from $3,000 to $8,500 for each well.

These figures cover the average cost of shallow and deep wells, without tankage. In some cases, however, they are put down cheaper, and in others, where accidents occur, such as caving in, tools, sticking, etc., they cost more money.

The time required for completing a well, ready for pumping, ranges from twenty-five to one hundred days.

From the reports issued by the Petroleum Producers' Association, 1871, we deduce the following interesting items in regard to the completion, abandonment and resumption of wells, and from these facts learn the average life of a well and the length of time required to drill one.

The facts given are as follows:

Date. 1871.	Wells Drilling.	Wells Completed.	Wells Abandoned.	Wells Resumed.	Number of Wells Producing.
January..........	132	90	83	14	2897
February........	173	57	63	31	3036
March...	240	64	69	9	2921
April	279	87	56	24	2994
May..............	356	99	30	29	3087
June.............	303	154	55	8	3177
July..............	329	147	54	39	3280
August..........	330	112	88	29	3369
September.......	439	128	69	38	3466
October..........	486	182	78	28	3606
November........	477	154	83	15	3692
December.........	394	191	131	21	3775
Total........	3938	1465	859	285	39,300

From which we deduce that 859 minus 285, the net number of wells abandoned during the year, was 574, or an average of 47.8 per month: the average number of wells producing was one-twelfth of 39,300, or 3,275; and the time necessary for all to become abandoned would be 3,275, divided by 47.8, or $68\frac{1}{2}$ months, or twice the average life of a well. Wells, therefore, (including all "dry holes" of which any record is had,) average to produce oil for $34\frac{1}{4}$ months.

The average time required to drill the 1,465 wells completed during the year is as many months as the number completed is contained in the total number reported as being drilled at the close of each month, or 3,938 divided by 1,465, equal to 2.69 months, equal to 81 days nearly; the extremes are, however, wide apart.

It would be interesting to ascertain, if possible, the average cost of all wells drilled, including the dry holes, and adding the cost of pumping to determine the average cost of each barrel of oil produced; an exact account it will be quite impossible to arrive at, but a close approximation is reached by the following method:

We divide the cost of a well into three parts; one a constant quantity, viz., the cost of labor and fuel expended in drilling the wells, for which $26 is a fair average daily expense, or for the 81 days $2,106. This for the 1,465 completed will amount to $3,085,290. Another part is the cost of the engine, boiler, rig, &c., necessary for each well; this is fully $3,000 for each successful well; but in the case of dry holes other wells can probably be put down with the same material for $1,000 each. To determine the total expenditure for this item it is necessary to ascertain the proportion of the dry holes to the successful wells; this proportion we learn from the experience of several of the most extensive operators, is five of every thirteen wells drilled, or 540 dry and 925 successful wells drilled in 1871, or an aggregate expense for engines, boilers and rigs of $3,315,000.

A third portion of the expense is that for tubing, rods, pumps, tanks and et ceteras necessary to work the well: this may be fairly estimated at $1,000 per well, or an aggregate of $925,000.

To reach the cost of the oil produced, we must add to this cost of the new wells, the expense of pumping those that have produced; this expense may be fairly averaged at $6.50 per day, or an aggregate for the 3,275 wells, of $7,982,812.

This makes a total expenditure for the year 1871, of $15,308,102, to produce 5,755,057 barrels of oil, or an average of $2.66 per barrel.

OIL SHAFTS AND DEEP WELLS.

Many, in the early history of oil developments entertained the idea of sinking shafts so as to obtain oil in vast quantity, and then, as it were, to tap the fountain at its head. Instances are recorded of such shafts being sunk to the depth of from two hundred to five hundred feet, in Burmah, which have yielded large quantities of oil for hundreds of years. In these Burmese shaft-wells, the mode of lifting the oil is not remarkably skillful, the entire work being accomplished by buckets. When it is necessary to clear the shafts, men are let down by means of ropes, and they often die from the effects of the gas. Life, however, is cheap in that country ; and there is no difficulty in keeping the wells clear at moderate cost.

The first oil shaft sunk in this country, was near Tarentum, in Allegany County, about twenty miles above Pittsburgh, which was finished in the latter part of 1859. The third sand rock in this locality is found at a depth not much less than two thousand feet, and as the shaft was sunk to a depth of only one hundred and sixty feet, it is needless to say that little or no oil was obtained. The salt wells of that section, which usually penetrate to a great depth, have always yielded more or less oil mixed with salt water.

On the south side of the Allegany river, opposite Tidioute, is a shaft, sunk in 1865 by the New York Enterprise and Mining Company. The aim of the company was to penetrate, if possible, the third sand rock, and then tunnel into it. The Tidioute shaft is the only one in this country which has penetrated the third sand rock. The shaft is twelve by eight feet in width, and a hundred and sixty feet deep. Upon striking the oil rock, holes were drilled at various angles, and quite a large amount of rock was removed and brought to the surface. The men worked in "towers" of eight hours each, and the shaft was kept supplied with fresh air by means of a powerful air blast. At the end of one of the towers the men came up to the surface, the engine was for some reason stopped, and the gas accumulated. The two gangs of men were seated on the curbings round the edge of the shaft, and Mr. Hart, the foreman, occupied a position on a plank directly over the mouth of the pit. As a preliminary to descending, one of the men dropped a lighted taper into the shaft, which was instantly followed by a powerful explosion. The men were thrown violently back from the curbing, and as soon as they recovered from the shock, they found that Mr. Hart had disappeared into the pit below. The body of Mr. Hart was found in a shockingly mangled condition, having been tossed from beam to beam on its way to the bottom. His death, more than anything else, put a stop to the operations,—at least no work was ever done after that.

A second shaft well was put down at Tidioute, about the date of the one described above. We have, however, been unable to obtain reliable data in regard to it.

Another shaft was sunk near the Hyde and Egbert farm, below Petroleum Centre. Work was suspended on reaching one hundred and sixty feet, owing to the large flow of gas, and the great cost of the undertaking.

In November, 1865, Mr. Jonathan Watson, of Titusville, conceived the idea of drilling a well beyond the third sand rock, in hope of reaching a fourth sand. Drilling on this well was prosecuted for upwards of two years without reaching a fourth sand rock. This well was cased with three and a quarter inch casing, to its full depth of two thousand one hundred and thirty feet, and pumped, but without any show of oil. This enterprise cost Mr. Watson upward of twenty-five thousand dollars.

Probably the deepest boring in the world, is to be found at Sperenberg, in Lusatia, Germany, at the salt region of that place, where a depth of four thousand feet has been attained.

EARLY FLOWING WELLS.

FROM 1861 TO 1864.

FOSTER FARM.

THE SHERMAN WELL.—In May, 1862, there was but one producing well on the Foster farm, and that was called "The Sherman Well,"—and that was struck in March of the year first-named. She started off at 1,000 to 1,300 barrels, and was for some months the largest well "on the Creek." "The Sherman Well" continued to flow until February 1864, gradually diminishing. Her daily product as a pumping well was quite large, and she held out for two or three years.

"The Sherman well" lease was the property of Mr. J. W. Sherman, now a resident of Cleveland, Ohio. He came to the oil region in the early days of petroleum, in somewhat straightened circumstances. He had some means, but not enough, as the sequel proves, to complete his first well. He obtained a lease upon the Foster farm, below Shaffer, on the Creek, and commenced the work of drilling a well. Passing over the trials and embarrassments encountered in getting ready to drill, he finally began his enterprise, employing "spring pole power. In the "first sand" he had a "good show" of oil, but long before he reached the "second sand rock," his money gave out, and he was compelled to shut down. The "spring pole" had become powerless to work the drill effectively, and a horse or steam-power was indispensable. Mr. Sherman waited for something to turn up, by which he could obtain either a steam engine or a horse. After many days of waiting, an interest in the well was disposed of "for an old horse,"—and the work proceeded. Two or three weeks of horse-power drilling, and the labor became too heavy for "OLD PETE." Another one-sixteenth was sold to two gentlemen who owned a small steam engine, and work was again resumed. Coal was an expensive item, and it could not be had without the "ready cash,"—and not one of the owners could muster enough to buy a single ton.! Another halt!

A week's delay and another interest was forced upon a reluctant purchaser, for "$80 in cash and a shot gun." Just before the last dollar of this money had been expended, the drill penetrated "a crevice," and the "Sherman" commenced to flow at the rate of 1,000 barrels per day! The fortunes of the plucky lessees were made—they had "struck ile," indeed. "The Sherman" continued to flow for two or three years, finally coming down to a pumping well. It is safe to say, the product of this well enriched its owners in fabulous degree, for its total receipt for oil sold is estimated at $1,700,000 !

Soon after "the Sherman" began to flow, a dozen or more wells were drilled upon this farm, but they were mainly non-productive. Mr. Frederick Crocker put down one well on the farm, which he pumped two months steadily, when she started off at the rate of five hundred barrels daily. The "Crocker well" had a short career, however. The surrounding wells let down the surface water and soon drowned her out. She produced for a year or more, and was then abandoned.

LOWER McELHENNY FARM.

This farm was purchased by Capt. A. B. Funk, in the fall of 1859, of David McElhenny, the original proprietor, for $1,500—McElhenny reserving one-quarter of the oil. In the spring of 1860, the work of development began, and the first well drilled was named "The Fountain." It was put down with spring-pole power, to the depth of 260 feet. To reach this depth required months of labor, running into the winter and spring of 1861. During the early months of the latter year, the "spring-pole" process was abandoned, and a small locomotive boiler and a stationary engine were obtained, and with this the drilling was completed. At this early day, few, if any wells, were drilled below the "second sand rock," obtained generally at about 160 feet. Capt. Funk was inclined to abandon this well at 260 feet depth—100 feet below the only oil-bearing rock yet discovered. His son, A. P. Funk, then and afterwards in charge of operations upon the farm, determined to sink it still deeper, persuaded that another oil rock could be found.

The well was completed in May, 1861—the drill having reached the "third sand," and perforated it to the depth of sixty feet before the slightest evidences of gas or oil were visible. Its entire depth was 460 feet, and the top of the "pebble rock" or "third sand" was struck at 400 feet depth. When the oil vein was reached, the drilling tools were "hammering away" at the bottom of the well, and the first intimation the drillers had of the presence of oil was the gradual rising of a foam, under which was a volume of water, bubbling and rushing over the top of the drilling-pipe. This continued for some moments, the column of water, mixed with oil, steadily rising to the height of eight or ten feet above the drilling-pipe, when it seemed to explode, and the oil followed in immense volume, rising to the altitude of the derrick and above it.

This was the first well put down to the "third sand rock" in the Pennsylvania Oil Region. It was, as before remarked, named the "Fountain Well," and produced (flowing) 300 barrels per day for about six months, and then stopped short—instantly, it is said, and never afterwards produced a barrel of oil. It was agreed upon all hands that the well was destroyed by paraffine, for the lead pipe from it to the tanks—two hundred feet distant—was completely filled up with it to such solidity that a sucker rod could not be driven through it with a sledge hammer. Only thirty feet of tubing was ever used in the well, and the hole doubtless presented the same appearance as to obstructions as did the lead pipe. The verdict therefore, was, "The Fountain Well" was destroyed by paraffine.

"The Empire Well," same farm, was put down by Bennett & Hatch, lessees, and was completed about the 20th of September, 1861. "The Empire" had the same sands and the same depth of "third sand" as that found in "The Fountain Well." When struck, she started off at 2,500 barrels! Six weeks after she began to "flow," 2,200 barrels was her regular daily product! She flowed nearly eight months, gradually falling off to about 1,200 barrels, when in May, 1862, she, like her predecessor, "The Fountain," stopped as suddenly and as mysteriously as did her consort, but not with the same fatality as to future profit. "The Empire" was soon afterwards cleaned out, and the pump applied to her, and for some months she produced 600 barrels per day, and then fell off, and for eight or nine months gave out about 300 barrels per day.

There is one incident connected with the history of "The Empire Well" which will bear repetition. While in the height of her "flow," one month's product of the well was sold to Bradley & Son, of Cleveland, Ohio, for five hundred dollars! Not less than 100,000 barrels of oil poured out of her during this thirty days' transfer, for which her owners realized not more than five cents per barrel!

The "Lower McElhenny farm" was among the most prolific of "flowing well" localities on "the Creek." After the "Fountain" and "Empire" wells were struck, the farm became rapidly studded with derricks, engine houses, and all the paraphanalia of an oil-producing locality. "The Davis and Wheelock Well," was struck in the fall of 1862, and daily poured out 1,500 barrels. "The Densmore Well, No. 1," struck about the same time, flowed 600 barrels per day. No. 2, same party and name, 400 barrels per day; and No. 3, same owners, about 500 barrels per day. These latter wells were all struck about the same time—in the fall of

1862, and were all put down upon a two-acre lease. "The Crocker Well" was struck about the same date, and flowed 1,000 barrels daily. This well was owned by Mr. FRED. CROCKER, of Titusville—since one of the prominent producers of the region, and now as anxious, and as industrious and determined to obtain good paying wells, as he was in 1861—thought he could hold the oil in the well, for better prices, as well as to save tankage expenses, and resolved to "plug it" below "the second sand-rock." His further object was to control the flow of oil to suit his convenience! In the "plug" was an inch-and-a-half hole, to permit the oil to escape into the tubing above; and upon the top of the tubing was placed a stop-cock of like dimensions. The idea was to turn off or on, the flow of oil at any moment, and thus secure a car, or a boat-load at pleasure! The contrivance worked well for a single day! But shutting down for the night, to remain idle for eight or ten hours, was fatal to it. The following morning, upon opening the stop-cock, little or no oil came from her. The "plug" was subsequently driven to the bottom of the well, for it could not be withdrawn, and the well became a "pumper" of thirty to fifty barrels per day, and finally after a brief life was abandoned.

To these noted "flowers" of "The Lower McElhenny farm," may be added "The Hibbard Well," struck in March 1863, and started off at 400 barrels. "The American Well," struck about the same time, and flowed 500 barrels. "The Canfield Well," struck in the summer of 1863, and flowed 400 barrels.

During the fall, winter and spring of 1862 and '63, the daily product of the Lower McElhenny farm, was between five and six thousand barrels. Oil was sold from this farm, during the years just mentioned, as low as 10 cents per barrel; the average price, however, being 25 cents per barrel, the purchaser furnishing his own barrels. In the spring of 1864, better prices were realized; oil being sold from the tanks on the farm at $5.00 per barrel.

THE ESPY FARM.

This farm, adjoining the Lower McElhenny, had some noted flowing wells, in the early days of petroleum development. "The Buckeye Well" was one of the most famous. She was completed in September, 1861, and flowed 1,000 barrels per day; while there were other good-producing wells upon this farm, they were small, compared to those upon the McElhenny farm. "The Buckeye" was a famous producer. The tanks to receive her oil, were set up on the hills above her, two hundred feet, and for a year the oil was forced through a lead pipe into these tanks from the well !

HYDE AND EGBERT FARM.

PETROLEUM CENTRE.

Dr. A. G. EGBERT, now an enterprising and wealthy resident of Franklin, Pa., purchased, or contracted to purchase the Davidson farm, of its owner and occupant, in 1860. Later, Mr. Davidson died, and some difficulty was experienced in obtaining a clear title to the property. Without further detail, we may add that all was cleared up, and in 1862, CHARLES HYDE, of Hydetown, became a purchaser of one-half the property from Dr. EGBERT, who, meantime, had effected a settlement with the widow Davidson, agreeing to pay her $2,625—"and one-twelfth of the oil," for a deed of the farm. This sum, $2,625, Mr. Hyde paid to Dr. Egbert for one-half his purchase from Mrs. Davidson.

Prior to this sale to Mr. Hyde, or in the spring of 1861, a well had been drilled upon the property. This was called "The Hollister Well," and when struck, "the oil flowed in great volume," flooding everything about the derrick. The lessees had contracted to deliver to the land-owners, their "one-half royalty," in barrels. Barrels could not be obtained in sufficient quantities, at any price, and $3.50 to $4.00 was demanded for all that could be procured. Oil was selling at 25 to 30 cents per barrel. The drillers, therefore, abandoned their enterprise, and the well was never tested! It continued to flow for some days, the oil running upon the ground. The lessees could not afford to barrel that portion going to the land proprietors, for it would cost them more than the entire product was worth!

"The Jersey Well" was one of the famous "flowers" of the Hyde & Egbert farm, and was the property of a company of Jerseymen, and was struck in the spring of 1863. It produced from the start 350 barrels per day, and this product was maintained with little variation for quite nine months.

"The Maple Shade Well"—working interest—was the property of an organized company, "The Maple Shade Oil Company," and was struck August 5th, 1863. Its product was 800 barrels per day, and continued at this standard for eight or ten months. It was a steady flower, and brought its owners a large amount of wealth. Dr. A. G. Egbert informs the writer that during its life its aggregate net earnings and clear profits were more than $1,500,000!

"The Coquet Well," Hyde & Egbert farm, was struck in the spring of 1864. An account of this well will be found in the biographical sketch of Mr. E. B. Grandin, and we omit its repetition here. "The Coquet" was pumped for ten or twelve days, when, upon drawing the sucker rods, in order to relieve her of an excessive quantity of gas, she began to flow largely, and for a few days produced 1,000 to 1,200 barrels. She finally settled down at 800 barrels, and continued for many months at this standard.

CHERRY RUN OR RYND FARM.

THE REED WELLS.

THE original "Reed Well" was struck on the 18th of July 1864, and flowed 280 to 300 barrels per day. The lease upon which this well was located consisted of *one acre of land*, and upon this small tract the lessees put down four wells, all proving abundantly productive. Mr. William Reed was the original lessee, and before he succeeded in getting down the first well—"the Reed"— he was joined by one or two parties, "with a little money." After the well was down to a proper depth, it had every appearance of "a dry hole." Several days were spent in pumping and testing it, when she began to flow at the rate of 300 barrels, and continued to produce largely for two or three years. The other three wells upon this lease added to the product considerably, and in the end netted handsome fortunes to the owners. One-quarter of the "land interest," belonging to Mr. Cresswell, who came into the enterprise plethoric with a *lack of greenbacks*, was sold soon after the first well was struck for $280,000, to the Mingo Oil Company, of Philadelphia, Mr. C. having previously realized from the product of the well $30,000! Mr. Reed, after realizing $75,000 from the sale of oil from the well, disposed of his one-half interest in the property to Bishop, Bissell & Co., for $200,000. Mr. Frazer, who owned one-quarter of the property, and who had received from sales of oil from the well, more than $100,000, subsequently disposed of his interest, to other parties, for $100,000! This in round numbers makes a total of $785,000 realized by the original proprietors, "for the working interest" of "The Reed well and lease," and all within "ninety days from the commencement of operations." The purchasers made money from their investment, for these four wells continued to produce largely, two or three years after they passed out of the hands of the original owners. Multiply $785,000 by two, and we have $1,570,000 as the grand total realized by the lessees and subsequent purchasers. Add to this princely sum the amount received by the "land interest," and it is safe to estimate the profits of this one oil operation at $2,000,000, and this is only "one of the many" having like fabulous history.

TARR FARM.

THE PHILLIPS WELL (No. 2), Tarr Farm, was struck on the 14th of November 1861, and commenced to flow at the rate of 3,000 barrels per day! Oil at this date was sold as low as fifteen, and even ten cents per barrel. Thousands of barrels of the product of this well, for want of barrels—worth then at the well $3.50 to $4.00 each—ran off into Oil Creek, or were allowed to waste in various ways. In December, one month after "the Phillips" began to flow, she produced by actual measurement three thousand nine hundred and forty barrels in twenty-four hours! She finally settled down to 2,500 to 3,000 barrels, and maintained this standard for months. The owners of interests in this marvellous well were accustomed to take their portion of the product by the hour! A rude trough, made of six-inch boards, was constructed from "the Creek" to the tanks, and as boats could be obtained, and sales made, the oil was "let on," and run, two, three, five, or more "HOURS" in each owner's interest.* If the boats, barges, barrels or tanks—supplied as fast as possible—filled up before the expiration of the party's allotted time, the oil ran into "the Creek," or upon the ground, and was thus wasted and lost.

Samuel Downer, Esq., later the proprietor of the Downer Oil Works at Corry, was one day standing at the discharge end of the leading trough, which had half an hour before been thrown from a flat boat just filled. The oil was running into the Creek in a volume as large as the trough would hold. "See here," said Mr. Downer, "don't you know you are wasting a hundred barrels an hour here?" "Yes," said the interested party addressed, "but what am I to do with it? You won't give five cents a barrel for it; and I can stand a loss of five dollars an hour rather than let you have it at that price!" Mr. Downer passed on up the Creek.

The lessees of this portion of the Tarr farm had obligated them-

* See Frontispiece.

selves to give "one half the oil, and deliver it to the land-owners in barrels!" For a few weeks after the well was struck, the flow was stopped by means of a stop-cock. The question of barrels was finally adjusted, and the flow began again, as above stated, in December 1861. The product of this well is variously estimated; some put it as high as 750,000 barrels, and others at 1,000,000 barrels. "The Phillips" flowed for a year or more, her product lessening, when the pump was applied, and she produced largely for twelve years, and was shut down as late as May, 1873, when her product was from seven to ten barrels per day.

During the "Stock Company epidemic" in 1864–5, all but the land interest of this well was stocked at a fabulous sum—one or two million dollars! While she had vigorous life and marvellous product, oil was sold from her immense wooden tanks as low as five cents per barrel, and as high as $13 per barrel!

The Phillips was 491 feet deep, and had 60 feet of oil rock.

THE CRESCENT WELL, Tarr farm, was drilled by N. S. WOODFORD during the summer of 1861, with a spring pole, and to the "First Sand." This was the first well put down upon the Tarr farm. She flowed thirteen months and twenty days, averaging 300 barrels per day, and "shut off in an instant," and never afterwards produced a barrel of oil. In 1871 efforts were made to resuscitate "the Crescent," but without avail. The well was cleaned out, drilled deeper, and pumped for several weeks; *but it was a dry hole!* During the life of this well, oil was sold at such low rates, that while there was little or no expense attending its running, beyond tankage, not a dollar of profit or dividend was ever realized by any one of its owners. She, however, paid for herself, but the land-owners claim to have lost money in outlays to save their portion of the product.

"The Woodford Well," Tarr farm, was put down by N. S. WOODFORD, in the winter of 1861. This well was located within a few rods of "the Phillips," and soon after she began to produce—two thousand barrels per day—the water flooded "the Phillips," and materially affected her flow. When these wells became "pumpers," neither would give out oil unless both were in motion. When the "Woodford" shut down, the "Phillips" produced only water, and *vice versa*. A compromise was subsequently effected by which both wells were to be operated at one and the same time, and each to have one-third of the product of the other well.

There were several large flowing wells upon the Tarr farm during the early developments there, from 1861, to 1863–4. "Phillips No. 1.," struck in June 1861, flowed two hundred barrels per day. "Elephant No. 1.," completed in December, 1861, was a bountiful producer,—six hundred barrels per day. " The Union," struck in April, 1862, measured out three hundred barrels per day. "The Eagle," started off, August 1862, at 100 barrels, and later, August and September, 1864, " The Cornwall," and " Sterling," each produced one hundred and twenty barrels per day.

The great flowing wells of the Tarr farm, in 1861 and 1862, were closely followed by others, at various points on " the creek." " The Van Syckel Well," on the widow McClintock farm, yielded one thousand five hundred barrels per day. "The Brawley Well," on the Buchanan farm, one thousand barrels per day. " The Blood Well," Blood farm, one thousand barrels per day. " The Noble Well," Farrel farm, two thousand five hundred barrels per day, and others which we have already mentioned in more detail, in this connection.

PIT HOLE IN 1865.

In January, 1865, the famous " United States or Frazer Well," was struck, on the Thos. Holmden farm, in a ravine on Pit Hole Creek, six or eight miles from its mouth, and almost as many miles from any other developments. "This intelligence," we quote an author, "who was himself of the moving mass," naturally created some excitement in the restless world of Oildom ; but the spring floods of that memorable year, as well as the attractions to enterprising operators afforded by more accessible localities, for several months prevented extensive developments in the direction of the new discovery. Indeed, the "town" could boast of only two buildings by the end of May, although the production of the United States Well had steadily increased, and was then fully eight hundred barrels a day. However, the beginning of June witnessed the striking of the " Grant Well," a " spouter " of twelve hundred barrels magnitude, and forthwith commenced the rush for the inviting hills and dales of Pit Hole, which resulted in the rise of a city in some respects the most wonderful the world has ever seen. Capitalists eager to invest their greenbacks, thronged in thousands to the spot. Labor and board commanded exorbitant rates ; every purchasable farm for miles around was immediately bought at a fabulous price ; hundreds of wells were begun with the least delay possible. New strikes continually intensified the excitement. Speculators roamed far and wide in quest of a source of wealth that promised to outvie the golden treasures of California. The value of oil lands was reckoned by millions ; small interests in single wells brought hundreds of thousands of dollars. New York, Boston, Philadelphia, Chicago, and numberless other lesser centres, measured purses in the insane strife for territory. Money circulated like waste paper, and for weeks the scene recalled the wildest fictions of the South-Sea Bubble or Law's Mississippi Scheme !

Everything conspired to favor the growth of the "city." The close of the war had left the country flooded with an inflated currency, besides throwing many thousands of energetic men upon their own resources, and hundreds of these flocked to the latest Oil-Dorado, which presented manifold inducements alike to the venturesome spirit, the active speculator, the unscrupulous stock-jobber, the needy laborer, the reckless adventurer, and the dishonest trickster.

Some time previously the Holmden farm had been purchased for $25,000, by Prather & Duncan, who surveyed the greatest portion into building lots, that found ready sale at figures varying from three hundred dollars to fifteen thousand, which latter sum was actually paid for the site of the Danforth House. Before the end of September the improved Chicago boasted a population variously estimated at from twelve to sixteen thousand, including the daily average of transient visitors. The post-office required seven clerks, and transacted a volume of business that ranked it third in the State, Philadelphia and Pittsburgh alone surpassing it. Hotels, theatres, saloons, public halls, and places of general resort could be counted by the score. A fire department was organized; stores and dwellings sprang into existence as if by enchantment; a railway to Reno was projected and completed almost the entire distance, and the unpoetic name of Pit Hole became familiar to every newspaper reader throughout the civilized world. Incredulous foreigners, unaccustomed to the "suddenness" of Americans, with unfeigned astonishment, learned that in the brief space of three short months, a dense forest had been transformed into a bustling city, possessing nearly all the conveniences and appliances of old-established towns—a city, the wondrous story of whose dazzling rise and unexampled fall sounds even now like the weird romance of ancient fable. What a rich field for a graphic sketch of fortunes lost and won in an hour, of strange vicissitudes and extraordinary reverses, of feverish excitement and unhealthy speculation, does the history of Pit Hole offer to some later Scott or Dickens.

But, alas! the youthful city was destined to decline as rapidly as it had risen. In October the production of the wells fell off largely; the laying of pipe-lines to Titusville and Oleopolis forced hundreds of teamsters to seek employment elsewhere; two destructive fires helped to accelerate the final disaster, and January dawned upon a comparatively deserted city, with scarcely anything more than long rows of empty buildings to indicate its former greatness and short-lived prosperity.

Many of the finest structures have since been removed to other places; not a vestige of the first wells is to be seen; the few hotels and stores that yet remain open, are no more crowded with liberal patrons. Occasionally a traveller finds his way to the spot, possibly impelled by an irresistible desire to behold again the scene of his disappointed hopes and buried greenbacks. Of the once busy city, the unused engine houses and derricks, the unoccupied tenements and unfrequented by-ways, are too often the only traces that still remain, silent, forsaken and alone, " to point a moral or adorn a tale!"

We add a single projected transaction of the Pit Hole furor in 1865, as an index to scores of others of like, or very like mammoth proportions.

In July, 1865, Mr. GEORGE J. SHERMAN, HENRY E. PICKET and BRIAN PHILPOT, then residing at Titusville, contracted with Messrs. Prather & Duncan, the owners, for the purchase of the Thomas Holmden farm, at Pit Hole, for $1,300,000! The farm contained about two hundred acres of land, and at the date of this contract, July 24th, 1865, was producing 3,500 barrels of oil per day, and had one hundred wells going down, at half royalty, and was besides part and parcel of the "city plot" of Pit Hole, upon which were building leases, netting $60,000 per annum. Dwellings, shops, stores and hotels were begun and completed every day. The contract spoken of, was for the purchase of this farm, just as it was, and thirty days were stipulated as the time in which to make the first payment of $300,000!

Mr. Sherman proceeded to New York city with his contract, survey, statistics, &c., to interest parties there in the scheme. A few days subsequent to his arrival in New York, he secured purchasers of his contract from Prather & Duncan, at $1,600,000! The preliminaries were all settled upon, and a committee of the purchasers was selected to visit Pit Hole, to make examinations and ascertain if the property was up to its representations. On the very day the committee were to leave New York for the oil region, the great Ketchum forgeries were announced, and as many of the gentlemen interested in this "Pit Hole oil scheme," were victims of Ketchum's rascality, the journey was abandoned, and subsequently the whole thing fell through!

In this dilemma, Mr. Sherman telegraphed to Mr. H. H. Honore, a wealthy gentleman at Chicago, giving him an outline of the property he had for sale, and urged him to meet him at Titusville in FIVE DAYS, prepared to close up the transaction. Mr. Honore, and a party of Chicago capitalists, made good Mr. Sherman's appointment, and after going over the property, reopened negotiations with Duncan & Prather, who, in lieu of the $400,000, cash, stipulated as the first payment, agreed to take that amount in real estate, situated in Chicago. To this end Duncan & Prather were to visit Chicago, examine the real estate,—which was to be priced by disinterested parties—and close up the sale. Delays followed—Messrs. Duncan & Prather were a week or more in reaching Chicago, and once there, they hesitated, and finally declined to receive real estate in Chicago as payment for their property. Among the many valuable properties offered was Honore's block, adjoining the Tremont House—at $175,000—since valued at $350,000!

The contract for the sale of the Holmden property had been renewed and the time extended two weeks! This extension had only about five days' life—and a Sunday intervened! Messrs. Duncan & Prather had left for St. Louis, intending to return home on the following Tuesday. Judge Beckwith, the attorney for the Sherman party, advised a tender of the first payment, $400,000, in greenbacks! This was late Saturday evening. On the Tuesday following, the $400,000 had been obtained, and Mr. Sherman, Joshua A. Ellis, President Second National Bank, Chicago, John G. La Moyne, and Mr. Honore, started for Titusville, with their treasure for a legal tender! The contract required that the first payment should be "made upon the Holmden Farm!" The party had reached Titusville with their valuables in safety. Pit Hole was twelve miles distant, and the country was just then infested by highwaymen of the meaner sort. Each of the gentlemen named provided himself with a pocket-pistol, and mounting horses—the $400,000 equally divided between them—they set out for Pit Hole. "It was the last day in the afternoon," and late at that, when the party reached Prather & Duncan's banking office. They entered, made known the object of their coming, and thereupon laid upon the bank's counter, in full view of Messrs. Prather & Duncan, $400,000 as a legal-tender for the first payment as required by the contract!

The tender was declined! Messrs. Prather & Duncan claimed that the life of the contract expired with the setting sun, of that very day, and they would listen to no further negotiations! Suit was soon after commenced in the United States District Court at Pittsburgh, which finally terminated in a compromise, by which the Honore party obtained title to seven-eighths of the Holmden farm property.

Soon after the tender was made, Mr. Samuel J. Walker, of Chicago, now one of the largest real estate operators in the west, became interested in the transaction, and he is now the owner of the seven-eighths interest in the Holmden farm, as also of the Copeland farm adjoining. The sum paid for the seven-eighths interest in the Holmden farm is not known to the writer of this.

Pit Hole, however, developed a great many flowing wells, a few of which we make brief mention of hereafter.

"THE UNITED STATES WELL."

This well was located on the Thomas Holmden farm, and was the property of the United States Oil Company, struck on the 7th of January, 1865. The well flowed at the rate of 650 barrels at the start, and continued to flow, gradually falling off for quite ten months. The well ceased to flow November 10, 1865. The same farm developed other great producers. Among the most noted were! the "Twin Wells"—800 barrels per day. "No. 54," 800 barrels per day. "The Grant well," 450 barrels per day. "The Eureka well," 800 barrels per day. None of these, however, "held up" their product beyond six to ten months. The daily product of the Holmden farm for some time during the season of 1865, was 3,685 barrels per day.

Upon the Rooker farm, adjoining the Holmden, were several large "flowers," during the summer of 1865. Among the most noted were the J. R. Johnson, "No. 110," which spouted out 800 barrels per day; "No. 15," the property of Pratt & Sumner, which produced 400 barrels daily, and "No. 108," 400 barrels a day. Nos. 18 and 147, each 200 barrels daily. The daily product of the Rooker farm for several months was 2,230 barrels. Leases of one acre upon this farm were sold as high as $3,000 and one-half the oil!

"The Homestead well," at Pit Hole, located upon the Hyner farm, was among the "great flowers" of 1865, in this prolific territory. This well started off at 500 barrels daily. "The Arletta" flowed 250 barrels per day, and "The Stevenson, No. 2," produced 175 barrels per day. But these wells, as "flowers," lasted only about three months.

"The Burtiss Well" was struck late in the summer of 1865. It was located on the Copeland farm, Pit Hole, and flowed from the start, and for months after, quite 800 barrels per day. Besides "the Burtiss," there were upon this farm three other flowing wells. "The Rice Well" gave out 300 barrels a day; "No. 1," 150 barrels per day, and the "Clara Well" ran up to 300 barrels per day for nearly four months.

FIRST FLOWING WELL.

The first "flowing well"—at least, the first we have any knowledge of—in the oil regions of Pennsylvania, was obtained in the summer of 1860, upon the Archie Buchanan farm, near Rouseville. It was called the "Curtis well," and was a little less than 200 feet deep. No tubing was then used, and only partial efforts were made to save the oil. The surface water was allowed to run into the well; and after a short season of flowing, "the Curtis" ceased to be.

OLDEST PRODUCING WELL IN THE REGION.

Near to the track of the Oil Creek and Allegany Valley Railroad at Rouseville, Venango Co., Pa., on the Buchanan farm, is situated the oldest well of the region, having now produced oil for a period of nearly fourteen years. This well was put down by Messrs. Rouse and Mitchel, the pioneer operators on the Buchanan farm. It was drilled only to the first sand, and pumped for several months at the rate of eight barrels per day, when it was sold to a Mr. Porter, who put it down to the third sand, and obtained a production of three hundred barrels per day, which lasted for several months, when it again declined. The well occasionally changed ownership, until in 1865, it passed through the hands of the Sheriff, into possession of the First National Oil Company, and was disposed of by them to Gould and Stowell. For some years, it produced from 4 to 6 barrels per day, up to the latter end of 1872. At present, it is producing some two barrels per day. It is unquestionably the oldest producing well in the region, and dates back to the earliest period of the oil operations following the success of the Drake well. This well, though of small average production, has produced upwards of a quarter million dollars' worth of oil!

EXTENT OF THE OIL REGION.

The area of a rectangle which will embrace all the territory of the Pennsylvania Oil Region, from which oil has yet been obtained is about 2,000 square miles; but the whole number of acres which have yet produced oil does not exceed 6,500, equal to ten square miles, or a two-hundredth part of what is known as the oil region.

PRODUCTION—AN ESTIMATE.

The region produced during the year 1872, 6,539,000 barrels. This amount of oil would fill 79,150 cars, making a train 446 miles in length. It is estimated that the total production previous to 1871 was about 33,500,000 barrels, or a total up to January 1st, 1873, of 45,789,000 barrels. This quantity of oil would fill 540,-548 cars, making a train nearly 3,507 miles in length.

VARIETIES OF PETROLEUM.

PENNSYLVANIA OILS.—QUALITY AND VARIETY.

THE oil found in the Pennsylvania Oil Region is, for the most part, of a greenish color, and by some considered of a rather unpleasant odor. The specific gravity ranges from .820 to .782, or from proof 40° to proof 48° Beaumé. The oil yields by distillation from seventy to eighty-five per cent. of illuminating oil, which, when properly manufactured, should not vaporize and inflame under a temperature of 110° to 116° Fahrenheit.

Of lubricating oil produced in the Franklin district, the specific gravity varies from .880 to .860, or from proof 28° to 32° Beaumé.

The Oil of Pennsylvania varies somewhat in color in the different districts and in the different sand rocks. The *black oil* district of Pleasantville, is so called from the fact that the oil is of a dark, inky, greenish color. This district extends from a little north-east of Pleasantville to the Story farm on Oil Creek, taking a north-east and south-west direction, and is in extent, so far as developed, about twelve miles in length and half a mile wide.

It is claimed by oil men that the rock in which the black oil is found is not the regular oil-producing sand rock; they term it a *stray* rock, as green oil is found in a lower sand on the same land.

On the eastern portion, or upon the lands of the Shamburg and Cherry Run Petroleum Companies, is the dividing line between the green oil and the black. The line is defined sharply, as if by a plummet. The rock in which the black oil is found is nominally thinner than the green oil sand rock.

The cause of this coloring of the oil is reasonably attributed to the metallic composition of the sand rock, it being largely impregnated with oxide of iron. In the Modoc and Millerstown districts the oil is of an amber color, with a very slight greenish tinge. As before stated, oil differs in color and, to a slight extent, in gravity in the different sand-rocks, which, we think, can be fairly accounted for by the presence of iron in the composition of the rocks.

———

THE FRANKLIN LUBRICATING OIL DISTRICT.

The Franklin lubricating oil district, lies in and around the city of Franklin, and is made up, for the most part, by the territory lying between and including Patchen Run, and Two Mile Run. The line or belt of the most important developments, is about one hundred rods wide, and so far as developed, two miles in length, and includes the following farms : Hyde and Blakely farm, Geo. P. Smith farm, McCalmont tract, Lamberton, Galloway, Dr. Fee, and Fee, Kunkle & Co's. farm.

Soon after the striking the Drake well, Mr. Evans, a blacksmith by trade, sank a well on the lot on which he resided, within the Borough limits. The well was put down at first for the purpose of obtaining water, and at the depth of seventeen feet, a vein of water was struck, which soon became covered with a thick scum of oil, so as to render the water almost unfit for use. On learning that Drake had obtained oil by drilling into the sand rock, he concluded to do likewise. Not having the means to procure the necessary implements, to carry his resolution into effect, he was obliged to seek for assistance, but for some time his efforts were in vain. Finally a merchant in Franklin, who became enlisted in the enterprise, sold him iron on credit, and he manufactured the tools himself. He then erected a derrick, and, by means of a spring-pole, bored the well to the depth of seventy-two feet, when he struck a heavy vein of oil. He then put down the tubing, and commenced pumping by hand, with a common pump, at the rate of twenty barrels per day, and which he readily sold for thirty dollars per barrel. The success of the well occasioned considerable excitement. A writer at the time says, " The town almost involuntarily poured forth its inhabitants to witness the natural curiosity. The attendants at court (which was then in session) went into a "committee of the whole," on the state of the oleaginous condition of the country, and adjourned to the Evans well. Attorneys, jurymen, and witnesses who were concerned in the various cases then pending in the Court of Common Pleas, suddenly became a self-constituted judicial tribunal to decide upon the merits of this *uncommon cause* of public excitement."

Mr. Evans, having raised in a few days, money sufficient to enable him to purchase an engine, he commenced pumping by steam-power. The yield of the well was variously estimated at from sixty to two hundred barrels in twenty-four hours. He was offered fifty thousand dollars for an undivided half-interest in his well, and refused the offer, as his income then was probably not less than two thousand dollars per day.

This well has earned the fame of giving occasion for the famous saying : " Dad's struck ile." The story is vouched for as true, and runs as follows : Mr. Evans had a daughter, who was courted by a young man living near by, and the course of their love ran smoothly enough, until the ill-starred day when the damsel's father reached the " third sand," and success in his well. On the evening of this day, the swain, not dreaming of anything less pleasant, than moonlight and love, called on his sweetheart, and was met coldly at the door, and promptly informed, that he needn't trouble himself to come there any more, for " Dad has struck ile !"

The quality of the oil obtained in this district, is not the common illuminating oil, but lubricating oil of nearly the best quality, being little inferior to the best West Virginia oil, which is twenty-eight degrees gravity, while that of Franklin ranges from 30° to 32°, and is now taking the lead in the markets of the world, as a lubricator.

At the present time, the production is estimated at · 900 barrels

per day. The largest production having been reached in the early
part of the present year, amounting to 1,250 barrels per day, which
was caused by the striking of a number of large wells on the
Galloway farm, one of which produced 150 barrels per day. The
number of wells now pumping, will reach about 150, many of which
produced between forty and fifty barrels each, per day. Not a few
of the wells, included in the above estimate, have been in operation
from three to ten years, quite a number of which produce a very
small quantity of oil, but such wells are only pumped " by heads "
—once, twice or three times a day.

That part of the Lubricating District, on Two-Mile Run, a
large portion of which was recently purchased by W. S. Mc-
Mullen, produces oil of 28° to 30° gravity, equal to the best
West Virginia oil. The gentleman referred to, is now making
preparations to refine the oil produced on his own territory.

The depth of the wells in this district averages from 260 to 700
feet. The oil-bearing sand rock is from 50 to 80 feet in thickness,
being an open pebbly rock.

The manufacture or refining of lubricating oils, has for years
been an important industry at Franklin. "The Eclipse Lubricating
Oil Works," are located here, and when fully completed will be the
most extensive of its kind in the world, and will have a capacity
of 1,000 barrels per day. At the present time the company have a
capacity of about 450 barrels, and generally keep the concern
run up to this point. A ready market is found for the oils in
England, Prussia, Austria and Russia. The company are now
making arrangements for supplying the governments of Prussia
and Russia for use on railroads, arsenals, navy, and other public
works. Oils were exhibited by the company at the Vienna Exposition, and were awarded a First Medal and Certificate for
Lubricating Oils made from Petroleum. We ought here to observe that the mode of refining adopted at these works is under
patent, granted to H. W. C. Tweddle, the general manager of the
company, which latter own the patents and use them exclusively.

An interesting feature of this company is, that the President, Directors and stockholders control the eight-tenths of the whole production. The stockholders are thus at the same time producers and refiners.

The manufacture of railroad axle oils is made a speciality.

Capital stock of the company is $200,000. The following are a list of the company's officers:

A. G. EGBERT, *President,*
CHAS. W. MACKEY, *Vice-President,*
H. W. C. TWEDDLE, *Genl. Manager,*
W. H. HOWARD, *Secretary,*
W. M. N. HAYES, *Treasurer,*
Hon. JOHN S. McCALMONT, *Solicitor.*

The second refinery of importance at Franklin is "The GALENA LUBRICATING OIL WORKS." This company has a capacity of 600 barrels per week, and they are well and favorably known for the excellent character of their oils. They have a quick market for their products in the east, as well as in the western states.

There are a number of smaller establishments at Oil City and Franklin, of which it is not necessary to give an extended notice.

DRILLING OIL WELLS.

EARLY AND LATER METHODS OF DEVELOPMENT.

In another part of this work we have made mention of the fact that the mode of operating or mining petroleum was borrowed from the plan adopted by salt miners in other parts of the country. To a great extent, the system of boring deep wells had been perfected long before the discovery of petroleum, so that the early operator had all the necessary appliances with which successfully to demonstrate the existence of oil in the rocks underneath the surface of the earth; had it not been for the knowledge of the plan of artesian boring, it is not unlikely, that the discovery of this most valuable article would have been indefinitely postponed.

Thus we see that the drilling tools and other mining apparatus used by salt miners, but in a more modified and simple form, furnished all that was requisite for testing and obtaining the rich deposits of oil that had lain hidden and almost unknown for so many years.

For some years operators were content with very shallow depths; indeed it was not necessary to go deeper, or take higher ground for sinking wells, as the few hundred feet to which they went gave as much oil as the markets, or the necessities of the demand required. Various kinds of power were employed. The most primitive and most laborious, was that of the "spring-pole," which has been described as follows:

"The spring-pole consisted of a green sapling, some forty feet in length and ten inches in diameter, with the butt end made fast in the ground, or attached to an upright pole. A second post, ten or fifteen feet from the butt, acted as a fulcrum, while the pole passed over the well, and about ten feet above it. The boring implements were attached to this pole, and the "power" adjusted near its smaller extremities. This was applied by the strength of two men throwing their weight upon the pole. Sometimes a small wooden staging, four feet square, was hinged by one of its sides to the derrick, and the other side suspended to the pole. In this case the two men stood on the staging, and brought down the pole by throwing their weight on the side next to the derrick. In either case the spring of the pole brought up the implements, while the downward motion of the pole permitted the stroke. The general term for this method of drilling a well, was "jigging it down," from its resemblance to the dance, so styled.

"Kicking down a well," another process used in the early days of the business, which was done at the expense of a great deal of human muscle. A short, elastic pole, ash or hickory, ten or fifteen feet in length, was arranged over the well, working over a fulcrum, to the end of which was attached stirrups, in which two or three men, each placed a foot, and by a kind of kicking process brought down the pole, and produced the motion necessary to work the bit. By this process the strokes were rapid.

Horse-power was used, of different patterns, suitable for one horse, and sometimes for two or three. They resembled in their general features the horse-power of a threshing-machine, the horses walking around the centre, and over a dumbling-shaft, that gave the necessary perpendicular motion. Water-power was used in many instances, at very trifling cost to the operator.

Steam-power next came into general use, which greatly reduced the labor and facilitated the work of the miners. Year by year, we might almost say, day by day improvements have been made on all things used in mining petroleum. The derrick has grown from 30 feet to 64, and even to 80 and 90 feet in height. Formerly it was built of rough poles, or hewn timber, the bottom being 10 to 12 feet square; the poles, four in number, being erected at each corner, converging toward each other, forming a square at the top, of two and a half feet, with girths and braces at suitable distances, to make the structure substantial. Derricks are now made of sawed boards, two inches thick and eight inches wide, the edges being spiked together, forming a half square, on each corner of the foundation, which is usually from 16 to 18 feet square. With the increased depth of boring, the derrick has grown stronger and higher, and in the same ratio the drilling tools have grown in weight from 150 pounds to 2,000 pounds! Formerly it was no unusual thing for the driller to take his set of tools in his hands and start out to look for a job—now it requires one and oftentimes two teams to haul the tools to the point of operation.

The pumping apparatus, valves, rods and the like, have been much improved during late years. And many devices, invaluable to the operator, have been brought into use. We would instance the process of "casing," which is now always used in wells, and obviates the necessity of breaking the seed-bag in drawing tubing. By the use of casing the well is never allowed to flood with fresh water, which flooding, experience has demonstrated to be of great injury to wells.

Early wells had a bore of four inches. At present the usual bore is from six to eight inches in diameter.

By the many improvements in the mode of drilling and pumping, the business of operating in oil has been shorn of nearly all its drawbacks. The industry is now reduced to a legitimate basis, and though it is said that more lose money, than make by it, we are convinced that no other enterprise in the country can show a less percentage of failures, and few can exhibit so many substantial successes.

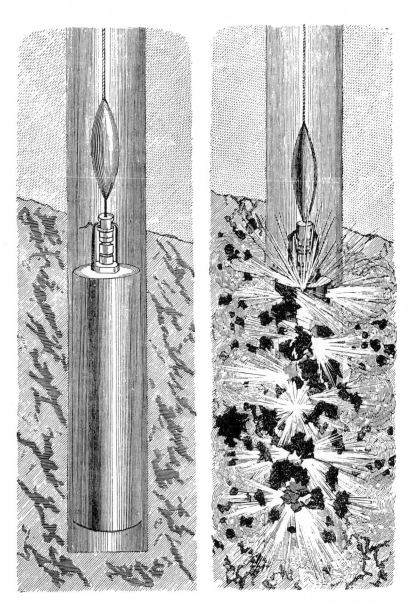

Before Explosion.   After Explosion.

THE ROBERTS TORPEDO.

TORPEDOES.

THEIR HISTORY AND POSITIVE VALUE.

"THE HISTORY OF PETROLEUM" would be singularly incomplete without mention of the Roberts Torpedo, for to this remarkable invention may be attributed, more than to any other agency, the success which has attended its prosecution. We propose, therefore, to present, as concisely as possible, a history of this invention, from its inception to the present time.

In 1862, Col. E. A. L. Roberts, then an officer in the volunteer service, and with his regiment in the Army of the Potomac, in front of Fredericksburg, conceived the idea of exploding torpedoes in oil wells, for the purpose of increasing the production. He made drawings of his invention, and in November, 1864, made application for letters patent. In the fall of the same year he constructed six torpedoes, and on the 2d of January, 1865, he visited Titusville to make his first experiment. Col. Roberts' theory was received with general disfavor, and no one desired to test its practicability at the risk, it was supposed, of damaging a well. On the 21st of January, however, Col. R. persuaded Capt. Mills to permit him to operate on the Ladies' Well, on Watson Flats, near Titusville. Two torpedoes were exploded in this well, when it commenced to flow oil and paraffine. Great excitement of course followed this successful experiment, and brought the torpedo into general notice. The result was published in the papers of the oil region, and five or six applications for patenting the same invention were immediately filed at Washington. Several suits for interference were commenced, which lasted over two years, and decisions in all cases were rendered declaring Col. Roberts the original inventor.

Notwithstanding the success of the first experiment, operators were still very skeptical as to the practical advantages of torpedoes, and it was not till the fall of 1865, that they would permit the inventor to operate in their wells to any extent, from fear that the explosion would fill them with rock and destroy their productiveness.

In December, 1866, however, Col. R. exploded a torpedo in what was known as the "Woodin Well," on the Blood farm This well was a "dry hole," never having produced any oil. The result of the operation secured a production of twenty barrels per day, and in the following month, January, 1867, a second torpedo was exploded, which brought up the production to eighty barrels. This established for the torpedo, beyond question, all that Col. Roberts had claimed, and immediately the demand for them became general throughout the region. We present below a tabular statement of the result of the first THIRTY-EIGHT wells torpedoed :

THE RESULTS OF THE TORPEDO.

NAME AND LOCATION OF WELLS.	Increase in Bills.	Pumping & Flowing.
Woodin Well, Blood Farm.........	80	Pumping
Two Wells for Mr. Archer, Tarr farm...	60	Pumping
Tarr Homestead, No. 1..	60	Flowing
Tarr Homestead, No. 2..	65	Flowing
Monitor Well, No. 2...	35	Pumping
Vogan	30	Pumping
Keystone Well......	185	Flowing
Sherman Homestead Well..	60	Pumping
Manhattan Well, Story Farm...	75	Flowing
Clara Well, Pit Hole, no increase, but made the Andy Johnson well flow	150	Pumping
Burnett Well, Tarr Farm...	65	Flowing
Gardner's Well, Pioneer Run.	8	Pumping
A. Aldrich, Tip Top Well, Tarr Farm..	35	Pumping
Smith Well, Tarr Farm..	10	Flowing
Hawkin's Well, Petroleum Centre...	20	Pumping
Anderson Well, Petroleum Centre...	90	Pumping
Monitor, Well No. 1, Tarr Farm. Two Torpedoes............................	10	Pumping
Mahaffy Well, Petroleum Centre..	4	Pumping
Ennis Well, Cherry Run..	35	Pumping
Hunter Well, Story Farm	20	Pumping
Hamburgh Oil Co., Story Farm...	30	Pumping
Morse Well, Blood Farm..	30	Pumping
Woodin Well, Blood Farm (second time)..	30	Pumping
No. 8 Well, John Rynd Farm...	75	Flowing
Hyde Well, Story Farm..	35	Pumping
Mitchell Well, Cherry Run.. .	10	Pumping
Parker Well, No. 1, Tarr Farm..	125	Flowing
Bakery Well, No. 1, Tarr Farm..	200	Flowing
Columbia Oil Co., Story Farm..	10	Pumping
Refinery Well, Blood Farm..	10	Pumping
Tarr Reserve Well, Tarr Farm..	35	Pumping
Blanchard Well, Blood Farm..	30	Pumping
Catskill Well, Cherry Run..	15	Pumping
Duff Well, Tarr Farm...	90	Flowing
Mahaffy, No. 2, Petroleum Centre..	10	Pumping
Hays' Well, Petroleum Centre..	30	Pumping
Briggs & Severence Well, Church Run...	40	Pumping
Anderson Well, Petroleum Centre (second time)............................	125	Pumping
No. 272 Well, Petroleum Centre (second time)...............................	200	Pumping

In 1865, immediately after operating on the Ladies' Well, a company was organized in New York for the purpose of prosecuting the business, with the following officers:

President, WILLIAM S. FOGG, 24 Fulton Street.
Vice-President, JAMES W. SIMONTON, 145 Broadway.
Secretary, W. B. ROBERTS, 47 Bond Street.
Treasurer, ERASTUS TITUS, 283 Washington Street.
Counsel, HON. GILBERT DEAN, 74 and 76 Wall Street
Superintendent, COL. E. A. L. ROBERTS, Titusville, Pa.

TRUSTEES: Walter B. Roberts, Wm. H. Dwinelle, M. D., A. G. Trask, Erastus Titus, Gilbert Dean, Wm. S. Fogg, Erastus Titus, Jr., Wm. H. Akin, James W. Simonton, Wm. H. Chapman, E. A. L. Roberts.

About the time the Woodin Well was struck (1866,) the wells of the region had materially decreased, and but little oil was produced. There was a general apprehension that the territory had been drained and would soon be quite exhausted, unless new belts were discovered. But the application of torpedoes immediately effected a revolution, and during the summer of 1867, the wells on Oil Creek were increased several thousand barrels. Immediately thereafter Col. ROBERTS introduced nitro-glycerine as an explosive for his torpedoes, and established a manufactory near Titusville, and during the last year (1872,) some twenty-five tons of this compound were used for this purpose alone.

The developments of Tidioute, Shamburg and other districts followed the operations of 1866, and the employment of torpedoes continued with the same striking success. And it may be safely stated that up to the present time nearly one-third of the oil production has been dependent upon the use of this invention.

In the summer of 1866, infringements commenced by different parties throughout the oil region, and suits were instituted by Col. R. against the parties and injunctions granted. In 1868, the Reed

Torpedo Company was organized, with several oil operators at its head, for the purpose of infringing and breaking down the Roberts patent. Suits were commenced by Col. R. against all parties and carried to a final hearing before Judge Grier of Philadelphia, and decisions given in favor of Roberts, and judgments rendered to the amount of about $10,000. Numerous other suits were commenced and final judgment rendered, among which was one against James Dickey, which was tried before Justices Strong and McKennan in Washington, in January, 1871. An elaborate opinion was rendered in this case in favor of Roberts. The case was regarded with great interest in the oil region, from the magnitude of the considerations involved, and the newspaper controversies upon the subject. Since the great Rubber suits, no patent-suit has elicited more general attention, involved so important considerations, or its termination more anxiously awaited. The sum of $50,000 had been subscribed among the producers, for the purpose of breaking down the Roberts Patent, and such a result was looked for with entire confidence. Few cases have ever enlisted higher professional ability, or been more earnestly contested. Messrs. Bakewell and Christy, of Pittsburgh, and George Harding of Philadelphia, conducted the case for Roberts, and Messrs. Kellar and Blake, of New York, were employed by the oil producing interest, for the defence. The decision was rendered in May, 1871, and was in favor of Roberts. It was made the occasion of a very elaborate and exhaustive opinion, which, as a matter of course, was received with general disapprobation on the part of the producers, and occasioned great disappointment.

Very many suits have since been brought for infringements, and over $100,000 have been expended by the inventor in protecting his legal rights. Thus far the Courts have uniformly sustained the Roberts patent.

OIL WELL RECORDS.

THE OIL WELLS AT BRADY'S BEND,—Lower Dist.

BY PROF LESLEY.

Original information on practical subjects is always useful, and we therefore register the following table and notes of thirteen oil wells sunk near the Brady's Bend Iron Works. These are situated on the Allegany River, at a remarkable oxbow curve of the river just above the mouth of the Red Bank, a few miles below that of the Clarion, and sixty-nine miles above Pittsburgh. Parker's Landing, a great oil centre now, is thirteen miles higher up than the iron works; Foster, another oil centre, is forty-eight miles above the works; Franklin, fifty-four miles; and Oil City, sixty-one miles.

Height of well mouth above Eng. No. datum.	Depth of well.	Depth below river, highest water mark.	First yield in barrels per day.	Present yield per day.
1... 96 feet	1,400			1 bbl.
2...232	1,111	1,268	5+bbls.	no sand rock.
3... 97.62	1,262	1,113		1 bbl.
4... 97.69	1,105	1,264	7 bbls.	abandoned.
5...100.31	1,290	1,105	5½ bbls.	2 bbls.
6...300.48	1,414	1,090	9 bbls.	4 bbls.
7...437.41	1,345	1,077	840 bbls.	8 bbls.
8...379.18	1,065	1,066	4½ bbls.	150 to 200 bbls.
9...101.38	1,300	1,066	1 bbl.	3+bbls.
10...330.27	1,200	1,070		abandoned.
11...111.13	1,212	1,189		powerful gas blow
12...216.50	1,402	1,095½	12 bbls.	13 bbls.
13...426.38		1,076	3 bbls.	2 bbls.

From the above table, it appears that all the oil-producing wells mentioned in it get their supply from one stratum lying in an undisturbed and horizontal position, varying in their actual depths below a fixed datum level from 1,113 to 1,066 feet, a difference of only forty-seven feet. This difference is due to three causes, viz: 1. The different depths in the oil-bearing stratum penetrated by the bottom boring of the wells; 2. The slight inequalities in the upper surface of the stratum; 3. And chiefly to a general slight dip of the rocks, both from the north-west and from the south-east, in toward the centre line or axis of the trough or basin which here crosses the Allegany River in its northeast-southwest course; and also to a still slighter and almost insensible decline of the axis of the basin itself south-westward.

The table also confirms what was proven years ago, long before the fact was acknowledged by oil men, namely, that it makes no difference whether a well is started in a valley bottom or on the hill-tops provided it goes down to the uniform and nearly horizontal oil-bearing sand rock. For some of these wells have their mouths at elevations more than 300 feet greater than others. Some on the river bank, and others high up at the heads of side ravines. The great No. 8 well was commenced at an elevation of (379— 96=) 283 feet higher than those on the river bank, which yield only from one to three barrels per day.

The following table shows the thickness of the third sand rock where it was passed entirely through:

No. 2.—No sand rock found and no oil.

No. 4.—Sand rock, twenty-six feet; hard fine white sand.

No. 5.—Sand rock, twenty-seven feet; fine pebbles.

No. 6.—Sand rock, sixteen feet; with slate partings.

No. 7.—Sand rock, twenty-seven feet; pebbles pretty coarse.

No. 8.—Sand rock? very coarse and open.

No. 9.—Sand rock? pebble very fine and close, very little gas.

No. 10.—Sand rock? ten feet; pebbles pretty fine, except in one thin streak.

No. 11.—No sand-rock, no oil, but great gas blow, doubtless from a fissure.

No. 12.—Sand rock, seventeen feet, all pebbles; steady flow of oil.

No. 13.—Sand rock, thirteen feet; coarse open pebbles; and a fair amount of gas.

No. 14. Sand rock, thirteen feet; large coarse pebbles; fair amount of gas.

Other noteworthy facts are as follows:

No. 1 well, on the river bank, one-half mile above the rolling-mill, begun March, 1865, finished 1866.*

No. 2 well, at the mouth of Cove Run, May, 1866—June, 1870.

No. 3 well, on the river above the mill, commenced August, 1868—pumping in September, 1872, one barrel a day.

No. 4 well, on the river above the mill, May, 1869—March, 1870. Cost $10,405. Record of strata given below.

No. 5 well, on the run above the mill, June, 1869—April, 1870. At 931 feet struck so powerful a gas vein, that the bore hole was deluged with water and abandoned for four months. In June, 1871, a three-quart nitro-glycerine torpedo was exploded without increasing the production of oil. The pebble-rock was almost as fine as sea sand.

No. 6 well, on Queenstown Run; August, 1870—April 5, 1871; drilled with the water cased out; all the previous wells were drilled in water; casing commenced at 357 feet; not much gas.

No. 7 well, on Queenstown Run; August 7, 1870—March 1, 1871; water cased at 512 feet; some gas at 1,050; commenced pumping about nine barrels a day, and has produced up to September, 1872, 4,133 barrels.

No. 8 well, on Queenstown Run; June 26, 1871—September 22, 1871; water cased out; first show of oil September 22, and

---

* The "Engineers' Datum" is an assumed level, 100 feet lower than a mark on the Brady's Bend Iron Company's warehouse, showing the extreme height reached by the flood of March 17, 1865.

began to fill up very slowly. At 12.35 A. M., September 23, struck a vein of gas and oil which spouted over the top of the derrick, and was fired by the night lamp hung in the derrick, burning the rigging down. The spouts occurred every two minutes. At 9 A. M., the fire was extinguished and the oil began to fill the tank at the rate of thirty-five barrels an hour, but gradually calmed down to about sixty barrels a day during the first month, and October 22, ceased to flow. Tubing and sucker rods were then put in, and it began to flow again at the rate of 150 barrels a day.

This well has been cleaned out many times to keep her in good running order. Immediately after any one such cleaning she produces from seventy to ninety barrels a day, and gradually falls off to about twenty to twenty-five, when it is understood that she again needs cleaning. In fifty weeks she has produced 9,505 barrels. There is not much gas except when flowing.

No. 9 well, on the river opposite Catfish; June 24, 1871—October 24, 1871; water cased out; cost $5,750.

No. 10 well, on Lower Campbell Tract; July 10, 1871—May 22, 1872; water cased out. After passing through third sand at 1,300 feet, put in a four-quart torpedo, which seemed to have very little effect. Sand-pumped for two days afterward, and found that she filled up with less than a barrel of oil per day, and therefore concluded it was useless to tube her. Not much gas at any time.

No. 11 well, on river half mile below the mill; August 24, 1871—June 24, 1872; water cased out at 437 feet. Struck very heavy vein of gas at 858 feet.

The gas from this well, by calculation, would supply fuel to run the rolling-mill and machine shop boilers, being therefore equal to 100 tons of coal per week.

The pressure of gas would sometimes lift the tools twenty or thirty feet in the hole, tools weighing 1,700 pounds and rope 300 pounds. The flow of gas is enormous and continuous.

No. 12 well, on Queenstown Run; December 9, 1871—April 12, 1872; water cased out at 394 feet. Struck heavy vein of gas

February 2, at 725 feet, which caused a flow of water until March 1, when casing was put in and the water stopped off.

Struck oil at the top of third sand April 4, at 1,183 feet, the rock being nearly all good pebble-rock; after passing through it (1,200 feet) drilled twelve feet into slate for a pocket; tubed well April 12; commenced pumping twelve barrels a day, and the well is now doing thirteen barrels. Much gas all the time. Cost $6,557.

WELL RECORDS IN DIFFERENT PARTS OF THE OIL REGION.

Name of Well.	Farm.	Township.	County.	Whole Depth.	1st Sand at	Thickness.	2d Sand at	Thickness.	3d Sand at	Thickness.	4th Sand at	Thickness.	5th Sand at	Thickness.
A. & G. W. Oil Co		Oil Creek,	Crawford	642					573	64				
Cadwallader & Warner		"	"	613			524		557					
C. G. Emery		Southwest,	Warren,	474					446		462			
Myers & Sherman		"	"	465	238		342		450	12				
E. Baum	Baum,			844							699		797	
William Wood	Zuver,	Oil Creek,	Venango	844	362	18	534	14	628	21	772	23	799	8
H. Davies	Steel,			526	313	31			475	20				
J. Watson		Perry,	Armstr'g	803			400		757					
"		"	"	900			400		452	37				
Murphy & McKenny	Hall,			559			382		520	23				
J. Watson	Ball,			868	587	77			834					
E. Poor & Co	Poor,			830							708	31		
Custer & Lowers	Bennehoff,			858									824	28
Dr. Egbert	Egbert,	Cornplanter	Venango	564	267		387	22	517	32				
"	Skinner,	"	"	791			605	24	764					
J. Watson	Weed,	Oil Creek,	Crawford	704	364		597		633					
B. D. Benson		Southwest,	Warren,	722					609		671		711	
"		"	"	596	241	27	325	19	369	13	418	16	482	
Whitman & Doubleday	McClintock,	Cornplanter	Venango	652	330		485	14	546	15	596			
F. G. Irving	Blood,	"	"	874					558	51	693	40		
Col. Roberts		Oil Creek,	"	823					779		794		813	
James Burns				634			472	20	597					
Reno Oil Co		Sugar Creek	Venango	882					668	15				
H. Joy & Co		Southwest,	Warren	522	221		342	10	469	19				
Benedict Estate		"	"	482					460	14				
Fee & Emery	Carp'ter Lot			560			372	8	509	35				
Watson, Williams & Co.	Barnsdall,			767	463		672		752	13				
A. R. Williams				579	275	44	429		560					
Redfield & Co	Sheridan,			778			542		688				706	
W. Barber		Cornplanter	Venango	518	208		342	35			489	24		
W. H. Marsden		"	"	477	200	31	333		364	7				
D. Kinney	Kinney,			591	247		500		531					
T. King	Withrop,			632	330	14	435	55	607	20				
C. D. Angell		Rockland,	Venango	660	365	5	469	59	635					
J. W. Brice				811	316	26			717	47			791	19
G. T. Ridgway	McMullen,			739			586		721					
E. Haines	Benedict,			495					480	10				
T. Chattels	N. Star Co.,			667					626	16				
Watson & Steele	Terrill,			796					620	37	775			
R. W. Throwbley	Shaw,	Cornplanter	Venango	789	497	43	614		763	23				
Linn, Pinkerton, & Co.	McClintock,	"	"	675	360	38	505	35	648	12				
Hess & Tarbell	Shoup,	Richland,	Clarion,	1,048	732		835		1034	16				

NOTE.—The sands, thickness, depth of wells, &c., in the Lower Oil Fields, will be found in the chapter devoted to a description of the Butler County and Parker's Landing District.

OIL COMPANIES.

A FEW SPECIALLY MENTIONED.

THE HARMONY OR ECONOMITE SOCIETY.

IT was not the intention to give special notice of the existence of individual associations and companies. We, however, make exceptions in a few instances. We do so because these present interesting and peculiar features.

In the case of the Economite Society their history is of marked interest to the public, because of their strange organization, their singular history, and their great success as oil operators. For the information herewith given we are indebted to their present Superintendent, Mr. William Merkle, to whom we tender our acknowledgments for his courtesy.

The Harmony Society was founded near the beginning of the present century, by George Rapp, and a colony of emigrants from Würtemberg, numbering over one hundred families. They belonged to that class of devout people in Germany, to whom, in the previous century, had been given in reproach, the name of *Pietists*. Dissatisfied with the state of religion in the established Lutheran church, they gave themselves to the more diligent private study of the Scriptures (a practice which they still retain), and to the edification of each other in social assemblages for conference and prayer. Two men arose among them, who by their force of character, became leaders. These were Michael Hahn and George Rapp. They were earnest, zealous men, who magnified their office by gathering together on the Sabbath, the people who sympathized with them, and administering to them words of instruction and encouragement.

Hahn had commenced the work of preaching at an earlier period than Rapp. He was a man of more literary culture, and made use of the press in the work of reformation. He was an humble farmer, with limited education, but he was a man of deep religious spirit, and great force of character. Feeling himself constrained to proclaim to others the religious convictions which filled his own heart, he soon gathered around him a number of followers, of kindred views. The work grew gradually until several hundreds looked to him as their leader. These movements excited the opposition of their more worldly neighbors, and especially of the clergy, whose ministrations they neglected. Hahn and his adherents, under the name of Pietists, still retained their connection with the established churches, giving at least occasional attendance on the ordinances as then administered. Like the first Methodists in England, they hoped to bring about a reformation within the church itself. They thus escaped in great measure the persecutions which arose against Rapp and his followers, who refused to attend upon the ministrations of the regular clergy. They were called Separatists, and although demeaning themselves as quiet, orderly citizens, and paying their dues both to church and state, they became objects of odium, and were denounced to the civil authorities by the offended clergy. They were persecuted with fines and imprisonment, and their appeals for redress were in vain. After long endurance, and after having made an ineffectual application to their own government for permission to form a settlement by themselves, they determined to emigrate. In the year 1803, George Rapp visited this country, in search of a location suitable for a colony. He purchased a large tract of land near Zelienople, in Butler county, Pennsylvania, and in the autumn of the ensuing year three ship-loads of colonists arrived.

Before they left Germany they had embraced some peculiar

views of religion and social economy, to which they were led, as they supposed, by their careful study of the New Testament. They had generally adopted the *Millenarian* theory of the personal and pre-millenarian advent of Christ, which they regarded as near at hand. They were also disposed to follow the example of the primitive Christians, in having all things in common. It was not, however, for some years later that this practice became a law among them. Soon after their settlement in this country marriage was prohibited, celibacy being strictly adhered to by members of the Society, for now, upward of fifty years.

The Society, for some reason, changed their location, having purchased lands in the Valley of the Wabash, in Posey County, Indiana, disposing of their property in Pennsylvania at a great sacrifice. After a residence of ten years on the Wabash, they again determined to change their location, finding the country unhealthy, and their neighbors ignorant, vicious, and turbulent. It is said unpleasant collisions occurred between them and the peacefully disposed Harmonists. These circumstances induced them to think of returning to Pennsylvania.

In 1825, they made a purchase of their present lands in Beaver County, Pennsylvania, on the Ohio river, about eighteen miles from Pittsburgh. They built a steamboat, and removed in detachments to their new and final place of settlement. Here they founded a town, to which they gave the name of ECONOMY, and from this circumstance, the popular name of the Society is derived.

A few years prior to the discovery of oil, the Economists' Society came into possession of a large tract of land, some six thousand six hundred acres, opposite Tidioute, in Warren County, Pennsylvania. This land was originally purchased as a lumber enterprise, for the supply of their own wants, and to meet the extensive demand for timber at their mills in Economy. This tract has since proved to be valuable oil territory, as is now well known. When oil was first discovered, it was a great and unexpected addition to the value of their lands, and they generously intrusted the management of it to the former proprietor, with such share of the profits as enabled him to retrieve his shattered fortune, and place his family in independent circumstances. They then took the whole business into their own hands.

In the early part of the summer of 1860, oil was found immediately below the river tract belonging to the Society, which created great excitement, and many persons sought to purchase or lease the land from them, but fortunately for them a law-suit was then pending for the possession of this tract, which prevented leasing or sales. In September of the same year they took actual possession of this tract, located five different wells upon it, made contracts for drilling them, and engaged workmen for the vigorous development of the property.

The first two wells drilled were entire failures. The third had a pretty good show of oil, but finally proved to be nearly worthless, after three months' hard labor, and the expenditure of a large amount of money. The prospect of success now looked gloomy, but they hoped and worked on, and unexpectedly, at the depth of 99½ feet, the drill struck a large crevice, and sank some eight or ten inches, and in a few moments large quantities of oil and water were thrown high above the derrick in a continuous stream. This well flowed steadily for six months, gradually decreasing in production, then stopped and flowed periodically, and at the end of nine months ceased to flow, and afterwards was pumped for a number of years. Immediately after the striking of the flowing well, a new well was struck near to it, which at first produced but little oil, but after some time proved a good well, and continued to produce for eight years.

During the spring of 1861, eight wells were finished, four of which were failures, and four of them good shallow wells of not more than one hundred and fifty feet in depth.

In March, 1862, the society had four good producing wells ; had erected a number of buildings, made roads, constructed wharves—a cooper shop; with a stock of several thousand barrels to send their oil to market; a blacksmith shop, carpenter shop, and all the necessary tools.

The history of the society's operations for the years following, up to 1868, is varied, meeting like all other operations, with successes and reverses, but still maintaining their position against all the drawbacks incident to the business.

In 1868, the society met with marked success, having sold during the year upwards of one hundred thousand barrels of oil, and in 1869, over seventy-five thousand barrels. They are now drilling their seventy-sixth well, having at the present time fourteen producing wells, yielding a little over one hundred barrels per day, which the Trustees have leased for a short term, to their superintendent, Mr. Merkle, and Mr. A. R. Moore.

From the records of their drillings, we find that the wells vary in depth from ninety-nine and a half feet, to five hundred and seventy-eight feet, and the oil-bearing sand rock from twelve to fifty-five feet in thickness. Their largest well was two hundred and fifty barrels per day.

It is something remarkable that the Economites never pumped their wells on Sunday. They declare that their wells do not suffer thereby, though of like character to that of their neighbors, being subject to water, which is presumed to injure wells by standing on the oil-bearing rock. We dare say the fact of the Economite wells being so long-lived is an argument in favor of their practice of cessation of pumping on Sunday.

In a few instances the Trustees have leased small portions of their oil territories, and in all such cases they have made binding, it being in the lease, not to pump on the Sabbath; their operators say that they have never known their wells injured by cessation of pumping on Sunday.

The society is represented in all its business matters by two trustees, one of whom formerly took especial care of their oil interests; we refer to Mr. R. L. Baker, now deceased, who was ably assisted by Mr. Jacob Henrici. Since Mr. Baker's death Mr. John Lenz has been selected by the society to fill his place.

There are no members of the Economite Society on their Tidioute property; the business is conducted by their agents, under the supervision of the trustees, who make frequent visits of inspection.

We ought not omit to mention that most of the land of the Tidioute property is valuable as timber land, on which the society has now two large saw mills in full blast.

Of the society's 6,600 acres, not more than 200 acres have been developed. The balance is supposed to be good oil territory.

The Economites are very few in number now; year by year death is thinning them out. Their adoption of celibacy, as a matter of course, will in a few years put an end to their existence. What is to become of their vast property? At present it is not known that they have made any provision for its disposal after their extinction. They expect, even yet, that the Lord will come in his glory before they are all gone.

SAGE RUN OIL FIELD.

The oil territory at the head of Sage Run, 2¼ miles south of Oil City, has of late proved valuable. The first strike was made early in the summer of 1869, which produced 40 barrels daily, located on the Schwartz farm, and owned by the McGrew Bros. of Pittsburgh. Land owners would not lease or sell on reasonable terms, and the excitement consequent upon the strike died out.

Prominent among the operators here is a widow-lady, Mrs. Sands, who, with a keen eye to future advantage, had several months before the McGrew strike, purchased a tract of some 200 acres, situated on the turnpike road, near the Schwartz farm; this land she bought at a nominal rate of eight or ten dollars per acre, from a Philadelphia Company. Mrs. Sands has now put down seven wells.   Her No. 1 was struck in September, 1870, and is said to have produced 60 barrels per day, gradually falling off, till at the present time, it is producing some three barrels.   The two following wells put down by this enterprising lady were moderately successful.   In the beginning of October, 1872, Mrs. Sands struck her No. 4 flowing well, which attracted much attention.   For days this well "spouted" not less than 500 barrels daily ; it gradually declined by filling up with sand and mud, and ceased to flow early in November.   At present this well is producing 12 barrels per day.

A few rods south of the Sands property is the first Green well, owned by the Green Bros. and Mead Bros.   It began pumping at the rate of 30 barrels per day in November, 1872, and is still producing at the rate of five barrels per day.   The second Green well was struck, 29th of January, 1873, and started up at the rate of 400 barrels per day, and then held out at the rate of 200 barrels for two months, when it decreased, and at 60 barrels it ceased to flow when pumping commenced, and at the time of our visit (August) it was producing 50 barrels per day.

A few rods south of the wells just described, is situated the Bly and Main well, which was struck July 12th of the present year, and continues to flow, starting up at 300 barrels, and now doing 200 barrels.   This well, like many of its kind, flows intermittingly, sending forth the oil with a tremendous rush every ten minutes, or thereabouts.

The territory hereabouts is from 1000 to 1045 feet in depth, the oil-bearing sand rock varying from 18 to 20 feet in thickness.

THE RENO OIL COMPANY.

The Reno Oil Company, now owners and operators, of what is known as the Reno district, was formed in 1867. The tract of land owned by this company comprises quite one thousand two hundred acres. The district lies on the north side of the Allegany river, extending back up the hill-side for about a mile, and so far as developed has proved valuable oil-producing territory.

The president of the present organization, is C. V. Culver, who was first identified with the oil region, as part owner of some wells upon the Clapp farm, early in 1861—he being at that time connected with the Citizens' Bank at Logan, Ohio. Gifted with a clear and far-seeing mind, he saw the opening which this region presented, and acted upon it at once, as in May of the above year he came to Meadville, and purchased the charter of the Bank of Crawford county, which was at that time in bad repute. He redeemed the old issue of notes, and established the bank on a sound basis. Comprehending fully, that for the development of the oil region a large amount of capital was needed, he conceived the financial scheme of organizing a chain of offices and banks throughout the district. He first opened an office of discount and deposit, at Franklin, in May, 1861. In October of the same year, he founded the Venango Bank, at Franklin, which institution afterward proved such a disastrous failure. The advent of this bank was followed by the opening of the Petroleum Bank at Titusville, and another of discount and deposit, at Oil City; and to accommodate Eastern capitalists who had invested in the oil region, he opened a house in Philadelphia, under the name of Culver, Brooke and Co., and in New York city, a banking-office under the style of Culver, Penn & Co. Although these offices facilitated the business of his country banks, they were not enough, and accordingly in the spring of 1864, he organized the Third National Bank of New York. With all these banks, he was not able to do the business which crowded upon him, and to meet the demands, he organized the first National Bank at Corry, the First National Bank at Titusville, the Second National Bank at Erie, and the Tradesmen's National Bank at Pittsburgh, thus completing the circle.

Mr. Culver organized the Venango Oil Transportation Company, for handling and storing oil, which company owned some of the finest store-houses and docks in New York city.

With the disasters of 1865-6, came the downfall of Culver, and the mighty fabric which his genius had reared, and with his fall came the ruin of thousands who had placed their savings under his care. We pass over this point with the suggestion generally received now, that the times, and not the man, contributed to the destruction of his grand scheme. Had the result been different, Mr. Culver might at this day have been a Petroleum king. As it is, he is generally acknowledged to be a gentleman of rare business qualifications, industrious, plucky, and sure to work out his financial salvation.

Mr. Culver was elected to Congress in 1865, in recognition of his great efforts to advance the interest of the Oil Region.

The Reno Oil Company was organized in the interest of the creditors of Culver, Penn & Co., and by judicious management and the good fortune of the company they have paid a number of dividends to the creditors.

The first well, known as No. 18, was struck on the 27th of May, 1870. It started off at the rate of one hundred and fifty barrels per day, and created quite an excitement at the time. It soon fell off, however, and after a time a torpedo was put in, which brought it back to its original production, when it caught fire and burned up the derrick.

The company have now some forty wells producing. No. 51 was finished early in 1872, and when completed did only fifteen barrels. It gradually increased, fluctuating from time to time, was torpedoed, and then commenced to flow two hundred barrels per day, and continued at this rate for a considerable time.

So far the company have found very few dry holes; in fact, only three during the present year. A number of the wells are small, but as there is plenty of gas to run them, they are made to pay by pumping them "by heads." The engine houses are models of neatness, with their cleanly swept floors and well kept boilers and engines. The oil from the wells is running into a large 10,000-barrel tank, from which it is pumped to the loading rack, as required.

Very little coal is used on any part of the farm, thereby saving an immense amount of cost, and making the profits of the company proportionately greater.

The sand found in the best wells is coarse and white, with large pebbles that look like bits of polished marble. It, however, changes its character somewhat in different localities. The oil in the wells near the river bank was reached at a depth of from 500 to 550 feet. Further back, No. 51, the elevation of the hill required about 200 feet more of drilling. This oil is taken to the New York and Cleveland markets, that to the latter place being shipped over the Jamestown and Franklin railroad, the remainder by the way of Meadville to New York. The prices and the freight are the same as those of Oil City, and the oil is in good demand on account of its quality and gravity, which varies from 43 to 47 degrees. The company have refused to lease any of their lands, reserving the entire development for themselves. No leases have been given at any time, with the exception of a few granted a number of years ago, before the creation of the present company. The wells on these leases are pumped by heads only, none of them being large enough to pay for more time and attention.

Quite a small village has sprung up upon the territory. The population numbers about 500—100 of whom are on the pay roll of the company, the remainder being generally the families of these. One rule rigidly enforced is that no liquor shall be sold in the place, and none drank by the men in the employ of the company, either when on or off duty, instant dismissal being the penalty for drinking even a glass of ale. The consequence of this is that perfect peace and quiet prevails, and the men are always able to attend to their work, in proof of which the company have never had a boiler burned since they began operations.

The officers of the company are an extremely gentlemanly and obliging set of men, always ready to afford information, or opportunities for inspection of the wells and machinery, to visitors. Their office is pleasantly located at the lower end of the village, and has from the windows fine views of the scenery both up and down the river.

*Conducted as it has been, the past of this company has been a success, and under the same efficient management there is no reason to expect a change from this in the future.

THE OCTAVE OIL COMPANY.

" The Octave Oil Company " was organized at Titusville in July, 1871, and consisted of an association of eight members, the majority of whom were members of the Mendelssohn Society of that city —all musically inclined—suggesting the title of the company— "OCTAVE." The association was formed for the purpose of producing, refining and transporting oil, and commenced with a capital stock of $80,000. The Octave territory is situated two miles north of Titusville, and embraces about nine hundred acres of land, including the Purtill and Hyde farms, which are owned in fee, also a large tract of leased territory on the Fleming, Pierce and Lamb farms, all on the west side of Oil Creek. On the east side, the company own one hundred acres in fee and one hundred and eighty-three acres by lease, on the Young and Noble farms. This territory stretches in a north-easterly and south-westerly direction, and from actual developments fully demonstrates " the belt theory." It is singular, however, that the Octave district is the only one of the many prolific tracts in the vicinity of Titusville which thoroughly proves the correctness of the belt theory.

The company have drilled twenty-one wells, the larger portion
of which have proved good paying wells.   One of them produced
200 barrels daily for a considerable time, and continues to produce
largely, though struck in November, '72.   The depth of the oil-
bearing strata varies from 460 to 900 feet, and the thickness of the
rock from forty-five to seventy-five feet.   Nearly all the rock is
white and coarse.   Operations can be carried on upon this territory
with much greater economy than on territory situated at great distances from the manufacturing centres.   The cost of wells averages
from $3,000 to $4,000 each.

The company possess peculiar facilities for economically conducting their business, as they pipe, transport and refine their own
oil, and sell their refined product to the exporter or the home
dealer.   The Octave Pipe Line is fifteen miles in length, with all its
connections, and furnishes cheap transportation to the oil farms in
the neighborhood of their operations.   The refinery of this company
is situated at Titusville, and has a still capacity, daily, of 606 barrels.

This company have extended their operations to the great lower
oil fields, having purchased 318 acres of the celebrated McClymonds farm at Karns City, Butler County.   On this farm they have
built one of the largest oil tanks yet constructed, it having a capacity
of 22,000 barrels.   The business plan of this Company may be
considered unique, as they produce, transport, refine and sell their
own oil and that of their neighbors, thus saving many profits for
their own pockets, and though the title of the association was at
first suggestive and appropriate enough, yet we think they have
earned for themselves the title of "Model" Oil Company.   The
present officers of the Company are as follows :

M. STEWART, President.

J. E. BLAKE, Vice-President.

D. O. WICKHAM, Secretary.

O. G. EMERY, Treasurer.

D. EMERY,   ⎫
L. EMERY, JR.,   ⎬ Directors.
                          ⎭

COLORADO OIL DISTRICT.

This territory, comprising what is now known as Colorado, is located on Pine Creek, in the south-west corner of Warren County, and in the town of South West—eight miles north-east of Titusville, or midway between Tidioute and Titusville. The extent of this territory may be stated, as at present developed, at about one mile up and down the course of Pine Creek, or lying near to that stream—embraced within a tract of 800 acres of land. The land was originally owned by the Holland Land Company. The title to four hundred acres of this tract, and first tested, we trace from the Holland Land Company, through several parties, until it reaches the possession of William and Benjamin McGee, who purchased it for lumbering purposes, years before the discovery of oil in this part of the State of Pennsylvania. About the date of the discovery of oil on the Watson flats, by Col. Drake, GODFREY HILL, of Pittsburgh, purchased these 400 acres for oil and lumbering purposes. In the fall of 1860, Mr. HILL put down the first well drilled upon the tract. This "test well" was very near what has since been demonstrated as the "Colorado Belt." He found the "third sand rock," and it was pronounced good. From it came a very little oil, but not in paying quantities. He soon after began a second and third well, but abandoned both before they were half down. This territory lay in its comparatively undeveloped state, until the fall and winter of 1869 and '70—the original purchaser, GODFREY HILL, having meantime died.

The remaining 400 acres, of which we have made mention, was purchased in 1860, by HENRY R. ROUSE,—whose tragic death by burning occurred at a well he was interested in, on the Buchanan farm,—for lumbering purposes. Just prior to Mr. ROUSE's death, he executed a will, by which, after certain bequests (see his will elsewhere), he devised the remainder of his estate to the County of Warren, to be expended in equal parts, for the benefit of the poor of the county and the improvement of the highways within its boundaries.

This property, after an equitable administration, was purchased by MYRON WATERS, of Warren, Pa.

In the spring of 1865, T. C. JOY, of Titusville, purchased Mr. WATERS' interest in this property, paying in round numbers $60,000 for it. In the course of the summer of the same year, Mr. JOY sold the north half of his purchase to "The Enterprise Oil and Lumber Company," represented by B. D. Benson & Co., of Enterprise, Pa. Soon after this sale, Mr. JOY began and completed his first well upon the south half of his purchase. This well exhibited a dark gray sand rock, but no oil. The following spring and summer, B. D. Benson & Co., put down a single well upon the tract purchased from Mr. JOY, called the "Metter well." This was also a "dry hole!" Little or nothing was done in drilling wells for a year or more thereafter. In 1867, "a JOINT WELL" was put down upon the east and west line of this tract—Mr. JOY and B. D. BENSON & Co. owning equal interests. This well produced about two barrels per day, but was soon after abandoned, as non-paying.

Shortly after this last development, the improvement in product and quality of sand rock, encouraged a number of Shamburg operators to try it once more. They sank the fifth well on the "Joy tract," and this was a dry hole!

In December, 1869, a "developing company" was organized, composed of F. W. AMES, Dr. W. B. ROBERTS, L. B. SILLIMAN, T. C. JOY and A. K. MURRAY of Titusville, and B. D. BENSON & Co., of Enterprise, and GRAHAM & HORTON, of Middletown, N. Y. In all previous developments a comparatively inferior "third sand" rock had been found, of forty to fifty feet in depth. This company was therefore organized to make still further searches for the "golden stream"—confident the territory would furnish it. The company drilled TWO wells, and both were dry! This concluded the operations of the "Developing Company," and soon after all its parties withdrew from the temporarily organized association, leaving Mr. JOY, and B. D. BENSON & Co., sole owners and operators.

Meantime, or in 1870, HENRY HILL, a son and heir of GOD-FREY HILL, put down the second well upon the first 400 acres, previously mentioned. This well made a promising "show," but it failed to produce in remunerative quantities—less than three barrels per day. It demonstrated, however, that oil existed there in some abundance, and that this last well was very near the deposits. The first good sand rock was found in this well.

Soon after this, B. McKinney and others secured a ten acre lease, and put down the third well upon the " Hill Tract," which produced ten barrels per day.

At this time the title to the property, was in dispute and in litigation. Terms of adjustment were, however, agreed upon, and out of the compromise, for such it was,—"The Colorado Oil Company " sprang—composed of the following gentlemen:—

T. C. Joy, F. W. Andrews, B. D. Benson, Marcus Brownson, R. E. Hopkins, David McKelvey.

In November, 1871, this company completed the fourth well on the first four hundred acres, or "Hill Tract," and within forty rods of the first well, put down by Mr. HILL in 1860. This proved to be a success. The well produced, after being torpedoed, one hundred and fifty barrels per day. This established the character of the Colorado Oil District, beyond cavil or doubt. The territory has been rapidly developed, since, and up to the date of this record, January, 1873, "The Colorado Oil Company," has put down eighteen wells, seventeen of which have been remuneratively productive—a few largely so.

During the two years' operation of this company, over 80,000 barrels of oil have been taken from their wells, and the line of development demonstrates that the "Colorado Belt" extends over both the "Joy" and the "Hill Tract," alike.

Messrs. D. B. Benson & Co., have developed other and larger tracts in the Colorado districts, and their operations began about the date of those above mentioned. We have not been able to obtain the information required, and therefore omit further mention of the Colorado district.

THE COLUMBIA OIL COMPANY.

This company was chartered by the Pennsylvania Legislature, May 1, 1861, with a capital of $250,000, divided into 10,000 shares of $25 each.

It commenced operations on the Story Farm, on Oil Creek, about seven miles from its mouth. During the year 1861, its production of oil was 20,800 barrels; in 1862 it had increased to 89,602 barrels. Its first dividend was declared July 8, 1863, but little over two years from the date of its charter, and was for *thirty per cent.* on its capital stock. This was followed on the 12th of August, by a second for 25 per cent.; another for the same amount September 9th, and still another on the 14th of October for 50 per cent.—making in all dividends to the amount of 130 per cent. on the capital stock within two and one-half years of the formation of the company.

In 1864, its production increased to 141,508 barrels. During the first six months of this year, it declared four more dividends amounting to 160 per cent. on its capital stock. At this time its capital was increased to $2,500,000, and it at once declared a dividend of 5 per cent. on this increased capital, and before the close of the year, five more—making in the aggregate 25 per cent. From that date to the close of the year 1871, its production of oil has been remarkably uniform; the minimum amount being 110,-655 barrels in 1867, and the maximum 142,034 barrels in 1871.

The whole amount of oil produced by the company during the ten years of its operations is 1,715,972 barrels, and the whole amount of its dividends $2,342,600, or 401 per cent. on its capital stock; and yet after ten years of active development, but a small part of its land has been touched, and the part where developments have been made is capable of receiving as many more wells as have been already sunk upon it; and although the fortunate owners of its stock have already received a princely return for their investment, there is no reason to doubt that these dividends can be maintained, or even increased, almost at pleasure, for years to come. The history of this company, as we have endeavored to give it, is one of figures, not words; but from these figures, some idea can be formed of the vast wealth still lying hidden in the crevices of the rock, only needing the hand of skill and the direction of prudent managers to bring it into the service of man.　Touch the point of a pencil ever so lightly upon a map of the oil region, and a larger proportion of the territory which is known to be oil-producing, is covered, than the whole property of this company.　Touch a pencil again upon a map representing the land of this company, and the spot covered by the mark may represent the developed portion of the land, and from this speck of ground, eight millions of dollars' worth of oil has been sold !

THE OIL MARKET FROM 1859 TO 1872.

Petroleum mining may be said to have commenced as a business in 1860. The demand previous to this was of a very limited character; yet the oil from the Drake Well commanded an average of 50 cents per gallon during 1859. In July, 1860, the price at the wells had declined to seven cents per gallon. In October it was ten cents per gallon, and from this time it continued to advance slowly to January 1st, 1861, when it was sold at 25 cents per gallon, and remained at this price till March 1st. A few days afterwards sales were made at 15 cents per gallon, and on the 18th of March it was ten cents per gallon. In June, 1861, the flowing well period commenced, and the production was suddenly increased from about 150 barrels daily, in February, to some three thousand barrels daily, in September, and more than 6,000 barrels daily, in December of the same year. The sudden and immense increase of production had now almost destroyed its value, as consumption was as yet very small. Thousands of barrels were allowed to run to waste for want of barrels and a market. In the summer the price had declined to five cents per gallon, and in August and September sales were made at fifty, twenty-five and thirty cents per barrel, when in July the price had receded to ten cents per barrel. Prices again took an upward turn, and sales were made during October, November and December at 35 to 40 cents per barrel; then again the price is quoted in January, 1862, at $2 per barrel. The spring of 1862 was signalized by a much larger production, and the price of oil went down to 40 cents per barrel. Excessive cheapness forced consumption, both in this country and abroad, with unparalleled rapidity, so that in the latter months of 1862, there occurred a large but spasmodic rise in the value of petroleum. The unremunerative price which had for some time prevailed checked production, causing all small wells to be abandoned. This state of the market continued, merging into a more even upward graduation of values through the year 1864, when crude oil sold at one time as high as $13.50 per barrel, at the wells. The following list of prices we take from the books of a firm engaged in the purchase of oil at Titusville, from 1860 to 1865.

December 7th, 1860,	Oil at Titusville,	23 cents per gallon.
" 24, "	" "	23 " "
January 4, 1861,	" "	25 " "
" 10, "	" "	26 " "
" 16, "	" "	26 " "
April 18, "	" "	12½ " "
July 22, "	" "	$2.25 per barrel.
August 12, "	" "	$2.25 "
November 17, 1862,	" "	$5.50 "
" " "	On the Creek,	$2.50 "
February 18, 1863,	At Titusville,	$2.00 "
March 27, "	"	$3.00 "
April 10, 1864,	On the Creek,	$5.00 "
June 7, "	"	$8.50 "
August, "	At Shaffer Station,	$9.50 "
Sept. 1, "	" "	$12.50 "
Sept. 26, "	" "	$8.00 "

The above quotation includes hauling from the wells to the place of delivery. The average price per barrel of crude on Oil Creek for the years named are as follows:

1862	.	.	.	$1.15 currency.
1863	.	.	.	3.25 "
1864	.	.	.	8.13 "
1865	.	.	.	6.71 "

In the Columbia Oil Company's Eleventh Annual Report, the average prices for the years named are given as follows:

1866	.	.	.	$3.78
1867	.	.	.	2.54
1868	.	.	.	3.95
1869	.	.	.	5.48
1870	.	.	.	3.82

OIL BROKERAGE—ITS COMMENCEMENT, &c.

BY ARNOLD BURGESS.

To a visitor in the oil regions, not the least interesting as well as surprising feature is the brokerage business. A stranger to the section and the trade, calling at the exchange, seeing the number of brokers and dealers engaged, and these augmented on the arrival of every train, the constant coming and going of telegraph messages, and listening to the terms used, finds himself in a *new world*, where thousands on thousands of dollars' worth of property changes hands with an indifference and ease which astonishes and bewilders him.

In 1868, brokerage was started by a few individuals, and for some time was chiefly confined to buying for refineries in Pittsburgh, Philadelphia and Baltimore. In most cases the brokers were paid by the buyers a commission of ten cents per barrel, and this was sometimes increased by an additional five cents per barrel by the seller. At the outset the business was a matter of experiment, but soon the attention of other parties was attracted thereto. New men entered the ranks, and the whole thing was placed on a legitimate basis by the formation of brokers' boards in the cities where the heaviest trade was carried on, and the establishment of regular rates of brokerage.

In 1869, Erie "cornered" the market, and by the large transactions in crude on the creek, influenced the refined markets of New York and Cleveland, thus bringing them in, as extensive buyers and opening these important points also to the brokers. Since that time, with each succeeding year they have assumed a more important position in commercial circles, till now all over the country, wherever oil is produced or shipped, either crude or refined, the greater part of the transactions are executed by brokers. A broker's business consists of buying and selling "spot," "regular" and "future" oil. As these are expressions which will convey to the uninitiated no idea of the particular trade

named, a few words may be appropriately given in explanation :
"Spot" is the term used when the oil is to be moved and paid for
immediately ; "regular" is where the buyer is allowed ten days
in which to put in his cars and take out the oil.   These are *parol*
contracts and without writings, the broker acting under orders
from his principal, whom he names to the other party, and he
being often the only witness to the trade ; but in the case of
"futures," this is not the custom, as the fulfillment of these takes
place at the expiration of the agreed time.   A written contract is
drawn by the broker and signed by him as such.   This is accepted
by both parties, and is equally binding, the one agreeing to *sell*
and the other to *take* a certain quantity of oil within a certain
period of time, at a price named in the contract, which also speci-
fies that the party, in whose favor the contract be drawn, shall
give to the other ten days' notice, within which he will move the
oil.   Regular future contracts are *buyers'* and *sellers'* options.   In
the first, the buyer has a right to demand the oil at any time he
sees fit ; and by the last, the seller can put it in whenever he
chooses, all within the time as named in the contract.   By these
contracts the buyer is also bound to take the oil or pay, or take a
difference in money according as the trade has proved in his
favor or against him, and this difference is that between the contract price and the regular market rate on the *last* of the ten
notice days.

I have spoken of these as *regular* contracts, because there is a
species of contract by which the amount of difference is named and
limited at the start.   These are called "puts" and "calls."

A "put" is where one party agrees to give a certain sum of
money—to be paid at once—for the privilege of delivering a
named quantity of oil at a price also named, within an agreed time.

A "call" is when the money is paid for the right to call on the
other to deliver the oil.   In these cases the prices of the "put" or
called oil is generally higher than the rate of regular contracts for
the same time.   This is because it is a one-sided affair—since

under no circumstances can the acceptor of the offer get more than the amount bid, while if the market goes against him, he is obliged to settle the difference at what may prove a heavy loss.

In all regular contracts the seller pays the brokerage; but in these irregular trades there is yet no established custom as to which of the parties it is due from. In New York the commission is three cents per barrel; in Cil City and on the Creek it is two-and-one-half cents. There is, however, one obstacle in this business to which brokerage in other commodities is not liable. I refer to the fact that in all futures the broker has to wait for the fulfillment of the contract ere he can collect his commission, and if either party fails, he loses his pay. By this he is actually made to insure the solvency of both parties to the amount of his brokerage, which is a manifest injustice. His business ought to end with the issue and acceptance of the contract, and though it is customary for brokers to attend to the taking or delivery of oil for their principals, they get no additional pay for this extra work.

In a business like this, the market is liable to great and sudden fluctuations. A combination is often formed to lower or raise the price of oil, and this is especially the case as the time approaches when a number of contracts mature. The bulls and bears are then rampant, and the talk is all of the " long " and " short " order. The close of the first, and last half of the year, is generally marked by some such struggle, and the brokers buy and sell thousands of barrels of " paper " oil, to effect settlements of the six months' contracts that are coming due.

Brokers are, by the nature of their business, very closely connected. Each has his correspondent " on the Creek " and elsewhere, with whom he shares the brokerage arising from the purchases or sales made through their joint exertions; and it often happens that each broker is obliged to call in the oil of another, till the commission is so divided and sub-divided that it will hardly pay the telegraph bills of the different parties. In fact, brokerage, is a hard-worked and poorly-paid profession, and yet there is an excitement about it that forms a great attraction. The men are a jolly, jovial set, free and generous with their money and kind offices, and as their's is a business where much is of necessity left to their honor, each man takes a pride in keeping his word on an equal footing with his bond.

Efforts are constantly being made to place brokerage upon a still higher and more responsible footing. During 1871, by the exertions of prominent men identified with the trade, regular exchanges were established both in Titusville and Oil City. The members of these are bound by the most rigid laws of equitable trade, and by a wise arrangement of arbitration committees, very much litigation is avoided. Within these halls of exchange, all possible aids to business are gathered.

Every facility is offered by telegraphic communication with both home and foreign markets for a thorough and accurate knowledge of the condition of affairs. Membership is not limited to brokers; but dealers, producers and consumers are admitted to the benefits thereof, and meet on the same footing. By such an arrangement, the best interests of all are consulted, buyers and sellers are brought together, and the brokers, through whom this is effected, take their rightful position before the world as a useful and honorable body of men.

OIL PIPE LINES.

EXTENT IN THE PENNSYLVANIA OIL REGION.

THE iron pipe lines for the conveyance of oil from the wells to railway shipping points play an important part in the transportation of the article. The difficulties experienced in conveying oil by teams in the early years of the petroleum discoveries and developments suggested the idea of using wrought-iron pipes for the purpose. Mr. Samuel Van Syckel, of Titusville, was, we believe, the first to reduce the idea to practice. The first pipe line, four miles in length, was put down by him in 1865, and extended from Pit Hole to the railway, at Miller's Farm. Like many other innovations, the idea was received with doubts and misgivings at first, but after a variety of changing fortunes, it at length worked its way into public favor, and was pronounced a success. From this small beginning has arisen a whole net-work of pipe lines, covering the entire oil-producing territory and introducing an entirely new system of transportation.

Soon after the completion of the Van Syckel line from Pit Hole to Miller Farm, Mr. Henry Harley had a line in successful operation from Benninghoff Run to Shaffer Farm, on the Oil Creek railroad. A fuller account of both these lines of pipe, is given in the sketches of Messrs. Abbott and Harley, and to these we refer the reader. Suffice it to say here, that "The Pennsylvania Transportation Company" own and operate nearly *five hundred miles* of pipe line in the upper oil region.

Messrs. Vandergrift and Foreman, of Oil City, are extensively interested in pipe line enterprises, and own and operate several in the upper as well as the lower oil field. They have one line from Pit Hole to Paxton, 10 miles; one from Fagundas to Trunkeyville, 10 miles; one from the Shaw farm to Oil City, 4½ miles; one from Sandy to Oil City, 11½ miles, and one from the Miller farm to Oil City, in one direction, and to Franklin in the opposite direction, 7 miles.

"The Rochester and Oleopolis Transportation Company," of Oil City, have a line from Oleopolis to Oil City—six miles—composed of *six-inch pipe*. This line and its capacity was found to be necessary during the great product of 1865 at Pit Hole—five to six thousand barrels per day.

Grandin Bros. & Neyhart own and operate extensive lines of pipe at Tidioute, Fagundas, and elsewhere in that vicinity.

Mr. Payne and Mr. Martin, of Petroleum Centre, are each operating a number of short lines for the convenience of producers in that locality.

The Cherrytree pipe line was one of the earliest constructed. It runs from Kane City to the Rynd Farm, is 15 miles in length, and is owned by John Wallace & Co., of Rouseville.

THE LOWER DISTRICT PIPE LINES.

The following is a very complete estimate of the Pipe Lines of the Lower Oil Field, comprising nine different lines, all of them gathering in the oil to places of shipment on the Allegany Valley Railroad. The extent and ramification of these lines is surprising, and testifies to the importance of the transactions in that direction.

The Emlenton Pipe Line is 50 miles in length, carrying the oil from Upper Turkey Run.

The Antwerp Pipe Line runs from Upper Turkey Run, and points adjacent, and is about 50 miles in length.

The Mutual Pipe Company's lines, consist of nine different

branches, running from the following points: Upper and Lower Turkey Run, Clarion District, St. Petersburgh, Antwerp City, and points as far as Beaver Creek, and is 100 miles in length.

The Grant Pipe Line runs from the Grant farm on the west side of the Allegany, coming out at Parker's Landing, and is 20 miles in extent.

The Union Pipe Line runs from the Bear Creek District, Sheakley, Argyle, Petrolia, Karns City, Millerstown and Modoc City, and is in all, 125 miles in length.

The Cleveland Pipe Line Co. (S. D. Karns, owner,) runs from Karns City and Greece City, and is 40 miles in length.

The Fairview Pipe Line, owned by Vandergrift and Foreman of Oil City, runs from Sheakley, Petrolia, Greece City, Millerstown and Modoc. Total length, 125 miles.

Relief Pipe Line runs from Story Farm and Armstrong Run, and is some 12 miles in extent.

The Butler Pipe Line runs from Greece City, Modoc, Millerstown, and intervening points, to the Butler Branch Railway, and is sixty miles in length.

With the lines now in operation, in course of construction, and those surveyed, soon to be commenced, the Oil Region of Pennsylvania will soon have upwards of 2,000 miles of pipe lines for the transportation of oil!

There are a few gravity pipe lines; but generally the oil is forced through the lines by pumping. The capacity of each line is about 1,500 barrels in every twenty-four hours. In some districts, the lines are run up to their capacity, while in others they do not exceed half that amount. The present cost of the pipe used is about 30 cents per foot; and the average cost of the lines, including pipes, tanks, pumps and boilers, is about $1,500 per mile. A large share of the production is purchased at the wells by the companies, and then transported on their own account. Many of the large producers, however, prefer marketing their own oil, and employ the pipe companies to transport it to the railway. The charge varies from 20 to 30 cents per barrel, according to distance. In such cases, the oil is usually pumped into the company's tanks, and from these, 42 gallons are delivered for each 43 gallons received at the well—the one gallon per barrel being deducted for wastage. The pipe lines are increasing with the development of new producing territory, and are proving a source of great benefit to the producers, as well as of profit to the companies.

EARLY AND LATER MODES OF TRANSPORTATION.

The early operator had many difficulties to contend with in the transportation of his oil to market, as the only mode of land-carriage consisted in hauling the oil from the wells to Titusville, Oil City, or other shipping points. The roads were bad, and when much used and in wet weather, they then became almost impassable. The author of *Petrolia*, writing on this subject, says:

"Oil Creek mud" attained a fame in the earlier and subsequent years, that will be fresh in the memory of those who saw and were compelled to wade through it. Teamsters and horse-men swore both loud and deep at it. Newspaper correspondents exhausted all their adjectives, epithets, and expletives in essaying to give a faint description of its demerits. Weary pedestrian pilgrims, like Bunyan's Christian, were inclined to part with their knapsacks after a brief experience; ministers of the Gospel and devoted laymen, earnestly desired sustaining grace while urging their weary beasts over and through it. Mud, deep, and indescribably disgusting, covered all the main and by-roads in wet weather, while the streets of the towns composing the chief shipping points, had the appearance of liquid lakes or lanes of mud."

The difficulties of moving the thousands of barrels of oil which it was necessary to transport, can be better imagined than described. It was indeed a huge task, and many were the mishaps attendant upon the rough and swearing teamsters, as was evidenced to any one passing along the line of roads leading to a large shipping point, as the way was literally strewn with broken wagons, dead horses, oil barrels, filled and empty. Some one conceived the idea of conveying the oil down Oil Creek to the Allegany in flat boats, to hold the oil in barrels or bulk, and the employment of *pond-freshets* to float the boats, when laden with oil. Flat-bottom boats were procured from the upper Allegany, and from all points where they were built. Arrangements were made with the mill-owners at the head-waters of Oil Creek, for the use of their surplus water at stated intervals. The boats were towed up the creek by horses—not by a tow-path, but *through the stream*—to the various points of loading, and when laden they were floated off upon a pond-freshet. The amount of oil brought down upon one of these pond-freshets averaged from 15,000 to 20,000 barrels—the largest quantity ever brought out of "the creek" upon a single freshet, would not exceed 40,000 barrels. The oil was transferred at the Oil wharves at Oil City to a larger and better class of boats, and floated down the Allegany to Pittsburgh.

At one time over 1,000 boats were employed on the creek and river, and in addition to these there were some thirty steamers, passenger and tow-boats engaged in the same traffic. This oil float furnished employment to about 4,000 men.

Collisions and "jams" were of common occurrence; a boat would by some mismanagement get aground, and thus swing round, by the force of the stream, when it filled with water and sunk. Against this obstacle the advancing boats dashed with great force, the weaker ones becoming splintered from the concussion, the stronger ones being wedged fast, in the order in which they came, and thus formed what is familiarly known on Oil Creek as "a jam." During the freshet of May, 1864, a "jam" occurred at Oil City, which resulted in the loss of from 20,000 to 30,000 barrels of oil.

The magnitude of the oil business, soon attracted the attention of railroads running near the Oil Region. The Atlantic and Great Western Railway, first built a branch road from their main line, at Meadville, to Franklin, thus opening an avenue of traffic to New York, Philadelphia and Cleveland. About this time, 1861-2, the Oil Creek road was projected and built from Corry to Titusville, thus connecting with the Philadelphia and Erie Railroad. The Oil Creek road was gradually extended down the valley of Oil Creek, to Shaffer farm. In the winter of 1865 and 1866, the Atlantic and Great Western, extended their branch road, from Franklin to Oil City. We need not here occupy the reader's attention with a particular account of railroad developments in the Oil Region. It is sufficient to say, that the country has ample accommodation in this respect at the present moment.

With the advent of railroads, the mode of doing business was revolutionized. Car tanks were brought into use, each car being mounted with two wooden tanks, having a capacity of about forty barrels each, and by the aid of pipe lines, were filled upon the railroad track with great ease, and at much less expense, compared with the old system. The wooden car tanks, have latterly given way to the iron cylinder-shaped single tank, which holds about the same amount of oil, as the two wooden tanks. These are used for transporting both crude and refined oils. A great many railroad companies own iron car tanks for the transit of oil. There are, on all the railroads that handle petroleum, about 2500 iron bulk cars, of an average capacity of eighty-five barrels to a car, giving a tankage capacity now on wheels of 212,500 barrels. Cars that carry oil in barrels are not included in this aggregate.

The expense of hauling by team, was an important and expensive item, and helped to reduce the gross price of oil. During the spring and summer of 1862, the price paid for hauling or teaming oil, from the flowing wells on the lower McElhenny farm, to Oil City and Titusville was, from half a dollar to a dollar and a half per barrel. Later in the fall of that year, three dollars to three dollars and fifty, and even four dollars per barrel was paid, for hauling from the Empire well (McElhenny farm) to Titusville. We have it from a reliable gentleman, BARNEY BOSCH, now a prominent citizen of Titusville, that he had in his employment, a teamster for a period of nine weeks during which time, this "oil forwarder" drew only money sufficient for the necessities of life and horse feed! The man slept in his wagon or under it, seldom washed, and during the nine weeks, never changed a single article of clothing. At the expiration of his nine weeks' services, he "thought he'd go home for a clean shirt, &c.," and called upon Mr. Bosch for a settlement. The amount standing to his credit was nineteen hundred dollars!

A Scene in the Lower Oil Field.

OPERATING WELLS BY GAS-LIGHT.

THE LOWER OIL FIELD.

—

ST. PETERSBURG, FOXBURGH, PARKER'S LANDING, AND BUTLER COUNTY, PA.

In the year 1860, Thomas McConnell, W. D. Robinson, Smith K. Campbell, and Col. J. B. Findlay, of Kittanning, purchased two acres of land on the west bank of the Allegany river, about ninety rods north of Tom's Run, from Elisha Robinson, Sen., and organized the "Foxburgh Oil Company," consisting of sixteen shares, and commenced putting down a well, which reached a depth of 460 feet, when an accident occurred to obstruct operations for a few days. In the interval the war broke out, and the excitement incident thereto, stopped all further work on the premises, and the well was abandoned.

Subsequently the same parties purchased 100 acres, known as the Tom's Run Tract, from Mr. Robinson, for which they paid $50 per acre. In 1865, the Company sold about thirty acres of this purchase to a number of gentlemen in Philadelphia, for the sum of $20,000. On these 30 acres the "Clarion and Allegany River Oil Company" put down their first well, which struck oil in October, 1865, the first to produce in that locality.*

Many were the scoffs and jeers and insulting remarks made

---

* Operations, during 1863–4, had been commenced and were successfully prosecuted on the Clarion River, near the Allegany, above Parker's Landing, by a Philadelphia Company or Companies, and the developments they made established the character of the surrounding country for oil purposes.

about these first operators. They were held up to ridicule by men of means, as well as by others, and were euphoniously called "crazy," having "oil on the brain," etc. The followers of these pioneers are now, however, numerous.

It is remarkable that the well of 1860, was put down on territory which has since proved dry; and had it then been finished would have undoubtedly been a failure, and possibly prevented all future development in this region. But it was not to be so. It was abandoned for a period of four years before the "Allegany and Clarion River" well, had been commenced, and which, proving a success, gave this field to the world at a time when Venango was rapidly declining in product.

The Pennsylvania Oil Region is divided into two grand divisions, termed the Upper and Lower region, or the Creek region, and River or Parker's Landing region. The tendency for operations has of late moved towards this lower division of the Oil field. Here, the great bulk of the present production is obtained, and here are to be found the leading men of this great industry. From the first developments in this section, operations have been conducted on the theory of a belt, or series of belts.

In 1868, some wells were struck at Lawrenceburg, situate on the hill, just above Parker's Landing. A well-known operator, Mr. Marcus Brownson, of Titusville, acting on data—the result of actual operations, projected a line or belt from this point north, 22 degrees east, and south 22 degrees west, in breadth about five miles, and in length as now developed, about 35 miles, and venturesome operators soon opened up a belt, the end of which has not yet been reached in either direction. Many were stimulated to "Wildcat," and it was found that north-east of this, which may be termed the central belt, is one extending up the Clarion river, out through Turkey Run, in Clarion County. Southward it passes a little to the west of Millerstown, Butler County. To the westward of the eastern, or central belt, is another, extending and developed from the Russel farm, opposite Antwerp, in Clarion County, to a point abreast of Lawrenceburg on the east, with the Robinson, Black and Grant farms, which may be termed the western belt. Then, there is a break in the development in this middle belt of some three miles, until the Stonehouse property is reached on the west line. This belt will pass not less than five miles to the left, or east of the borough of Butler, Butler County. Between the central and east belts the distance is about one mile; between the central and west belt about three miles. The eastern belt is of the greatest importance; in fact, the middle belt appears to run into it, after they cross Bear Creek, east of Lawrenceburg. (We ought here to say that many operators differ in their opinions, from this last suggestion.) Actual developments north-east of Lawrenceburg show that this eastern belt runs through Parker's Landing, Foxburgh, St. Petersburg, Antwerp, etc. Below, and just across Bear Creek, are the Say, and the Stonehouse farms. Then follow, lying end to end, the Fletcher and Campbell farms; the Martin and Hutchinson, the Gibson and Turner farms; the Robert Campbell and Marcus Brownson farms; the Mayville tract; the James Campbell and Ward farms; the Canada Oil Company and A. L. Campbell and Wilson farms; the Blaney and Dougherty farms at Petrolia City; the McClymonds, Wilson and Bank farms; the J. B. Campbell and Adams farms; the Story and Riddle farms at Karns City; and last, the Moore and Hepler farms, now known as the Angell Oil Company's tract, which consists of two hundred and seventy acres, and lying in the Millerstown belt. From Bear Creek to the Angell tract, the distance is about ten miles, and the average width four miles, giving an area of forty square miles developed or in process of development.

Prominent among the first wells of this region, and perhaps we should say the belt just described, was the noted one on the McClymonds' Farm, at the date of its completion, one mile and a-half in advance of other developments. It was drilled some fourteen hundred feet, fifty feet deeper than the wells thereabouts, and the owner, fearing the rock had run out, sold it as a dry hole to Morehead, Tack and Preston, who purchased it solely on the strength of their belief in the existence of belts. Three hours after they had purchased it the drill entered the sand, and before the well could be tubed seven hundred barrels of oil flowed out of it ! Their success encouraged others, and the intervening territory was rapidly developed. A town called Karns City, a compliment to Mr. S. D. Karns, a well-known operator, was soon built up.

The success at Karns City greatly emboldened operators, and the line of the western belt was then projected in the direction of Butler, seven miles out. Several wells were started on the Jamieson Farm, in the latter part of last year. Four of them produced two hundred barrels per day for a considerable time. This point has since been called Greece City, and now has a population of 4,000 to 5,000. This is the largest jump ever taken, there being some four or five miles of undeveloped territory in the rear. On the Angell Oil Company's territory, lying on the line of the central or eastern belts, successful strikes have but recently been made, and thus all the territory near or between it and Fairview, a distance of some two miles, has been opened.

The St. Petersburg district, upon the east side of the Allegany River (it may be said to be from one to eight miles from it, and its limit has not yet been reached) is north-east of the Butler oil field. The general direction of the belt, like that just described, is northeast and south-west. Developments in this district were commenced in the summer of 1869, and the first well was struck in September of that year. This was the "Mead Well," south of the Clarion River and near its mouth. Soon after the Elephant Well, near the first named was struck. Parties then began to extend developments north of the river, and in October following, a well was struck there, and operations continued on up the Allegany, to where Foxburgh now stands. The wells on this portion of the new field were not extraordinarily large. The pioneer who pushed developments back from the Allegany to St. Petersburgh, was Marcus Hulings, who struck the Hulings' or "Antwerp Well," in November, 1871. This well started up at the rate of 100 barrels per day, and maintained this production for some time. It is still producing oil, and pumps about seven barrels per day. Within two weeks after the Hulings' well was struck, twenty-five wells were commenced on the J. J. Ashbaugh and Dan Ritz farms, in and around the borough of St. Petersburg. The farms between the Allegany River and St. Petersburg were soon taken up by active operators, and derricks were rapidly reared all along the line. These farms are the Frederick Rupert, Whitting, Shoup, Collins, Foust and Keating. Then commenced the building up of St. Petersburg, which is now one of the important villages of the lower oil field. Early in the spring of 1872, building went on rapidly, (it had previously been a farming settlement and centre,) as there was a large influx of population. Houses were put up at the rate of twenty-five per day. Shortly afterwards it was incorporated as a borough. The distance from the Allegany to St. Petersburg is two miles. A quarter of a mile beyond St. Petersburg stands the village of Antwerp. St. Petersburg contains a population of 2,500 to 3,000.

The thickness of the oil-bearing rock in this district is twenty-five feet nine inches. The average depth of the well is 975 feet. The deepest is the Fountain Well, which is 1241 feet. The shallowest, the Antwerp or Hulings' Well, which is 790 feet. These two wells are half a mile apart.

The lower oil field proper, is varied and beautiful in scenery. The land is rolling, fertile and fairly cultivated. The homes of the old settlers bear the marks of peace and plenty. The hills and valleys contain rich deposits of coal, and their bowels, rivers of oil.

Recent developments have centred at points lying between and including Petrolia and Fairview on the north, and Millerstown on the south, and Karns City on the east, and Greece City on the west, all in Butler Co., Pa., which comprise the best producing oil territory of late years. At no period in the history of petroleum developments has there existed such a large number of flowing

wells as are to be found at the present writing, and at no time has
there been so large a production of oil.

The country has been chiefly developed by combinations of individuals, some ventures being divided into sixteenths. The most
prominent gentlemen engaged in developing this territory are S.
D. Karns, C. D. Angell, Parker, Thompson & Co., Lambing Brothers, Campbell Bros., Fisher Brothers, Tack Brothers, Moorhead
and Ripley, Robert Leckey, H. W. Scott, F. F. A. Wilson, Marcus Brownson, Dimick, Nesbitt & Co., Jno. Preston, Jno. L. and J
C. McKinney, and Jno. H. Gailey, Vandergrift & Foreman, Phillips
Bros., H. L. Taylor, Jno. Satterfield, Tarbell & Hess, and others
not known to the writer.

THE MODOC DISTRICT.

Having in the early part of this chapter noted the lines of development, and marked its progress, we will now proceed to chronicle
the more recent developments with a brief sketch of the progress
of the work.

We have already made mention of the striking of the Troutman
well, on the Troutman farm, in that portion of this oil field, known
as the Modoc City district. This famous well, which began flowing enormous quantities of oil on the 23d of March, 1873, is situated on a tract of land of some fifty-five acres, upon which a French-
man named Troutman settled some four years since. A party of
capitalists, known as "The Hope Oil Company," purchased the land
about one year ago. This well created great excitement among
operators, and soon extensive operations were commenced in the
neighborhood. The amount of oil produced for the first few days
is variously estimated at from 800 to 1,000 barrels per day. From
the 23d of March to the 10th of September of the present year,
1873, this well produced, according to the Pipe Line Co.'s receipts,

85,413$\frac{44}{100}$ barrels! Add to this 3,000 barrels, lost for want of tankage—the first few days of its production. The production on the 10th of September was 308 barrels,* after flowing five months and seventeen days! "The TROUTMAN" is 1,440 feet in depth, and was sunk as a test well, the success of which soon attracted operators to the locality. Surveys were made, and leases given out. The following shows the results thus far of this wonderful territory:

	BBLS.
The Troutman Well, Troutman farm, struck 23d of March, 1873, now producing daily,	308
Capt. Grace Well, John Starr farm, struck in July, '73, now producing daily,	300
Boyer Well, John Starr farm, struck in July, '73, now producing daily,	300
Capt. Grace No. 2, John Starr farm, struck in July, '73, now producing daily,	300
Percy & Beck Well, John Starr farm, struck in July, '73, now producing daily,	250
Brawley Well, Jerry Starr farm, struck in July, '73, now producing,	300
Captain Jack Well, Harper farm, struck in July, '73, now producing daily,	200
Dean Well, Harper Farm, struck in August, '73, now producing daily,	300
Modoc Well; Troutman farm, owned by Hope Oil Co., struck in August, '73, now producing daily,	300
W. W. Thompson Well, Morrow farm, struck in August, '73, now producing daily,	500
Seep Well, McClurg farm, struck in August, '73, now producing daily,	350
Fleming Well, No. 1, Ralston farm, struck in Sept. 6th, '73, at first produced at the rate of 700 barrels, and is now producing daily,	500

* The production, as reported above, is the actual *flow* of the well on Sept. 13th, 1873.

Fleming, No. 2, Ralston farm, struck Sept. 6th, 1873, BBLS.
    flowed at the same rate as No. 1, and is now doing daily,   500
Tip Top Well on the Troutman farm, started up at 700
    on Sept. 7th, '73, and is now producing daily,   500
Phillips and Vanansdall well on the Harper farm, struck
    1st Sept. '73, and started at 500 barrels, is now doing daily,   300
Phillips well on the Sutton farm, struck on the 1st Sept.
    '73, commenced at 500 barrels and is now doing,   300
Miller well, on the Troutman farm, struck Sept. 3, '73,
    commenced at 500, and is now doing daily,   350
Gordon well, on the McClelland farm, struck 28th Au-
    gust, '73, commenced at 350, and is now doing,   200
Columbia well, on the Columbia Oil Co.'s tract, struck
    7th Sept., '73, commenced at 350, and is now producing
    daily,   225
The Markham and Jock well, struck 11th Sept., '73, and
    producing,   500
Capt. Grace well, No. 3, on the Starr farm, was finished
    on the 4th Sept., '73, the well filled with salt water,
    which was cased off and pumping commenced, which is
    now producing daily,   150

These wells are all situated in "THE MODOC DISTRICT,"
and all, with one exception—the Grace well, No. 3—are flowing,
making a grand total daily production of 7,033 barrels!

The Gas well, on the Banks' farm in the district, is supposed by
many to have a greater flow than the famous Newton gas well,
near Titusville. It was struck about the 5th of August, and a
volume of gas has ever since been pouring forth, with a noise like
"the rush of mighty waters." An attempt was made to lower the
tools into this well, but the tremendous force of gas forced them
out of the hole!

There are about seventy-five wells going down in this district,
principally on the Troutman, Ralston, Starr, Sutton, Harper,
Grover, McClurg, and Brown farms, embracing an area of one

thousand acres. All these wells flow through the casing, not one of them having been tubed. No sooner do the tools strike the sand rock than the oil spouts forth. A contrivance on the top of the casing having been provided, the oil is allowed to flow undisturbed.

The depth of the wells in this district average 1500 feet. The oil-bearing sand rock is from twelve to fifteen feet in thickness, being pebbly and porous.

The Starr farm is now owned by Phillips Bros., of Parker's Landing, Pa., having been recently purchased by them for the sum of $100,000

GREECE CITY DISTRICT.

The first well struck in the Greece city district was the Morrison well, on the 24th August, 1872, on the Jameson farm, which flowed at the rate of 250 barrels per day for four months. It then gradually fell off and is now producing 40 barrels per day. S. D. Karns struck the next well, "The Dogley," on the 25th December, 1872. This well is situated half a mile below the Morrison well. It flowed liberally for several weeks, and is now pumped, producing in moderate quantities. A third well was put down by the same gentleman, with what result we could not ascertain. The fourth well, owned by John Preston, was struck on the 12th January, 1873, and began flowing at the rate of 130 barrels per day, and is at present (Sept. 1873,) yielding oil in paying quantities. Numbers of others followed in close succession during the latter part of February, and through March and April. Some of these started off very largely; a fair percentage never yielded above twenty-five barrels per day, while all fell off materially after the first "spurt."

The greatest number of wells producing at this point at any time did not exceed thirty-five. The largest daily production never

stood above 1200 barrels daily, and this occurred during the
months of February, March, April and May of the present year.
At present the production is about 300 barrels per day. No new
wells are going down here.

———

MILLERSTOWN DISTRICT.

The first well struck in the Millerstown district, was the Shreve
well, on the Stewart farm, in March of this year, which produced
at the start 150 barrels per day, and is now doing 125 barrels per
day. Next came the Dr. James well, on the Barnhart farm,
which was struck in May, and started off at 150 barrels, and is
now doing 130 barrels per day.

The Lambing well followed, and produced 100 barrels per day,
and is now doing fifty barrels. This well produced a large amount
of gas, and is situated on the Barnhart farm.

The Howe and Clark well, on the McDermot farm, next followed, and produced at the rate of 125 barrels, and continues to
produce 75 barrels per day.

The Green well, on the Johnston farm, never penetrated the
third sand. About the 1st of August, the Wolf well commenced
flowing at 150 barrels per day. It is situated on the Barnhart
farm, and continues to produce at the above rate.

The Carlien and Mosier well, on the McDermot farm, began at
the rate of 150 barrels on the 21st of August, and continues about
the same rate. The Preston well, on the McKinney Bros. &
Gaily tract, was struck on the 10th of August, and on striking the
oil rock was burned down. It is now pumping at the rate of 100
barrels per day.

The Parsons well, on the McKinney Bros. & Gailey tract, was
struck on the 1st of September, and flowed at the start, 250 barrels
per day, and is now flowing 200 barrels.

Dr. J. McMichael or Salsbury well, on the McDermot farm, started at 125, and is now doing 75 barrels per day.

The Farquar well, on the Farquar farm, one mile south-east of Millerstown, was struck on the 20th of August, and is now flowing 250 barrels per day.

The Salsbury well, No. 2, owned by Dr. McMichael, was struck on the 1st of September, and started at the rate of 300 barrels per day, continuing to do about the same rate.

The Dubenspeck well, on the Dubenspeck farm, adjoining the McDermot farm, struck on the 1st of September, 1873, and flowed 300 barrels per day. No perceptible reduction in the produce is noticeable at the present writing.

A new well on the Abidiah Barnhart farm, was struck on the 10th of September, 1873, and is flowing 100 barrels.

The Kepler well, on the Kepler farm, struck on the 10th of September, 1873, is flowing 200 barrels per day.

The Hulings well, on the Barnhart farm, is in the sand, and flowing in large quantities. (Sept. 11.)

The Shidemantle well, on the Dubenspeck farm, began flowing (11th Sept. 1873) at the rate of 250 barrels per day

---

PETROLIA DISTRICT.

The first wells in the Petrolia district, were put down by Messrs. Dimick, Nesbitt & Co., in November and December, 1871, on the Sheakley farms, which attracted considerable attention from operators. These wells proved quite remunerative, but it was not till April of 1872, that the first great strike was made at Petrolia, then a rural district. Early in April, 1872, Dimick, Nesbitt & Co., finished the " Fanny Jane Well," which yielded liberally for a considerable time. This successful venture was attended with the usual result, and forthwith began a regular rush for the latest Oil-Dorado.

The Blaney Farm was purchased by Fisher Bros. for $60,000, and other tracts in the vicinity were either bought at extravagant prices or leased at high rates; houses multiplied rapidly, and ere long the infant settlement presented all the bustle and activity characteristic of new oil towns. Large numbers of wells on the Blaney, Wilson, Jamieson and neighboring farms produced in abundance. The wells on the farms above named are but short-lived, as with few exceptions they declined materially within a short period, gradually falling off, many of these ceasing to yield in paying quantities six months after their completion.

At present a large number of these wells are shut down in consequence of the low price of oil. At present there are only some five wells going down in this section. No new strikes of consequence to be recorded within the last few months.

_____

KARNS CITY DISTRICT.

The success attending the operations at Petrolia induced operators to extend developments further south, and in May, of 1872, the Cooper Bros. began on the McClymonds Farm. This property is situated on a branch of Bear Creek, Fairview township, one mile and a-half from Petrolia, and the same distance from Fairview. In June the Coopers fearing it destined to be a failure, disposed of a well partly down to S. D. Karns, who drilled a few feet deeper and struck a hundred barrel well. The next was completed by the Coopers, and for several weeks this well flowed 200 barrels a day. Other wells soon followed, on the J. B. Campbell, Story, Riddell and Kincaid farms. On the 9th of January the famous Salsbury Well, on the J. B. Campbell farm, began to flow at the rate of over five hundred barrels per day. This point became at once the centre of developments, and soon the nucleus of a town was built, which, as before mentioned in an early part of this chapter, was named after S. D. Karns, and called Karns City. At present operations are at a stand-still, only five wells drilling in this section and forty-three wells producing; operators rushing to newer and for the present more productive fields. We ought here to remark that a *fourth sand* has been reached in two wells near Karns City with good results, which discovery is likely to give rise to the deepening of all the small wells on the line of development in the neighborhood of Karns City and Petrolia. The opinion of operators in regard to these two wells differ. Some believe that no oil exists in the fourth sand, while others are of opinion that the reckoning in these two wells is correct.

Passing through Karns City, one and a half miles to the south, is the Moore, Hepler and Myers farms. The first well struck in this section, was on the 31st of January of the present year, which flowed some 200 barrels per day. Another was struck soon afterwards, which proved a good well. Mr. C. D. Angell, the owner of the Moore & Hepler farms, has now five producing wells, doing 500 barrels per day, and five new wells going down.

A new town has sprung up, on this property, called in honor of the owner, ANGELICA. Extensive operations are in progress at Fairview and Angelica.

Having now taken a cursory glance, at the present state of developments, and given the initial operations at different points of interest, we will bring this chapter to a close.

The importance of this lower oil field, must be evident to all conversant with the history of its steady development; but we would here say that with this rapid progress of developments, a large and very extensive scope of territory has been left behind untouched and undeveloped. Oil men pushed ahead, only desiring to make great conquests in the way of flowing wells. That portion of the field which has been left in the *rear*, has only been skimmed, but will most assuredly be once again opened up, and again become the scene of fortunes made and fortunes lost.

At the present moment there are in this lower oil field no less than twenty-eight flowing wells, producing daily the enormous sum total of 8,833 barrels, giving an average production to each well, of 315 barrels!

DISCOVERY OF THE FOURTH SAND-ROCK.

We have already mentioned, briefly, the discovery of a FOURTH SAND, or oil rock, in the vicinity of Karns City. It was first found by Mr. CHARLES STEWART, a native of Butler County. He purchased an abandoned well on the Scott farm, near Karns City, and after a month's vain effort to make it pay as a pumping well, resolved to sink it deeper. He began this labor about the middle of June last, and after nearly eight weeks' drilling in daylight, struck the first flowing well in the fourth sand! This well is known as the Banks' or Stewart well, and averages four hundred barrels daily. The entire oil community, including the shrewdest operators, scouted the idea of this being the "fourth sand," and claimed that the well had not originally been drilled deep enough. Messrs. Tack & Morehead, however, on the adjoining farm, commenced to drill one of their abandoned wells, known as the McCleer No. 1, about four weeks ago, and struck the fourth sand on the 15th September, with a good show of oil. They continued to drill until the 18th of September, when the well commenced flowing at the rate of 700 barrels per day! The most experienced operators claim that the fourth sand is only prolific at these points, where a spur branches out from the main belt, and this is evidently the spur of the Modoc belt. This theory will be very fully studied within the next sixty or ninety days, as there is a determination on the part of everybody owning an abandoned or non-paying well to try it.

The "FOURTH SAND," thus far developed, is from 65 to 75 feet below the "third sand," and is of excellent quality.

STATISTICAL INFORMATION.

PRODUCTION.

The following shows the average DAILY product of the Pennsylvania oil region district in 1867, during the months indicated. No reliable monthly reports were published prior to this date:

1867.

September..	9,700
October...	9,600
November.........................	9,800
December..	10,400

MONTHS.		1868.	1869.	1870.	1871.	1872.
January	bbls.	8,700	10,192	12,634	15,477	16,286
February	"	9,200	9,967	11,917	14,391	17,012
March	"	8,621	9,891	12,385	13,457	15,506
April	"	8,537	11,067	12,974	13,308	16,308
May	"	8,790	10,153	14,165	13,987	18,345
June.....	"	10,102	11,334	14,817	14,806	17,749
July..............	"	10,693	11,697	16,969	17,261	18,513
August...........	"	11,981	12,157	17,777	18,161	18,816
September......	"	11,033	12,645	19,489	17,648	16,561
October	"	10,133	13,071	20,158	16,068	14,309
November	"	10,276	13,317	18,012	16,651	23,275
December	"	9,737	12,844	15,214	16,703	22,054

The total production in 1872 was 6,539,103 barrels of forty-three gallons, a daily average for the year of 17,925 barrels against 15,800 barrels in 1871, showing a daily average increase in 1872 of 2,115 barrels, and a total increase of 671,975. The daily average in 1870 was 15,350 barrels, in 1869, 11,560 barrels, and in 1868, 10,180 barrels. The average in 1865 was between 6,000 and 7,000 barrels daily.

The annexed table gives the production of Pennsylvania oil region each year since 1859:

BBLS.

			BBLS.
Production in	1859		* 87,000
"	"	1860	500,000
"	"	1861	2,118,000
"	"	1862	3,056,000
"	"	1863	2,631,000
"	"	1864	2,116,000
"	"	1865	2,497,000
"	"	1866	3,597,000
"	"	1867	3,347,000
"	"	1868	3,715,000
"	"	1869	4,215,000
"	"	1870	5,659,000
"	"	1871	5,795,000
"	"	1872	6,539,000
	Total bbls		45,840,000

* In all published statements of the product of Petroleum for 1859, this is the amount given. It is palpably wrong. Col. Drake's well was struck in August, 1859, and produced not more than ten or twelve barrels per day. *And this was the first well in the Pennsylvania Oil Region.* The second producing well was struck in February, 1860—the Barnsdall well—fifty barrels per day. It is questionable, therefore, if the entire product of 1859 would reach 3,000 barrels.

PRODUCT OF AMERICA FOR THE YEARS GIVEN.

The production of America in 1872, and previous years compare as below :

BBLS.

			BBLS.
Total product Pennsylvania Oil Region in 1872			6,539,000
"	" of West Virginia, Ohio and Kentucky Oil Regions in 1872		325,000
"	" " Canada in 1872		530,000

BBLS.

			BBLS.
Total product of America in 1872			7,394,000
"	"	1871	6,638,000
"	"	1870	6,535,000
"	"	1869	4,917,000
"	"	1868	3,965,000

The daily average product in America in 1872 was 20,271 barrels against 18,100 barrels in 1871, and 17,900 barrels in 1870.

In Canada the yield is estimated at 530,000 barrels for the year. At one time there was a production there of more than 2,000 barrels daily. In West Virginia and Ohio the product is given at 325,000 barrels.

AVERAGE PRICES.

The following were the average monthly prices of Crude on the Creek—of barrels of 43 gallons—and of Crude and Refined *per gallon* in New York for 1872 and 1871 :

MONTHS.	Crude—In Bulk.		Refined. Standard White. In Barrels.		Naphtha. In Barrels.		——
	Highest and Lowest.	Average Price.	Highest and Lowest.	Average Price.	Highest and Lowest.	Average Price.	Crude On Creek.
January	12¾ @ 13½	13.11	22¾ @ 23½	23.29	12 @ 13	12.44	$4.05
February	12½ @ 13¼	13.01	21¾ @ 23	22.22	11 @ 12½	11.66	3.85
March	12⅝ @ 13⅜	13.06	22 @ 23	22.58	10 @ 11	10.28	3.67½
April	11⅞ @ 13	12.31	21¾ @ 22½	22.17	10 @ 12	10.39	3.55
May	13 @ 14	13.34	22⅞ @ 24	23.52	12 @ 18	15.25	3.95
June	12½ @ 13½	12.92	22⅝ @ 23¾	23.04	16 @ 18	17.50	4.10
July	12 @ 12¾	12.33	22 @ 23	22.37	14½ @ 16	14.75	3.75
August	11½ @ 12½	11.88	22⅛ @ 23¼	22.55	14¼ @ 15½	14.86	3.42½
September	11⅛ @ 12¼	11.71	23¼ @ 24½	24.17	15 @ 17½	16.14	3.25
October	13 @ 14¾	13.74	24½ @ 26⅞	25.97	17½ @ 20	18.94	4.25
November	13½ @ 14½	14.10	26½ @ 27½	27.15	17½ @ 19½	18.42	4.50
December	11 @ 13	12.08	24 @ 27½	*26.00	16½ @ 18	17.07	3.62½
Av. for 1872	....	12.89	....	23.75	....	14.81	3.75
" 1871	....	14.04	....	24.24	....	10.01	4.00
				* Nominally			

Monthly average of prices of Crude and Refined, at New York, for the years 1871, 1870, 1869, 1868, 1867, 1866 and 1865 :

YEARS.	Crude.		Refined. Standard. White.	Naphtha.
	Bulk.	Barrels.	Barrels.	Barrels.
1871	14.04	18.09	24.24	10.01
1870	13.93	18.45	26.35	9.83
1869	18.25	23.25	32.73	10.33
1868	14.40	19.66	29.52	18.91
1867	12.17	17.43	28.41	23.75
1866	....	25.78	42.45	37.84
1865	....	38.37	58.87	50.37

THE CONSUMPTION OF THE WORLD.

The total consumption of crude in 1872 was 6,663,000 barrels, an increase over the previous year's consumption of 662,000 barrels, or eleven per cent. The rate of increase in consumption in 1871 over 1870, was over two and one-half per cent. greater than the rate of increase in 1872 over 1871. Among the causes that led to the falling off in the rate of increase in 1872, was the increased manufacture of shale oils, and the prices demanded by the refiners of Petroleum in America.

The following table shows the consumption throughout the world in 1872 :

Production, 1872, bbls...............		7,394,000
Stock Jan. 1, 1872, bbls...............	3,269,000	
Stock Jan. 1, 1873, bbls...............	3,849,000	
Deduct increase Jan. 1, 1873, bbls.	580,000	
Deduct losses by fire, &c., in 1873.	150,000	730,000
Total consumption 1872, bbls. crude.		6,664,000
Consumption in 1871, bbls.............		6,002,000
Increase in 1872, or about ten and eight-tenths per cent...............		662,000

The average daily consumption in 1872, was nearly 18,500 bbls.

PETROLEUM TRADE OF PITTSBURGH.

We give below the statistics of the petroleum trade of Pittsburgh for the last fourteen years. The figures do not include lubricating oils, the quantity and value of which it would be impossible to ascertain.

The following are the receipts of crude oil from 1859 to 1872, inclusive.:

	BBLS.		BBLS.
1859	7,063	1866	1,253,326
1860	17,161	1867	727,494
1861	94,102	1868	1,061,227
1862	171,774	1869	1,028,902
1863	175,181	1870	1,050,810
1864	208,744	1871	1,149,493
1865	630,246	1872	1,186,501
		Total barrels	8,746,756

RECEIPTS AT PHILADELPHIA FROM 1865 TO 1872.

The following are the receipts of Petroleum by railroad, at Philadelphia, for the past seven years :

	BBLS.
1872	1,165,613
1871	1,329,250
1870	1,476,564
1869	1,049,516
1868	1,064,702
1867	970,798
1866	743,504
1865	640,019

EXPORTS OF REFINED OIL FOR PAST EIGHT YEARS.

	BBLS.		BBLS.
1865	298,111	1869	596,475
1866	424,848	1870	811,158
1867	498,226	1871	733,943
1868	724,991	1872	743,610

THE PETROLEUM TRADE.

PRODUCTION, CONSUMPTION, AND EXPORT FOR 1871 AND 1872. VALUES, DECREASES, &c.

The following interesting and reliable statistics we gather from " The Trade and Commerce Reports " of 1871 and 1872 :

	1872. Barrels.	1871. Barrels.
The total production of Pennsylvania oil region equaled...	6,839,103	5,795,000
West Virginia, Ohio, and Kentucky.	325,000	401,000
Canada	530,000	442,000
	7,694,103	6,638,000

DAILY AVERAGE PRODUCTION :

Pennsylvania	17,917	15,800
Daily average of America	20,271	18,100
Exports from the United States of Crude and refined	150, 385,869	155,674,741

Some 4,683,922 gallons less than preceding year.

	1872.	1871.
Consumption of the world equaled	6,644,000	6,002,000

An increase of about 108.10 per cent. The daily consumption estimated at 18,500 barrels against 18,000 barrels in 1871.

Stock in America January 1, 1873, 2,316,000 barrels, against 1,600,000 in 1872.

World's stock January 1, 1873, 3,849,000 barrels. Stock of the world has more than doubled in three years.

The value of production of Pennsylvania, West Virginia, Ohio, Kentucky, and Canada equals $28,516,250, against $30,570,500 in 1871.

	1871.	1872.
Value of the world's consumption	$27,009,000	$24,990,000
Value of daily consumption	73,800	69,395
Value of world's stock, December 31	14,710,500	14,433,730
Value of United States stock, Dec. 31	7,155,000	6,903,550
Value of exports of crude, refined, and Naphtha....	$38,077,501 24	$33,174,182 52
World's value amount on hand Jan. 1, 1872, and consumed in 1871		$41,719,500 00
On hand Jan. 1, 1873, and consumed in 1872		30,423,750 00
Decrease in value of refined exports		5,294,978 23
" crude		113,225 89
" naphtha		278,433 62

Crude oil and naphtha show a gain in exports of $391,659,51, and refined, which should have been the greater, fell off $4,903,318.72.

DECREASE IN VALUE OF PRODUCTION, CONSUMPTION, STOCK AND EXPORTS :

Increase in value of production..		$2,054,250 00
"	" World's consumption....................................	2,019,000 00
"	" Daily consumption......	4,405 00
"	" World's stock...	276,750 00
"	" United States...	251,450 00
"	" Exports.......................................	4,903,318 72
Value of world's consumption during the year, and stock in hand		
Jan. 1, 1871...............		41,719,500 00
Do. 1872..		39,423,750 00
Decrease in value...		$2,295,750 00

VALUE OF EXPORTS FROM THE UNITED STATES DURING THE YEAR AND
STOCK ON HAND :

	Jan. 1, 1871.	Jan. 1, 1872.
Value of exports.....................................	38,077,501 24	33,174,182 52
stock.................	7,155,000 00	6,903,550 00
Total.........	$45,232,501 24	$40,077,732 52
Decrease of exports and stock 1872 under 1871...		5,154,768 72

The production, consumption, and stock in hand were greater than the preceding year, but the value was less. The exports of crude and naphtha were increased; but the refined oil decreased in commercial value, and shows a large falling off.

The falling off in most cases occurred among the largest consumers of American petroleum in Europe heretofore.

The facts show that refined oils sent upon the markets of the world of late years have been of poor quality and dangerous, and the public have been seeking other illuminators, and putting up with some inconvenience to be safe. This has stimulated the manufacture of coal oil, from the cheap and almost exhaustless shales of Wales and Scotland, and the increase has been unparalleled in the last few years; all the abandoned works of 1865, have been rebuilt, with new improvements and new machinery.

The future consumption must govern the demand of this article as in all commercial products, and must meet the wants of consumers both in quality and price. This, we are glad to know, will be the results of the petroleum trade in the future. Several of our State Legislatures, Pennsylvania among the number, have enacted laws requiring a manufactured article of refined oil at such fire test as to render it absolutely non-explosive.

EXPORTS FROM THE PORT OF NEW YORK FROM 1868 TO 1872.

	GALLONS. 1872.	GALLONS. 1871.	GALLONS. 1870.	GALLONS. 1869.	GALLONS. 1868.
To Liverpool	1,388,419	1,866,538	1,836,675	877,667	1,291,200
London	1,372,263	1,457,628	2,047,118	872,118	947,311
Glasgow, &c	....	....	... .	....	....
Bristol	556,261	414,322	248,132	410,605	184,070
Hull	65,814	392,919	83,119	....	....
Falmouth, E., &c	1,021,079	....	551,649	367,233	92,210
Grangemouth, E	....	....	....	....	....
Cork, &c	3,141,436	5,328,811	4,689,283	2,648,865	2,272,534
Bowling, E	....	....	....	....	... .
Havre	4,139,619	2,832,134	1,417,851	4,275,096	2,925,413
Marseilles	1,399,830	2,549,793	2,508,468	2,410,308	3,269,600
St. Nazaire and Rouen	....	....	309,522	....	149,450
Cette	226,300	....	108,743	....	....
Dunkirk	850,886	762,369	288,231	831,398	369,501
Bordeaux and Bayonne	852,292	557,639	455,677	428,306	184,600
Nantes and Rouen	229,828	....	118,772	346,650	78,539
Dieppe	....	....	....	....	....
Antwerp	6,489,132	4,747,167	9,977,114	8,202,931	7,052,177
Bremen	11,822,831	12,356,572	10,162,399	11,374,282	8,578,026
Amsterdam	678,914	....	....	....	....
Hamburg	5,776,354	5,866,532	4,456,226	4,333,982	2,458,557
Rotterdam	1,897,546	6,987,302	5,305,299	2,115,838	1,695,235
Stockholm & Gottenberg	783,702	....	....	....	....
Cronstadt, &c	3,433,905	5,997,362	7,227,273	4,163,320	1,523,387
Ancona	397,799	216,047	216,942	....	150,028
Konigsburg and Stettin	5,644,478	5,650,978	2,645,677	4,594,363	2,537,086
Arendel	143,864	....	....	....	....
Lubec, &c	294,229	186,260	97,242	138,570	....
Danzig	873,889	1,177,776	767,999	810,596	374,671
Copenhagen, Elsinore, &c	3,858,708	2,967,345	894,422	341,572	118,492
Borga, Finland	....	73,321	121,540	....	....
Sodertolje	100,230	169,023	189,148	....	....
Syria, &c	490,520	985,250	287,500	168,220	....
Venice	....	472,201	610,110	....	....
Cadiz and Malaga	1,068,555	608,487	1,101,049	436,058	380,581
Tarragona and Alicante	35,000	50,760	135,500	66,038	518,260
Barcelona	774,723	786,685	571,462	530,029	470,929
Gibraltar and Malta	8,023,509	7,397,196	7,982,173	2,774,547	4,289,017
Oporto	74,590	71,690	210,759	362,708	251,704
Naples and Palermo	520,945	870,113	379,912	1,064,943	1,032,209
Genoa and Leghorn	1,425,261	3,159,142	2,515,926	1,774,223	2,229,928
Trieste	2,131,130	2,601,290	2,816,655	1,413,743	900,161
Smyrna, &c	1,684,482	1,463,882	1,045,376	748,494	398,873
Alexandria, Egypt	911,532	411,660	228,394	....	223,000
Lisbon	310,302	140,729	451,582	194,812	43,194
Canary Islands	61,230	187,365	18,234	16,353	16,461
Constantinople	738,218	1,492,905	1,508,240	602,180	603,012
Bilboa, Seville and Vigo	1,385,671	2,233,671	2,136,551	1,498,682	417,210
Palma, Spain, &c	1,138,408	592,915	935,207	330,221	199,163
China and East Indies	1,353,030	457,290	451,610	207,180	120,300
Japan	200,000	....	....	....	....
Africa	169,990	169,980	99,272	30,200	24,560
Australia	1,318,328	1,794,993	1,633,663	619,649	959,959
Otago, N. Z	133,820	319,680	....	43,680	37,500
Sydney, N. S. W	433,614	337,280	231,080	139,280	224,526
Brazil	2,713,409	1,036,943	1,364,294	835,299	804,390
Mexico	382,542	559,809	243,022	169,541	155,576
Cuba	1,850,051	1,534,751	1,566,547	1,144,378	988,955
Argentine Republic	828,573	374,950	396,403	101,000	169,200
Cisalpine Republic	529,779	534,050	417,580	109,120	91,000
Chili	270,750	266,160	174,884	193,990	168,000
Peru	233,490	181,629	305,673	142,780	233,956
British Honduras	12,462	8,072	5,049	9,027	4,220
British Guiana	50,897	37,150	79,543	36,106	40,700
Dutch Guiana	....	11,322	8,235	....	....
British West Indies	397,693	489,227	586,492	298,997	236,805
Br. N. Am. Colonies	69,969	34,930	38,598	54,221	47,215
Danish West Indies	27,121	10,596	10,058	16,473	12,255
Dutch East Indies	....	277,517	....	....	....
Dutch West Indies	48,061	19,823	30,267	40,698	17,463
French West Indies	14,600	88,701	86,600	73,138	77,260
Hayti	19,377	40,399	19,654	16,678	8,066
Central America	15,465	17,916	8,273	1,858	2,848
Venezuela	132,764	76,620	68,251	77,266	57,911
New Grenada	110,478	98,509	78,186	60,312	64,219
Porto Rico	103,379	93,346	46,934	36,492	34,228
Sandwich Islands	....	....	3,000	....	....
Total	90,027,726	94,955,850	87,667,299	65,933,690	52,803,202

TOTAL EXPORTS FROM ALL PORTS IN THE UNITED STATES.

Annexed are the total exports of Crude, Refined, Naphtha, and Lubricating Oils, since 1862, from all ports in the United States :

FROM	1862.	1863.	1864.	1865.	1866.	1867.	1868.	1869.	1870.	1871.	1872.
New York........galls..	6,720,273	19,547,604	21,335,784	14,626,090	34,501,385	33,834,133	52,599,483	65,933,690	87,667,299	94,955,850	90,027,726
Boston......................	1,071,100	2,049,431	1,696,307	1,511,173	1,591,694	2,264,113	2,367,865	2,117,939	1,790,271	2,185,096	1,717,689
Philadelphia............	2,800,978	5,395,738	7,763,148	12,552,882	28,811,853	29,437,429	38,484,157	33,445,552	49,889,736	55,901,590	56,645,350
Baltimore................	175,100	915,866	929,971	973,117	2,483,419	1,515,454	2,587,207	1,251,423	1,731,321	2,570,528	1,995,104
Portland..................	120,250	342,082	70,762	11,088	12,100	900	705,107	....	....	....	....
Cleveland................	....	....	80,000	81,173	30,000	....	270,000	....	159,528	....	....
New Bedford	....	....	....	50,000	....	....	....	....	....	....	....
Total........galls..	10,887,701	28,250,721	31,872,972	29,805,433	67,430,451	67,052,029	97,013,819		141,238,155	155,612,974	150,385,869
Equal to bbls. of 40 gallons..	272,192	706,268	796,824	745,138	1,685,761	1,676,300	2,429,498		3,530,068	3,890,326	3,759,646

The above figures fairly represent the rapid increase in the consumption of the article abroad.

EXPORT FOR 1873.

The figures upon the opposite page fairly represent the rapid increase in the consumption of the article abroad. It will be noted that the export of 1872 was more than double that of 1867.

The following table of the quantity shipped from leading ports, from January 1 to June 1, 1873, will show how greatly the foreign demand has augmented within a few months. The table is compiled from the issue of the New York *Commercial and Shipping List*, reliable authority on all matters pertaining to shipments or imports to or from the United States. The exhibit is as follows :

FROM JANUARY 1st TO JUNE 1st, 1873.

	GALLONS.
From New York	46,224,596
" Boston	987,368
" Philadelphia	22,437,417
" Baltimore	1,221,438
Total exports from United States	70,870,819

In addition to the above, the exports from New York alone aggregated nearly three million gallons during the first three days of June, an increase of over twenty-seven million gallons since the first of January as compared with the same months of 1872. The coal famine in England is rapidly undermining the prejudices existing against kerosene as an illuminator, and the petroleum trade with that country is fast acquiring vast proportions as the result. Thus Liverpool imported 255,708 gallons last year, up to the first of June, against 1,150,877 this season ; London, 353,433, against 1,741,551 ; Bristol, 136,534, against 781,852; and other ports in like proportion. The exports to Ireland have more than doubled ; those to France quadrupled, and the demand from Germany and Belgium and other European countries is enormously increased.

These facts and figures are at once interesting and suggestive, indicating, as they do, in unmistakable terms, an enhanced value of petroleum at no distant date. With so great an enlargement of the foreign demand, a production certainly no greater than the markets of the world require, the chances of its application in immense quantities to new purposes, and the continuous increase of home consumption, it is difficult to believe oil will not advance in price till it reaches a figure at which the average operator will be able to carry on his business, if not at a large profit, at least without positive loss, as has been the case in too many instances during the depression of the last few months.

NUMBER OF WELLS DRILLING AT VARIOUS DATES.

No reports have been made up since December, 1872, upon this subject, and we are left to conjecture as to the number of wells drilled from January 1st, 1873, to July 1st, 1873. A fair estimate would be about 225 to 250 during the first six months of 1873.

MONTHS.	1872	1871	1870	1869	1868	1867
January	394	167	364	378	182	.....
February	369	173	388	341	150	.....
March	313	159	395	334	160	.....
April.................	302	231	433	292	193	.....
May...................	336	247	412	312	217	.....
June..................	391	306	463	345	257	.....
July	359	386	349	305	299	.....
August...............	392	353	319	310	327	.....
September.........	301	364	306	315	331	.....
October.............	.....	426	305	331	370	.....
November	354	481	206	360	435	255
December	318	490	191	346	401	232

STATISTICS OF REFINING.

REFINING CAPACITY OF THE UNITED STATES.

Statement showing the Refineries in the *Oil Region of Pennsylvania*, with their respective daily Still Capacity, for Crude:

NAME.	LOCATION.	DAILY CAPACITY. BBLS. OF 43 GALS.
Porter, Moreland & Co.,	Titusville, . . .	1,213
Bennett, Warner & Co.,	" . . .	856
Octave Refining Co., .	" . . .	606
Pickering, Chambers & Co.	" . . .	512
Easterly & Davis, . .	" . . .	496
R. M. & J. W. Jackson,	" . . .	288
M. N. Allen, ...	" . . .	251
Decker & Co., . . .	" . . .	190
A. H. Lee, ...	" . . .	185
J. A. Scott, ...	" . . .	139
Cadam & Donohue, . .	" . . .	68
John Johnson & Co., .	Miller Farm, . . .	308
Dudley & Co., . . .	" . . .	250
A. R. Williams, . . .	" . . .	243
Z. Chandler, ...	Gregg Switch, . . .	187
H. De Zebala, . . .	Pioneer, . . .	127
Patterson Refinery, . .	Petroleum Centre, .	292
Hermann, Cornell & Co.,	" . . .	198
Bartlett & Newton, . .	" . . .	47
Doe & Frazer, . . .	Rousville, . . .	117
Producers Oil Works, .	" . . .	100
Levi Kerr, ...	Tarr Farm, . . .	227
Imperial Refining Co., .	Oil City, . . .	1,385
Standard Oil Co., . .	" . . .	418
Economy Refining Co.,	" . . .	321
Solar Oil Works, . . .	Oleopolis, . . .	171
L. D. Galligan, . . .	Tidioute, . . .	36
	Total daily still capacity,	9,231

REFINING CAPACITY OF NEW YORK.

Statement showing the daily refining capacity for the city of New York and vicinity.   Furnished by Peter Schmid :

WORKS AND LOCATION.	OWNED OR RUN BY.	DAILY CAPACITY. BBLS. OF 43 GALLS.
Kings Co. Oil Works, Newtown Creek, Green Point, L. I.,...	Sloan & Fleming, 159 Front St., N. Y......	1,700
Pratt's Oil Works, Brooklyn, E. D..........	Chas. Pratt & Co., 108 Fulton St., N. Y.....	1,500
Empire Oil Works, Hunter's Point, East River, L. I. City, Queens Co Oil Works, Newtown Creek, Long Island City, Franklin Oil Works, Newtown Creek, Brooklyn, E. D.,	R. W. Burke, 181 Pearl St., New York............	1,500
Olophine Oil Co., Greenpoint, L. I........	Olophine Oil Co., 322 Broadway, N. Y.....	1,000
Brooklyn Oil Works, Greenpoint, Brooklyn, E. D........	Wm. A. Byers, 181 Pearl St., N. Y.......	600
Central Oil Works, 66th St., N. River......	Lombard, Ayres & Co. 58 Pine St., N. Y.........	600
Hudson River Oil Works, Bull's •Ferry, N. J.....	I. H. Wickes, 120 Maiden Lane, N.Y...	400
Locust Hill Oil Works, Newtown Creek, Long Island City......	I. Donald & Co., 124 Maiden Lane, N.Y...	140
Union Oil Works, Brooklyn, E. D.........	T. Meyer, 126 Maiden Lane, N.Y...	105
Washington Oil Works, Newtown Creek, Brooklyn, E. D.....	Thomas McGoey, 143 Maiden Lane, N.Y...	215
Wallabout Oil Works, Brooklyn, E. D.........	S. Jenney & Son, Kent Av., foot of Rush St., Brooklyn, E. D.........	280
Vesta Oil Works, Gowanus Creek, Brooklyn,......	W. & G. F. Gregory, 125 Maiden Lane, New York.........	200
	Geo. Sommer I., Jersey City, Cor. Warren & 1st St...	175
Peerless Works, Brooklyn, S. D., Foot of 25th St......	Denslow & Bush, 128 Maiden Lane, New York.........	175
Long Island Oil Works, Long Island City.....	Long Island Oil Co , 140 Pearl St., N. Y.......	1,200
	Total daily Capacity Bbls.....	9,790

REFINING CAPACITY OF CLEVELAND.

Statement showing the Refining Capacity of Cleveland, Ohio, and vicinity :

NAME OF OWNER.	LOCATION.	NUMBER BBLS. CRUDE. DAILY CAPACITY.
Standard Oil Company, . .	.. Cleveland, ..	10,000 Estimated.*
Hanna, Chapin & Co., . . .	. . . " . . .	732, 40 Gall. to Bbl.
Scofield, Squire & Teagle,	. . . " . . .	675, 42 " "
Bishop & Heisel, ...	. . . " . . .	300, " " "
W. H. Doan, ...	. . . " . . .	825, " " "
Corrigan & Co., ...	. . . " . . .	200, " " "

* The Standard Oil Company has a Capacity, it is said, of over 10,000 barrels per day. We estimate it, therefore, at this amount. The balance of the statement is given by the parties named, and may be relied upon.

REFINING CAPACITY OF PITTSBURGH.

Statement showing the daily refining capacity in Pittsburgh and vicinity:

NAME OF REFINERY.	OWNERS.	CAPACITY PER DAY.
Central,*	Central Refining Co.,	1,165
Penn,	H. S. A. Stewart,	130
Standard,	Standard Oil Co.,	650
Iron City,	H. S. A. Stewart,	75
Vesta,	R. S. Waring,	335
Nat'l. Ref. & Storing Co.,	Nat'l. Ref. & Storing Co.,	330
Keystone,	P. Weisenberger,	65
Petrolite,	Wormsen, Myers & Co.,	130
Cosmos,	Braun & Wagner,	260
Lily,	Brooks, Ballantine & Co.	100
Citizens' Co.,	Citizens' Oil Co.,	400
Riverside,	Elkins, Bly & Co.,	110
Fairview,	Alonold Hertz,	110
American,	L. Irwin & Co.,	330
Crystal,	Livingston Bros.,	200
Brilliant,	Lochart, Frew & Co.,	670
Model,	Model Refining Co.,	260
Liberty,	J. A. McKee & Sons,	200
Star,	Ralston & Waring,	130
Empire,	D. P. Reighard,	60
Nonpareil,	Warden & Oxenerd,	80
Hutchison,	Hutchison Oil Ref. Co.,	200
		6,090

* These works are in course of completion, and will have a capacity as stated. This company has absorbed eight refineries or firms.

REFINING CAPACITY OF PHILADELPHIA.

Statement showing the Refineries in Philadelphia and vicinity, with their respective daily Still Capacity :

NAME OF WORKS AND LOCATION.	OWNED OR RUN BY.	DAILY CAPACITY. BBLS. OF 43 GALS.
Atlantic, Point Breeze · ·	} Warden, Frew & Co., .	665
Point Breeze, Point Breeze · ·	} Stewart, Matthews & Co ,	266
Franklin, Gibson Point . . .	J. L. Stewart, . . .	200
Phœnix, Gibson Point . . .	M. Lloyd, ...	133
Harkness, Gibson Point . . .	N. W. Harkness, . .	100
Monumental, Hestonville . . .	} Taber, Harbut & Co., .	100
Belmont, Hestonville . . .	} W. L. Elkins, . . .	165
Reliance, Hestonville . . .	W. D. Heston, . . .	100
Excelsior, Hestonville . . .	W. King, ...	100
Greenwich Refinery, Greenwich . . .	Greenwich Oil Co., . .	100
Stephen Carr, City	Stephen Carr, . . .	66
Victoria, City	Carson & Conlin, . .	66
		2,061

REFINING CAPACITY OF BALTIMORE, MD.

Statement showing the Refineries of Baltimore, Md. with their respective daily Still Capacity:

NAME OF OWNERS.	NAME OF REFINERY.	DAILY CAPACITY. BBLS. OF 43 GALS.
Merritt, Jones & Co., . .	Canton, ...	650
Sylvia C. Hunt, . . .	Monumental, ...	90
Robert Read, ...	Baltimore, ...	30
Brown, Hamill & Co., .	Standard, ...	35
C. West & Sons, . . .	Crystal, ...	133
Newbold & Son, . . .	Belvidere, ...	60
Carswell & Son, . . .	Rising Sun, ...	40
Christopher & Co., . . .	Patapsco, ...	60
		1,098

REFINING CAPACITY OF ERIE, PA.

Statement showing the Refineries of Erie, Pa., with their respective
daily Still Capacity.　Furnished by M. B. Parsons :

NAME OF OWNERS.	LOCATION.	DAILY CAPACITY. BBLS. OF 43 GALS.
Ira G. Hatch, ...	Near Phila. & Erie R. R. on 10th St., . .	305
Brown Bros., ...	Sixth Street, ...	430
O. C. Thayer & Co., . .	Mill Creek, ...	155
Wallace & Vaughn, . .	Mill Creek, ...	160
M. V. Dawson, ...	Mill Creek, ...	98
I. W. Watkins, . . .	Mill Creek, ...	20
		1,168

BOSTON, MASS.

The Refining Capacity of Boston, Mass., and vicinity, is estimated at 3,500 barrels per day.　We have been unable to obtain
the names of the Refineries or their proprietors, although making
every effort to do so.

BUFFALO, N. Y.

THE STAR OIL WORKS, owned and operated by Thayer & Riddell, No. 385 Hamburg St., Buffalo, has a Still Capacity of nearly
two hundred barrels per day.

Dudley & Co., Buffalo, have a refining Still Capacity of 251
barrels per day.　A portion of this labor is done at Miller Farm,
on Oil Creek, before it is shipped to them at Buffalo.

PORTLAND, ME.

PORTLAND OIL WORKS.

Portland has one refinery, originally built and used for the manufacture of Coal Oil, with a capacity for working ten thousand tons of coal annually. It was one of the best and most perfect Coal Oil works in the United States, and was among the most extensive. It was the last to give up the manufacture of Coal Oil.

The Still Capacity of these works is 350 barrels Crude daily.

Wm. Atwood, Sup't.

JAMESTOWN, N. Y.

MARVIN & CO. OIL WORKS.

This is a small refining locality, the works being owned by Messrs. Marvin & Co. Their Still Capacity is about 50 barrels per day.

BINGHAMTON, N. Y.

Binghamton has a small refinery, the Capacity of which we have been unable to obtain. It is said to be not more than fifty barrels per day.

THE ORIGINAL "DRAKE WELL."

This was the first Artesian well drilled in the Pennsylvania Oil Region. It was located upon the Watson Flats, below Titusville, and was 69 feet 6 inches in depth—struck August 28th, 1859—and produced twelve barrels of Oil per day.

SKETCHES.

PIONEER AND PROMINENT OPERATORS.

COL. E. L. DRAKE.

BETHLEHEM, PA.

THE subject of this memoir, whose useful life will leave the mark of its individuality upon the events of the nineteenth century, was born on the 29th of March, 1819, at Greenville, Green County, New York. His parents were poor, but respectable and intelligent people, and earned their living by farming. EDWIN L. was the eldest of two sons—their only children. The brother died in the far West about the time Mr. DRAKE'S name was heralded to the world in connection with the first oil well. When the oldest of their sons was about eight years of age, the parents removed to the vicinity of Castleton, Vermont, where they gave their children the benefit of the old-time, New England common-school education—no mean advantage.

Passing an uneventful childhood, there was, perhaps, but a single incident so indicative of his future useful career as to leave any impression on his own mind, or to be worthy of remark in a sketch of his life,—and that incident was a dream. It is, of course, only singular in so far that with the superstitious, it is capable of prophetic interpretation; but one can hardly be said to have had any childhood who has not had strange dreams. It was a day-dream—not a waking dream, however. He sat upon the wide old porch

323

that shaded the entrance to their plain abode. The autumn sun shone down upon his head; and the autumn breezes, heavy with the fragrance of the fields, lulled him to sleep, and sleeping he dreamt. With his brother—in fancy—he raked the dry stubbles of the wheat field. Together they tugged and toiled, and after infinite labor they had raked a great stack of straw into a corner, nearly half a mile from the house. Then for a bon-fire! While his smaller brother watched with gleeful anticipation, EDWIN touched a match to the pile. They watched it a moment in ec-stacies—but their mirth was turned to horror when, their stack consumed, the ground continued to blaze and burn! They exerted all their strength to quench it, but in vain. The devouring flames rose higher and higher. The fire burned deeper and wider. It followed their receding footsteps; and now, completely terrified, they turned and fled to their mother. When they reached the house, EDWIN, breathless and guilty, buried his face in her lap and confessed the deed. She led him gently to the door, and after watching the flames a moment, she said calmly, and without re-proach: "My son, you have set the world on fire!"

Nearly thirty years later these words of his mother were recalled by the burning of his oil tanks a few weeks after the first well be-gan to produce. When the tanks burst, and the creeping flames spread over the surface of the creek, he may possibly have enter-tained a momentary suspicion that his mother's words were about to be fulfilled. The incident recalled the dream.

At the age of nineteen he left home to seek his fortune—which meant to go West. Like the majority of emigrants in that lati-tude, his ultimate destination was Michigan, where he had an uncle living. At Buffalo, however, he obtained a situation as night clerk on the steamer Wisconsin, plying between that port and De-troit, where he remained until the season closed, when he went to his uncle's, near Ann Arbor, and worked on a farm for about a year.

He then procured a situation as a clerk in a hotel at Tecumseh.

This was a type of the western hotel of the day, and around the hospitable log fire upon the broad hearth, it is not unlikely that DRAKE caught that droll and happy faculty of story-telling which has ever since been among the genial characteristics of his manner. In this situation he remained two years, acquiring something of that western "push" which was not developed until brought out by the difficulties which beset his labors years later on Oil Creek.

After leaving Tecumseh he returned to visit his parents in Vermont, and was persuaded to remain in the East.

He next went to New Haven, Connecticut, where he served three years as clerk in a dry-goods store. They were three uneventful years, and in the hope of bettering his prospects he gave up his situation, and obtained a position in one of the retail dry-goods stores on Broadway, New York. While here he married a young woman, whose home was in Springfield, Mass., and soon afterwards falling into a lingering sickness, it became advisable to seek country air, and they went to Springfield. While there Mr. DRAKE was offered the position of Express Agent on the Boston and Albany Railroad, at a salary of fifty dollars a month, which he accepted, and held the position till 1849, when he resigned it to accept the office of Conductor on the New York and New Haven Railroad, then just opened, which he held nearly ten years, with entire satisfaction to the superior officers of that corporation, and only resigned it to take charge of the developments on Oil Creek in Pennsylvania, as described in the opening chapters of this work.

The position he held on the railroad gave him the opportunity of forming an extensive acquaintance, which his inclinations prompted him to improve. In 1854 his wife died, leaving him one child, two others having already died; and he broke up the comfortable little home he had provided in New Haven, and went to boarding.

It was about this time that he made the acquaintance of Jas. M. Townsend, a banker in New Haven, into whose society he was thrown at the Tontine Hotel, where, at the time, both made their home. A few years afterward, when the prospects of the Pennsylvania Rock Oil Company were under a shadow, Mr. Townsend, who, amidst the allurements of social intercourse, kept an eye upon business, induced his friend DRAKE to invest a little balance of two hundred dollars which he had in bank, in stock of that corporation, and sold him a part of five hundred shares, which he himself held. This was the beginning of his connection with the business which has rendered his name famous. About the first of the year 1857, he married Laura Dow, of New Haven, a young woman of most excellent character, who has ever been to him a friend and guide in prosperity, and a staff and a light in the gloomy days of adversity and want. During the summer of 1857, Mr. DRAKE was compelled by debilitating illness to give up work on the railroad for a couple of months ; but at the same time he was not prostrated, and having at least an "inquiring" interest in the Pennsylvania Rock Oil Company, he began to investigate its prospects, and the subject of Petroleum generally. He had leisure for conversation with the directors, of whom his friend Townsend was one, and also president of the board.

The new idea of developing the property by artesian wells had been suggested some time before, and found in Mr. Asahel Pierpont, an intelligent and persistent advocate. Business complications forbade the thought of his going to attend to a matter so far away from home, and perhaps the growing dissensions of the company discouraged the hope of efficient action in a legitimate way. The board of directors consisted of five members, three of whom were residents of New Haven, as required by the by-laws adopted, and though representing only a third of the whole stock of the company, they controlled the management of its affairs. From what followed—all of which has been minutely described in the opening chapters of this book—it is indisputably clear that the New Haven stockholders were determined to secure to themselves the advantages of this new idea.

In December of the year 1857, Mr. Townsend, then president

of the board of directors, engaged Mr. DRAKE to proceed to Venango County, as has been previously stated. He finished his business and returned, enthusiastic to embark in the enterprise which they had projected.

On the last of the month, the New Haven members of the board —a majority and a quorum—met and executed a lease of the lands to Mr. Bowditch—one of the largest New Haven stockholders— and Mr. DRAKE, the terms of which were remarkably advantageous to the lessees, but which it was found necessary to change before the other members would permit them to go on.

When all was satisfactorily arranged with the old company, a new corporation was formed called "The Seneca Oil Company," of which Mr. DRAKE was the nominal president, and in which he appeared as the principal stockholder.

In the published articles of association the stock was subscribed as follows :

		SHARES.
W. A. Ivis,		2680
E. L. Drake,		8926
J. F. Marshall,		394

But of the 8926 shares which were in his name, DRAKE, according to a previous understanding, transferred all but 656 to the other members of "The Pennsylvania Rock Oil Company," and it then stood as follows:

Asahel Pierpont,	3334
James M. Townsend,	2785
William A. Ivis,	2680
Edwin E. Bowditch,	1630
E. L. Drake,	656
Henry L. Pierpont,	521
J. F. Marshall,	394
Total,	12,000

This comprised all the New Haven members of The Pennsylvania
Rock Oil Company, of which the largest stockholders in the new
company, Pierpont, Townsend and Ives, were directors.

In the following spring Mr. DRAKE set out for Titusville with
his little family, and until a house was prepared boarded at the
American Hotel.   Himself, wife, and two children and a horse,
were boarded for six dollars and-a-half per week, where a few
years later they would only have been entertained for about twice
that amount per day.

Shortly after arriving he bought a tract of twenty-five acres of
land in Titusville, of Jonathan, Watson, through the centre of
which Drake Street now runs.

He unfortunately sold this in 1863, realizing about ten thousand
dollars by the bargain.*   It was shortly afterwards sold for ninety
thousand dollars, and must now be worth not less than treble that
amount.

Nothing perhaps better indicates the condition of the little village, than the fact that a few weeks after his arrival, being in want
of a couple of picks and spades, he found there were none to be
had short of Meadville or Erie.   Though his life at the well was
crowded with incidents, they were incidents now too common to be
any longer interesting.

After oil was struck there was some difficulty in obtaining
a market for it—a difficulty which indeed continued to increase,
until in a couple of years' time, it was for a season nearly impossible to sell at any price.   There was no room for delay, and relying
upon his integrity to shield him from the imputation of improper
motives, naturally counting something on his services to the company and his own interest in that company's welfare, he hastened
at once to Pittsburgh and contracted to furnish about a third of the
oil to S. M. Kier, and arranged hastily with Mr. Geo. M. Mowbray
for the disposal of the rest on commission.

In 1860, Mr. Bissell proposed a division of the lands in lieu of the

* See the Sketch of Dr. Atkinson for a detail of this transaction.

twelve cents per gallon, royalty, and the Seneca Company thus obtained in fee simple one-third of the island which DRAKE afterwards sold for them, for enough to clear them of all indebtedness, though it is doubtful if they made a dime by the whole transaction. Indeed they declare they did not.

In 1860, Drake was elected Justice of the Peace for Titusville, an office worth about three thousand a year at that time, when every man was rushing to sell or buy leases, the documents for which he mostly drew and acknowledged. At the same time he bought oil for Shiefflin Bros., of New York, and thus increased his income to about five thousand a year.

In 1863, he sold his property, and left the oil region forever, taking with him between fifteen and twenty thousand dollars, and united himself with some Wall street broker, in oil stocks. It was a very unfortunate, not to say short-sighted move, for a man with his total ignorance of the manipulation of stocks and with so limited a capital.

His little fortune was soon engulfed. His health, already impaired by his labors on the Creek, gave way, and his noble wife now cast about to secure the future. She removed the family to a cheap and quiet abode in Vermont, and hoarded to the last the little she had been able to save from the wreck.

But his illness lingered and his strength failed, and his physician advised him, if possible, to seek the sea air. A friend kindly offered the use of a cottage on the Highlands-of-Never-Sink, near Long Branch, New Jersey, and thither they removed. But their funds were now exhausted, and their misery began indeed.

His disease was most agonizing; neuralgic affection of the spine, which constantly threatened paralysis of the lower limbs. He needed constant care, and his wife, surrounded by a family of four helpless children, attempted to keep them in bread by her needle. Sewing she could obtain in plenty, when she could tear herself from other absolute duties, to go after it, tramping through wet meadows, and chill and choking sea-fogs that roll in on that dreary point. But with all her noble and uncomplaining effort to keep them in bread without begging, she found it impossible. Medicines were out of the question, and with the greatest difficulty she got together the price of his fare to New York and back— eighty cents—and he struggled up to the city to get a situation for his eldest son among some of his old acquaintances. Before returning in the afternoon he was met and recognized in the street by Mr. Z. Martin of Titusville, who noticed his wretched appearance, and drew from him the story of their misery.

Mr. Martin, after providing him with a warm dinner, of which he stood sorely in need (for above the money to pay his fare he had not enough to pay for a cup of coffee, and he was weak from hunger), gave him twenty dollars, and cheered him with the hope of raising a fund for him in the oil region. No sooner was his distress made known, than with a generosity for which they have ever been famed, the citizens of Titusville, with some aid from individuals throughout the region, raised four thousand two hundred dollars for his relief; which wisely enough was committed to the management of Mrs. Drake, who has frugally hoarded it, and yet continued to meet a part of the family expenses with the wages of her needle.

In 1870, on the advice of his physician, she removed her invalid husband and three smallest children, to Bethlehem, near Allentown, in this State, where they are still living, beloved and respected by a large circle of friends who have gathered about them.

In conclusion, we have the pleasure to record the fact that the Legislature of Pennsylvania, at its session in 1873, deemed it proper to pass a law which grants to Col. DRAKE a pension of fifteen hundred dollars a year during life, or that of his wife. This is not charity, but simple justice.

Woodburytype. A. P. R. P. Co., Phila.

Capt. A. B. FUNK.

CAPT. A. B. FUNK—*Deceased.*

TITUSVILLE, PA.

AMONG the many noted, and successful pioneer operators, in the Pennsylvania Oil Region, Capt. A. B. FUNK, merits distinction and prominence. He was a man of superior intellectual acquirements and yet a fair type of the hardy settlers of the wilds of that portion of the Commonwealth of Pennsylvania. When the oil developments at Titusville became a reality, Capt. FUNK was among the first to enter into the new enterprise, lending to it the whole force of his character and ample wealth. We regret we have not the data at hand for a completed history of his eventful life; for few men, active as he was, in the earlier years of the development of petroleum in the Pennsylvania oil fields, deserve so generous a remembrance. Such facts as we have, however, we make use of, more for the purposes of a tribute to his memory, than a detailed sketch of his life.

Capt. A. B. FUNK was a native of West Newton, Westmoreland County, Pa., born in 1811, and grew to man's estate in his native town. His earlier years were devoted to commercial pursuits, in which he earned for himself an enviable character for integrity, and an unblemished repute for uprightness and honesty. Later in life he engaged extensively in the lumber trade on the Youghiogheny River—building and running small steamers upon its waters—and here, we infer, he obtained his title of " Captain."

In the spring of 1848, he had superintending charge of the construction of a lock and dam, known as "the Upper Lock and Dam," on the Youghiogheny Slackwater Improvement. During the same summer he began the construction of a large side-wheel steamer, intended for the Youghiogheny river trade. This vessel, called " THE FARMER," he completed in 1850, but when launched, she proved to be of too heavy draft, and he was compelled to run her on the Ohio and Mississippi rivers. The management and care of this steamboat enterprise required his almost constant attention, and necessarily, he was, during the season of navigation at least, away from his home in West Newton. But the enterprise was abundantly successful, and he continued his connection with it, until the fall of 1851, when he disposed of his vessel, having previously determined to purchase timber lands in Western Pennsylvania, and engage in the manufacture of lumber. In pursuance of this predetermined, and we may add, well considered enterprise, he soon after the sale of his steamer, in 1851, purchased of Judge Warner, of Allegheny City, a large tract of timbered lands, located in Deerfield Township, Warren County, Pa. In March, 1852, with his family, he removed to his new home, then in the wilds of this portion of the commonwealth, and entered industriously upon the work before him. He continued his lumbering operations until the spring and summer of 1859—meantime largely increasing his capital and his products as well. When oil was discovered at Titusville, in 1859, he was among the largest lumber manufacturers of that region, and his enterprise has been abundantly successful.

In the fall of 1859, Capt. FUNK purchased from the original proprietor, David McElhenny, his farm of less than one hundred acres, paying him $1,500 for it—McElhenny reserving one quarter of the oil! In the spring of 1860, the first well, "The Fountain Well," was commenced upon the property, "spring pole" power being used, until a depth of 260 feet had been attained. It was late in the fall of 1860, and winter and spring of 1861, when the "spring hole" was abandoned, and a small boiler and engine, procured to complete the drilling. "The Fountain Well," was completed in May 1861, and started off at 300 barrels per day! THIS WAS THE FIRST WELL IN THE PENNSYLVANIA OIL REGION, DRILLED TO THE THIRD SAND ROCK!

The development of this farm, always known as "The Lower

McElhenny," was rapidly prosecuted by Capt. FUNK, and by lessees, under him, and it is needless to say it was among the most bountiful producers of that early day. Its wonderful product during Capt. FUNK's ownership, and the handsome sum received for the property in 1864, netted a princely fortune to its owner. Very many of the lessees and operators realized large fortunes from their investments, and retired to more inviting homes.

Early in 1864, Capt. FUNK sold his oil lands and property, including the lower McElhenny farm, to " The McElhenny Oil Company," for $100,000 ! This, after realizing from it, in profits, during his four years' ownership, more " hundreds of thousands of dollars," than we care to mention. The sale of this property practically terminated his career as an oil producer. During 1863, he disposed of his lumber-lands, steam-mills, &c., in Deerfield Township, and had removed to Titusville, Pa., where he subsequently built a substantial residence, which he lived in until his death, and which his widow still owns and occupies.

The later years of Capt. FUNK's life were spent in doing good. He was ever a noble-hearted, generous man ; always ready to aid those who deserved it. We might give many instances of his liberality and generosity to individuals and to communities, but we must content ourselves with a rehearsal of only a few.

At " Steam Mills," Deerfield township, where Capt. FUNK first settled in 1852, are monuments of his liberality and enterprise that speak volumes for the goodness and kindness of his heart. In 1861, he built a substantial church edifice, paying every dollar of its cost himself. Did his neighbors supply timber, lumber, stone, or labor, he insisted upon the payment of their bills, and would allow no one to be called upon for assistance to complete it. This church, when it was ready for use, with five acres of land surrounding it, he presented by warranty deed, to the Methodist Episcopal Society of Deerfield, and it is held and occupied by them to this day. Later, he built a school-house, in the same locality, and supplied it with an extensive library—these two bounties cost him in all, not less than $10,000. Nor did he confine his generosity in this respect to his immediate neighborhood. He gave bountifully to churches and public schools about him, and seemed always anxious so to do.

During the winter of 1859, Hon. JNO. FERTIG, now Mayor of Titusville, and a producer of prominence, was employed to teach the district school in Capt. FUNK's locality. The District paid him "$18 per month and board." Capt. FUNK generously added $18 per month more, and towards spring presented Mr. FERTIG a sizeable lease upon the Lower McElhenny Farm, upon which that gentleman, in 1861, put down a well, the product and profit from which became the basis of his subsequent success and later ample wealth.

Capt. FUNK died on the 2d day of August, 1864, universally mourned by all who knew him. His record was that of a good man. He was kind and generous to the poor, liberal in aid of his less fortunate neighbors and friends, and ever ready to assist those struggling with adversity and misfortune. But above, and beyond all, he was AN HONEST MAN—true to his destiny, true to his fellows, true to himself, and faithful and devoted to his family. His religious convictions were the results of a life-time of eventful experiences, and these were "a shield and buckler" to him in his dying hour. He passed away quietly and peacefully, beloved and lamented by a large circle of devoted friends and bereaved relatives and kinsmen, in the fifty-third year of his age.

Woodburytype. A. P. R. P. Co., Phila.

HENRY R. ROUSE.

HENRY R. ROUSE, *Deceased.*

ENTERPRISE, WARREN CO., PA.

THE tragic death of this gentleman, which occurred at an oil well he was part owner of, upon the Jno. Buchanan farm, at Rouseville, in 1861, has lent an interest to his short career in the oil region at once remarkable and melancholy. The likeness herewith given of him was taken some years previous to his death, but those who knew him while he lived will recognize its truthfulness.

HENRY R. ROUSE was a native of Westfield, Chautauqua County, New York, where he was born, on the 24th day of August, 1824. Very little is known of his earlier years, beyond the fact that he was kept at school until he was twelve years of age. Thence he was sent two terms to the select school at Jamestown, N. Y. Returning to Westfield, he entered the academy there, and continued his studies a year or two, and before he attained his majority commenced to read law, in the office of a Mr. Dixon, then, as now, a practicing attorney of note in Westfield. He remained here preparing himself for his chosen profession, two years. Laying this aside temporarily, he engaged in teaching a district school in the town of Ripley, in the same county, for one winter. At the close of his school he received $100 for his services, and seemed at this time to have changed his determination. It is said a slight impediment in his speech—absolutely overcome when he became interested—determined him to abandon the law as a profession. He did not return to his law studies, but in company with a friend, named Brigham, resolved to go into mercantile and lumber pursuits in the wilds of western Pennsylvania. Procuring a number of letters of recommendation from friends at Westfield, with his friend Brigham, and $100 in cash, he proceeded to the city of New York and purchased a considerable stock of dry goods, groceries, &c., and in a few months after, we find him located at Enterprise, Warren Co., Pa., five miles from Titusville, in the full tide of success as a busy, industrious merchant and lumber dealer. This was about the year 1844–5. He was not then far from twenty years of age. He continued in the lumber and mercantile trade for a few years, when he disposed of his store and goods and gave his undivided attention to his large lumber enterprises. He prospered beyond his sanguine hopes, and was known as a clear-headed, comprehensive and successful lumber man.

In the fall of 1858, he was nominated by the republican party of Warren County for representative in the General Assembly, and was elected by a flattering majority—serving in the legislatures of 1859, and 1860—for he was re-elected in 1859. His career in this responsible position was marked by the same vigor, industry and integrity as had characterized him in the conduct of his private affairs. He was incorruptible in his legislative action, and supported or opposed every measure presented to him upon his own convictions of right. He attained prominence among the leading members of the legislature of that period, and was known and acknowledged by all, as an honest man, a high-toned gentleman, and an incorruptible law-maker.

His election was opposed by Mr. Jonathan Watson, of Titusville, with considerable bitterness and effect. After the session had closed, Mr. Watson met Mr. ROUSE, and proffered him his friendship, desiring to have the "old score" obliterated. "I can forget and forgive all but one thing," said Mr. ROUSE. "What's that?" said Mr. Watson. "Your story during the election," said ROUSE, with a knowing wink to the bystanders—"that I was a crazy spiritualist! I'll never forgive you that, *if I find that you set the story in motion!*"

The rumor was in circulation, we are told, but Mr. Watson was not responsible for its currency, and the old score was cleared off, and Mr. Watson was ever afterwards a warm personal friend and admirer of HENRY R. ROUSE.

When in 1859 and '60, the oil excitement burst upon the quiet of "the Creek," and the country adjacent, Mr. ROUSE was yet an extensive land owner and lumberman at Enterprise. He was not long in taking the inspiration of the hour, and with his partners, Sam. Q. Brown and John Mitchel, of Franklin, Pa., in September, 1859, secured leases of the two Buchanan farms at Rouseville. In October following he had completed one well upon the Barnsdall farm near Titusville; and another upon one of the Buchanan farms, near Rouseville. This was the commencement of his career as an oil producer. Fully convinced of the value of the discovery, he, with his partners before named, began a general investment in oil lands, and as the result proved, were soon the owners of large tracts of the best oil territory then, or years later, developed. Wealth poured in upon him in fabulous volume; and as yet, the territory he was interested in, was but partially developed. His connection with the early developments of oil in western Pennsylvania began late in 1859, and terminated with his terrible death on the 17th of April, 1861—about eighteen months in all. During this brief period he had established a character for energy, industry and boldness in his operations, that won for him universal admiration and general prominence. It has been remarked by one who knew him well, that had he lived, "he would have been a giant or a bankrupt in the oil business." With him, however, as with Cardinal Richelieu, there was "no such word as fail." During his brief career as an oil operator and producer he had laid foundations of a great fortune, and up to the day of his death his accumulations continued to increase with wonderful rapidity.

The facts and detailed circumstances attending his death, we obtain mainly, from Mr. GEO. H. DIMICK, at that time his confidential clerk and cashier, and a relative of the family. We give them mainly in Mr. DIMICK's own language:

"Just after supper on the evening of April 17th, 1861, Mr.

ROUSE, Mr. PERRY, Mr. BUEL and myself and others were in the sitting-room of Anthony's Hotel, (now Cherry Run Hotel,) discussing the fall of Fort Sumter, (the nearest railroad point at that time being Union Mills and Garland, and news from the outside world requiring from two to three days to reach us,) when a laborer on the fatal well hurried into the room to say that a monstrous vein of oil had been struck and barrels were wanted to preserve it.

All ran to the well with the exception of myself, and I not seeing the man who attended to the distribution of barrels, started in the opposite direction for teams to haul the necessary packages. I had completed my errand and was on a full run for the well with less than twenty rods to make, when an explosion occurred which nearly took me from my feet. On the instant an acre of ground with two wells and their tankage, a barn and a large number of barrels of oil were in flames, and from the circumference of this circle of fire could be seen the unfortunate lookers-on of a moment before, rushing out, enveloped in a sheet of flame which extended far above their heads, and which was fed by the oil thrown upon their clothing by the explosion. Scenes followed each other, and occurred simultaneously, beggaring both description and imagination. One poor wretch struggled out of the fire, believing himself to be in the hands of the evil one. His charred and naked figure was speedily placed in a blanket, and he was borne from the place. He lamented his supposed arrival in ———, in piercing tones of agony, which proceeded from lips burned to a cinder, and hence powerless to give proper accent to his language. He bemoaned his own fate, and calling the names of various friends warned them of his own terrible punishment. Death ensued in four hours.

Above the well and against the foot of the hill had been rolled two long tiers of barrels. One of the victims it would seem had been standing on these barrels near the well when the explosion occurred; for I first discovered him running over them away from the well. He had hardly reached the outer edge of the field of fire, when coming to a vacant space in the tier of barrels from which two or three had been taken, he fell into the vacancy, and there uttering heart-rending shrieks, burned to death with scarcely a dozen feet of impassable heated air between him and his friends.

So numerous were the victims of this fire and so conspicuous, as they rushed out, enveloped in flame, that it would not be exaggeration to compare them to a rapid succession of shots from an immense Roman candle.

Before speaking of ROUSE, a word about the well. It had produced oil for some time from the first sand rock, but failing there, was being drilled deeper in search of another oil-bearing formation. Its location was near the upper line of the John Buchanan farm on the east side of Oil Creek, and back at the junction of the bottom land, with a steep hill. A few rods up the hill, and a little south of the well issued a spring, which had formed a small ravine in running down, and created something of a swamp at the bottom and around the well. The well must have commenced flowing (as measurements of other later wells would prove,) at the rate of three thousand barrels per day, and although but eighteen or twenty minutes of flowing preceded the explosion yet the little swamp was covered deep with oil, excepting several small elevations on which the astonished spectators were standing when the shock occurred. Had this catastrophe happened even ten minutes later it is safe to say its victims would have been quadrupled; for word of the strike had spread with the rapidity of thought, and hundreds were running in breathless anxiety to behold it.

Mr. ROUSE standing probably within twenty feet of the well and among the very nearest of the spectators did not lose possession of his mind for an instant. He remembered the ravine, and dashed toward it. In the breast-pocket of his coat was a book containing valuable papers, and in the pocket of his pantaloons a wallet containing a large sum of money. These he jerked from their places and threw far outside of the fire, where they were afterwards found in safety. He had accomplished but half a dozen steps when he stumbled and fell, still being within the circuit of fire. He buried his face in the mud to prevent inhalation of the flames; then recovering himself bounded again up the ravine, falling a second time completely exhausted at a point where two men barely endured the heat long enough to seize and drag him forth. He was taken to a shanty near by, placed upon the bed of a workman, and gasped through five hours of excruciating agony before death gave relief. His body from the top of his head down the back and legs to the knees was burned to a crisp. The front of his person being less exposed was less seriously injured, but the face and feet were the only portions so far escaping, as to remain in any degree natural. The former was partially protected by the ground when he fell, and the latter by high-topped boots. Of his clothing, which was very heavy, but a handful of shreds remained.

ROUSE during the period of consciousness, which lasted up to within an hour of his death, maintained a coolness of manner most astonishing, and neither by word or action betrayed his terrible bodily suffering. With the precision and unconcern of a man without a care he dictated a will concise in terms and correct in language, and this, too, while being obliged to have water given him with a spoon not only at the end, but in the middle of every sentence. The bulk of his estate was bequeathed in trust to the Commissioners of Warren Co., one half of the proceeds to be applied to the improvement of the public roads. The proceeds of the other half for the benefit of the poor. Suitable bequests were made to all his relatives, and some of his intimate friends also found themselves remembered. We publish herewith a copy of this remarkable Will, certified by the Register of Warren county.

LAST WILL AND TESTAMENT OF HENRY R. ROUSE.

IN THE NAME OF GOD, AMEN:

I, HENRY R. ROUSE, being as I believe near my last moments, but sound in mind, do make this my last Will and Testament.

1st. My Executors to be George H. Dimick, Samuel D. Rouse and Samuel Q. Brown.

2d. I bequeath to my father, Samuel D. Rouse, Five Hundred Dollars per year, during his lifetime.

3d. Rouse & Mitchell hold the notes of A. Skinner and Allen Wright for Twenty-five hundred dollars. My half I bequeath to them; they are having hard enough times, without having to pay the notes.

4th. All the Lessees of Rouse & Mitchell, and Rouse, Mitchell & Brown, I want to have their leases at one-half the oil, and I bequeath to them all of my share of said rents over the one-half the product of the wells as now stipulated to be paid in their respective leases.

5th. I bequeath to George H. Dimick Two Thousand Dollars, for the use of himself and his mother, to be paid out the residue when my estate is settled up.

6th. To John Mitchell I bequeath my black mare.

7th. I have the Sheriff's Deed of the Store and Dwelling House occupied by Thos. Morean. I bequeath said property to his two youngest children, Eva and Maggie. Their father to have the use of it until they come of age.

8th. I bequeath the residue of my estate, after making some other bequests, to the Commissioners of Warren County, the interest of it to be expended on the roads of said County, after I make some other bequests.

9th. I have a little namesake, Harry Rouse, in East Granby, Connecticut. I bequeath to him Five hundred dollars. I cannot think of his name. His mother is the daughter of Joel C. Rouse; his name is Harry Rouse Victs.

10th. David H. Taylor, I bequeath to him Five hundred dollars.

12th. I bequeath to my aunt, Clara C. Hart, Five hundred dollars.

13th. I bequeath to Miron Waters, Five hundred dollars, to be paid when my estate is settled up.

14th. I also bequeath Five hundred dollars to my hired boy, Miron Dunham, to be paid when my estate is settled up.

15th. I wish to change the object of the bequest contained in No. 8, so as to give the benefit of one-half of it to the poor of Warren County. It is given in trust to the County Commissioners for that purpose.

16th. To Almedia Arnold I bequeath Two Hundred dollars.

17th. To Joel C. Rouse, of Saratoga, N. Y., I bequeath Three hundred dollars.

18th. I bequeath to Mrs. Morean, wife of Thomas Morean, Three hundred dollars.

19th. Two gentlemen carried me out of the fire. I bequeath them each one hundred dollars.

20th. Let my funeral be without display. No funeral sermon to be preached. Bury me by the side of my mother at Westfield.

21st. I have a beautiful picture, an engraving, in Herstfield's store, at Pittsburgh. I bequeath it to William Hirst, of Meadville.

22d. I bequeath my library to my father.

23d. I bequeath my wardrobe to Mrs. Thomas Morean.

I have nothing more to add at present. I authorize all who are here present to witness the foregoing as my last Will and Testament.

In testimony that the foregoing is the last Will and Testament of HENRY R. ROUSE, and at his request, we the undersigned hereby sign the same in his presence, in Cornplanter Township, Venango County, this seventeenth day of April, A. D. one thousand eight hundred and sixty-one.

his HENRY R. ⋈ ROUSE. mark. Physician in attendance,	N. F. JONES, ALLEN WRIGHT, S. S. CHRISTY, Z. MARTIN, W. B. WILLIAMS, W. H. KINTER.

Rouse was a man of peculiar religious views; while entertaining exalted ideas of the Creator and Ruler of the universe, he yet deprecated the popular forms of worship. Fear of the present or of the life to come had no place in his heart. After the completion of his will and but little more than an hour this side of the grave, a preacher friend standing in the little group watching his fading life, desired to administer religious consolation. Mr. Rouse replied :—"My account is already made up. If I am a debtor, it would be cowardly to ask for credits now. I do not wish to discuss the matter."

The well burned three days before it could be extinguished, which was finally done by smothering it with manure and earth. Its appearance while burning was grand. From the driving pipe, six inches in diameter, to the height of sixty or seventy feet arose a solid column of oil and gas burning brilliantly. Above this hovered an immense cloud of black smoke, which would seize sections of the ascending flames, and rolling over and over, first exposing to the view cloud, and then flame, would rise a hundred feet higher before the flame would fade out. From the main column below, millions of individual drops of oil would shoot off at an angle and then turning the arc of a circle drop burning to the ground, presenting all the hues of the rainbow making a scene like enchantment. The whole accompanied by a roar hardly inferior to that made by Niagara Falls.

Of the little group of eight sitting in the hotel at the time word of the well was brought, Mr. Dimick was the only one escaping injury, and Rouse was the only one burned to death.

The disaster, although so mournful, was not wholly destitute of ridiculous incidents. One woman in an agony of fear, rushed nearly across Oil Creek through water waist-deep, ere she realized the situation; and a man of strong religious convictions, climbed one hundred and fifty feet of the hill in rear of the well, before he could understand that it was not the day of judgment which he was making such good time in getting away from.

The evening of the fire was damp and murky, and the gas, issuing from the well crept along the ground like fog in a valley. All fires in the immediate neighborhood had been extinguished, and ignition undoubtedly occurred from a boiler eight or ten rods distant. The amount of gas escaping from the well, and the condition of the atmosphere, made explosion almost a certainty, and it is believed that fire anywhere within a distance of thirty or forty rods up or down the bottom land, would sooner or later have been reached with a result much more terrible.

Rumor has often charged Mr. ROUSE with smoking a cigar which caused the explosion. This we know to be false, from his own statement of the direction from which the explosion originated, and the extreme care which he invariably took in banishing cigars and pipes from the immediate vicinity of the wells.

Nineteen persons lost their lives by this fire, as follows: Henry R. Rouse, Enterprise, Pa.; W. S. Skinner, Wattsburg, Pa.; James Walker, Butler Co., Pa.; George Hayes, Chautauqua Co., N. Y.; Albert Gardner, Pontiac, Mich.; Judson Mason, New York State; G. W. Bentley, Harlansburg, Pa.; and Philander Stevens, Chautauqua Co., N. Y. Badly burned: Levi Walker, Butler County, Pa.; S. Houston Walker, Pittsburgh, Pa.; John Restling, Chautauqua County, N. Y.; —— Easton (fatally;) Whiteson, Oneida County, N. Y.; Constant Burnell, Erie County, Pa.; James Perry, Utica, N. Y.; (fatally:) Smith Cushing, Sherman, N. Y.; Thomas Page, Mercer Co., Pa.; J. G. Stratton, Crawford Co., Pa.; James Smith, Venango Co., Pa.; James Johnson, Mercer Co., Pa.; Archibald Montgomery, Venango Co., Pa.; Willis Benedict, Warren County, Pa.; —— Lockwood, Michigan; Augustus Cummings, orphan boy, fatally; —— Buel, Utica, N. Y.; Jos. Floyd, Utica, N. Y.; (fatally;) J. A. Kent, Chautauqua Co., N. Y.; John Glass, Butler Co., Pa.; Geo. Glass, Henry Chase, Mr. Burly and a few others slightly burned. In all nineteen persons lost their lives, and eight or ten were disfigured or maimed for life.

Thus perished HENRY R. ROUSE, by a fearful disaster which swept from the habitations of men, nineteen human beings, half of them without note or warning, and almost in a twinkling of an eye. The explosion was instantaneous, and death to a majority of its victims was sudden and painless. The large number who survived the catastrophe, and who yet live, bear the marks of the terrible conflagration about their persons.

HENRY R. ROUSE was a little more than 38 years of age when this fatal accident came upon him. From a feeling tribute to his memory, written by a kinsman, and published in the Warren (Pa.) *Mail*, of May 4, 1861, we transcribe the following truthful estimate of his leading characteristics:

" Mr. ROUSE was distinguished for many noble traits of character. Foremost among these were energy and decision. He was rather small in stature, and of light frame, but the energy of his will, carried him forward to the accomplishment of his object, through difficulties and over obstacles that would have deterred common men. To this he owed his success in life. He never knew fear. To resolve was to execute, and his business capacities were of the brightest order. He was emphatically the architect of his own fortune. He was a man of a noble public spirit. There was no enterprise by which the public were to be benefited, the resources of the region developed, facility of intercourse increased, in which he was not interested. He spent much time, labor and money opening and improving roads, constructing bridges, and helping on every work which tended to develop the wealth of the new region about him. How dear to him was the accomplishment of these is evident from his will, which appropriated one-half of his property, after the payment of legacies, to the construction and improvement of roads and bridges in Warren County.

He was a man of much literary taste and culture, and a great reader. The impulse which he received in this direction in his academical course was never lost. He kept himself informed of the affairs of the nation, and had collected a large library of standard works in history and general literature.

He was a man of warm impulses and strong personal attachments. He loved his friends with his whole heart, and never forgot them; and he greatly enjoyed friendly and social intercourse with them. Having no family of his own, he was continually showing kindnesses to families and children of his neighbors, making them presents, taking them to ride; and he surprised a number of his little friends by legacies in his will. Many of his early friends were thus remembered; and to each of the persons who picked him up, when insensible, and carried him to a place of safety, he gave a legacy of $100.

To the poor he was proverbially liberal, dispensing his favors with a lavish hand. Many are the anecdotes current of his timely and sympathizing aid. He never forgot that he himself was once poor, and he had a strong sympathy for indigent merit, and always lent with an open hand to its encouragement; and in his will he showed the liberality of his heart by giving one-half of his princely fortune for the support of the poor of Warren County. He was loved by all who knew him; and in his death Warren County lost a most valuable and public-spirited citizen, his fellow-townsmen an energetic and liberal-minded business man, the cause of virtue and sound morals a firm supporter, and the poor a sympathizing and most helpful friend."

GEORGE H. BISSELL.

NEW YORK CITY.

AMONG the earliest of the early pioneers of the Western Pennsylvania Oil Region, GEORGE H. BISSELL, of New York city, must take a leading and a prominent place. We show clearly in succeding sketches and by historical data, that petroleum was gathered upon the Watson Flats near Titusville, and at McClintockville, just above Oil City, as early as 1840, and so on down to 1856–7–8 and '59. It however remained to Mr. BISSELL to give force and effect, and final triumph in developing this world-renowned benefaction.

The facts, as we give them below, are obtained from reliable data, and are given without fear of contradiction. But first of Mr. BISSELL's early history.

GEORGE H. BISSELL was born at Hanover, New Hampshire. He is descended from a family of Norman-French origin, which came from Somersetshire, England. His mother came of Belgic and Holland descent. One of his ancestors was the first settler at Windsor, Connecticut, in 1628. The late Governor Clark Bissell, of Connecticut, and Governor William H. Bissell, of Illinois, were relatives.

About the age of twelve years his father died, and GEORGE was thrown upon his own resources for support. He has gained education and fortune, but never by the aid of a dollar from any one. While at school and college he supported himself by teaching and writing for magazines and papers. In the business struggle, it has been his own energy and talents which have won the victory.

Some two years were spent at the Military School at Norwich, Vermont; another period at Kimball Union Academy, at Meriden,

Woodburytype. A. P. R. P. Co., Phila.

GEORGE H. BISSELL.

New Hampshire, and he was graduated at Dartmouth College in 1845. For about two months he held the professorship of the Greek and Latin languages at the University at Norwich, but resigned on account of the inadequate salary.

Going to Washington, D. C., he was employed during the winter of 1845–6, as correspondent of the Richmond *Whig*. In the spring of 1846 he went to Cuba, and thence to New Orleans, where he became connected with the editorial department of the New Orleans *Delta*. For several years thereafter he contributed largely to the columns of the different papers of that city.

In 1846, on the organization of the High School, Mr. BISSELL was elected its first Principal, over many competitors. Subsequently he was chosen Superintendent of the Public Schools in New Orleans. His remarkable administrative ability, and high qualifications as a scholar, were of great service in his onerous position. The schools reached a discipline and prosperity before unknown. Amid the pressure of official and editorial duties, he still found opportunity to study law and several of the modern languages. In the summer of 1853 impaired health compelled him to come to the North.

It was during this year that Mr. BISSELL's attention was first called to Petroleum. He saw, at the office of Professor Crosby, of Dartmouth College, a bottle of petroleum, given Professor Crosby by Doctor Brewer, of Titusville, Pennsylvania, found upon his (Doctor Brewer's) land on Oil Creek. He became greatly interested in the product, and, about six months after, sent to Titusville, Mr. J. G. Eveleth, who was then, and had been previously, his partner in other business. They bought together, what were then thought to be the principal oil-lands of Pennsylvania. The lands were in extent one hundred acres in fee simple, and one hundred and twelve acres on lease for ninety-nine years, on Oil Creek, about two and a half miles below Titusville, for which they paid five thousand dollars. In 1854 they organized "The Pennsylvania Rock Oil Company," which was the first petroleum company in the United States.

This Company was organized under the laws of New York, with a nominal capital of $500,000, most of the stock being owned and retained by Messrs. EVELETH and BISSE L, who were its officers.

The Company proceeded to develop the lands by trenching them, and raising the surface oil and water into vats. The supply was very limited, amounting to perhaps, a few barrels in the course of a season, which was sold at one dollar and fifty cents per gallon, to parties who retailed it for medicinal purposes. In the spring of 1855, Professor Silliman, of Yale College, was employed to analyze the oil, and Messrs. BISSELL and EVELETH furnished him with all useful apparatus for his experiments, and paid the entire cost of the analysis. Professor Silliman's report, published in the fall of 1855, attracted attention in New Haven, and led to the reorganization of the Pennsylvania Rock Oil Company, with that gentleman as President.

The work of trenching the lands was continued until 1858, when the question of boring an artesian well was discussed, and advocated strongly by Mr. BISSELL, it having been suggested by the fact that Mr. Kier of Pittsburgh had obtained a small quantity of oil from one of his salt wells near Pittsburgh, at a depth of about 400 feet.

The New York and New Haven stockholders were not harmonious, and finally, after much discussion and difficulty, a contract was concluded between the Pennsylvania Rock Oil Company and some of its members, by which the latter agreed to lease the lands for a term of years, and pay the parent Company a royalty of 12 cents a gallon on all oil raised. They then organized in New Haven a new Company, based on the lease aforesaid, and employed one of their number, Mr. E. L. Drake, as Superintendent, and furnished him with the necessary capital. He proceeded to Titusville, and after many delays and obstacles, on the 28th day of August, 1859, the first vein of oil was struck, and the first petroleum obtained from an artesian well drilled on Oil Creek, Venango county, Penna., and this was accomplished under the auspices of "The

Seneca Oil Company," lessees of "The Pennsylvania Rock Oil Company," the organization of which, and the first purchase and development of oil lands under it, were mainly due to GEORGE H. BISSELL.

Soon after the completion of the "Drake well," Mr. BISSELL and Mr. EVELETH began the purchase of large tracts of oil lands along "The Creek," investing between two and three hundred thousand dollars in the enterprise. Thenceforward they engaged in the production of the oil by drilling wells at various points on "The Creek," at Franklin, Petroleum Centre, &c., doubling and quadrupling their investments in munificent pecuniary returns. We have not the details of all the operations of these pioneer operators, and it is perhaps, needless that we give them. It is enough to know that GEORGE H. BISSELL's name is identified prominently, and, we may add, honorably so, with all the early struggles and later triumphs in connection with this great national blessing, and that his name and fame is "a household word" among oil men from end to end of the continent.

From 1859 to 1863, Mr. BISSELL was a resident of the oil region, his home being at Franklin, Venango county. He erected a large barrel factory at Franklin, and continued this industry for some years. In 1866, he established a banking-house at Petroleum Centre, which has withstood the vicissitudes and disasters of the oil region from year to year, and is to-day regarded as one of the soundest and most substantial banking institutions in the oil country. Mr. BISSELL still continues his connection with it, and this fact is a sufficient guarantee of its stability and unquestioned soundness.

In 1863, Mr. BISSELL removed to the city of New York. In 1864, he represented the oil dealers of Pennsylvania, and the Petroleum Board of New York, at Washington. He made a powerful and effectual argument before the Committee on Ways and Means in opposition to the tax on the crude material, which would have proved ruinous.

In addition to conducting an immense petroleum business, he was at one time carrying on three banking institutions, building a railroad in the oil country, and was president and director of various companies in New York. With a majority of these he is still actively connected. He has recently been prominent in the organization of the New York Loan and Indemnity Company. He is also President of the Peruvian Petroleum Company and of the Peruvian Refining Company. These companies supply most of the petroleum used on the Pacific coast of South America, and have made large shipments to Australia, England, and other countries.

Mr. BISSELL was admitted to the bar of New York in 1855—to practice in the United States Courts in 1857, and to the bar of Pennsylvania in 1861. He was married in 1855, to Miss Ophie Louise Griffen, of New York city, who died suddenly in the spring of 1867. He has been a liberal donor to various institutions. Dartmouth College is indebted to him for a gymnasium which cost twenty-four thousand dollars.

Mr. BISSELL has an erect, well-proportioned figure, an active step, and an intellectual head and face. His head is long, towering to a round, high brow, while the other features are not less significant of mental force and the purest character. His eyes are sharp, and look forth with much directness. The mouth is closed, having the expression of decision and energy, which are the leading characteristics of the man. The peer of his contemporaries in the walks of business, he is endowed with scholarly accomplishments which fit him for any of the most exalted positions of life. Friendly and honorable in social intercourse, he is, also, one who is to be admired for the successful application of varied and brilliant talents in all other relations with his fellow-men. A man of fine mental accomplishments and of commanding business talents, he has distinguished himself in many widely different fields of effort, and his life affords events of much popular interest.

Woodburytype. A. P. R. P. Co., Phila.

CHARLES HYDE.

CHARLES HYDE.

PLAINFIELD, NEW JERSEY.

AMONG the number of successful pioneer oil producers, the subject of the following sketch is perhaps one of the most notable. His successes were not in all respects the results of " good luck" either, but were rather the offspring of judicious investments.coupled with clear-headed business tact. With his good fortune, he united in himself a sagacity and discernment that led him, as wealth poured in upon him, to gather up his accumulations and hold them for other profitable ventures; the sketch we give herewith will warrant us in these preliminary suggestions, and without further remark we proceed.

CHARLES HYDE is a native of the town of Eagle, Allegany County, New York, and was born on the 27th day of February, 1822. The place of his birth was known as Hydeville, his father being one of the early settlers of that locality. The town or village is now known as Eagle Village, and is a point of considerable note in Allegany County, N. Y. Charles is the third of a family of four children—three sons and one daughter. At the age of eleven, his father removed to Nunda Valley, then in the same county, but now a thriving town in the county of Wyoming. In 1837, or when CHARLES had attained the age of fifteen, the family removed to Western Pennsylvania, settling upon a partially cleared up farm, two miles south of Titusville, Pa. This farm adjoined the Stackpole farm, and has lately been developed into valuable oil territory, by the " Octave Oil Company "—a history of which we give elsewhere in these pages. The price paid for the property by Mr. Hyde, Sen., was $3.33 per acre.

Charles was schooled winters and assisted his father the balance of the time, until he was nineteen years of age. At twenty, in company with his father and two brothers, W. C., and E. B. Hyde, he purchased a small tract of twenty-five acres of land near Centreville, Crawford County, Pa., and in connection with the grocery and hardware trade, engaged to a limited extent in the manufacture of salts and saleratus, from ashes gathered or brought to their place of business. The firm was known as E. Hyde & Sons. Three years were spent in this enterprise, the business steadily increasing, demanding greater facilities and larger capital. Success in a marked degree followed these ventures, and industry and frugality characterized all the operations of this firm of father and sons. To their other interests they added the manufacture of lumber, which was run out of Oil Creek into the Allegany River, and thence to Pittsburgh. At the end of three years successful business, CHARLES purchased his father's and brothers' interests, and thenceforward assumed its entire responsibility. Soon after he became sole proprietor, his ashery burned, and he abandoned the idea of rebuilding it, and resolved to confine himself exclusively to the manufacture of lumber. The "Hydetown Mills," were at this date the property of JOHN TITUS, who had become hopelessly involved in debt, and was soon after compelled to compromise with his creditors. Mr. Hyde, Sen., who was one of Mr. Titus' largest creditors, finally purchased the property upon a fair valuation, paying Mr. Titus the difference, over and above his indebtedness. Soon after the conclusion of this transaction, CHARLES again entered into co-partnership with his father and brothers, and removed to "Titus Mills,"—assuming charge of the property and business. This was in 1846, and from this time on "Titus Mills," was succeeded by "Hydetown," and by this name has the little village since been known—now come to be a point of considerable importance, and boasting a thrifty growth, with many handsome residences, which villages of greater pretensions might point to with pride.

This last partnership continued for two years, and was measurably successful. At its termination—the father purchasing the interest of the sons—CHARLES spent a month or more, in exploring the great lumbering regions of the Clarion river, traversing the country for a hundred miles or more, on foot. He, however, returned to Hydetown, and the following year purchased the Hydetown Mills—in which he had retained a small interest—and at once entered largely into the manufacture of lumber. This property he still owns and operates, and it is a source of considerable revenue to him. The property has meanwhile, however, been much improved, the capacity of the mills greatly increased, and the facilities for manufacturing materially added to. To his lumber business, he united soon after his purchase of the property, a large mercantile trade, all which he successfully conducted for many years, and altogether he may be set down as a well-to-do merchant-lumberman, long before the oil excitements of 1859–'60, and later, came to the knowledge of the residents of that locality, or startled the public mind from end to end of the country.

As early as 1840–'41, Mr. HYDE knew of the existence, and in quantities, too, of petroleum oil below Titusville. With his father, he had many times visited these " oil springs," to procure supplies for lighting their mills and other purposes. Their mode of obtaining it was by digging trenches, or excavating from four to six feet in depth and diameter, into which the oil would run upon the surface of the water in such quantities, that with cloths, they were enabled to gather and " wring out" pails full of the odorous fluid. This was also true of the McClintock farm near Rouseville. Mr. HYDE, from 1840, down to 1858–9, frequently visited the McClintock farm, and purchased barrels of petroleum, which was taken thence to Titusville and elsewhere, and sold at $1.00 per gallon, for illuminating and for medicinal purposes.

When in 1859, the old " Drake Well" was commenced, Mr. HYDE was still a resident of Hydetown, driving his lumber and mercantile interests, as only he knew how to drive them. Col. Drake became a customer of his, in a small way, purchasing some of his supplies, small tools, shovels, etc., at his establishment in Hydetown. He was not, therefore, an uninterested observer of this new enterprise. Col. Drake visited Mr. HYDE often, and thus he was enabled to keep himself thoroughly informed in regard to the progress of events. When later in the summer of 1859, the well was down and pumping ten to twelve barrels of " Seneca oil, worth one dollar a gallon," Mr. HYDE did not craze himself over the event, as many of his neighbors and acquaintances did. An additional demand was made upon his well-filled lumber yards at Hydetown, for derricks, engine houses, dwellings, tank lumber, etc., etc., and these he supplied as promptly as he was able to, and during the fall and winter and spring of 1859, and '60, his business was amply remunerative, and he was satisfied with its profits, and its rapid increase.

During the winter or early spring months of 1860, however, Mr. HYDE was invited by his friend, SAMUEL GRANDIN, of Tidioute, to become a share-owner in " The Tidioute and Warren Oil Company," then about organizing at that point. He left his home a day or two after receiving Mr. GRANDIN's invitation, and proceeded to Tidioute, where, after a careful canvass of the matter, he became a purchaser of one share of this property, paying $1,000 for it. The property was divided into ten shares of $1,000 each. This was his first investment in oil territory.

[A fuller detail of " The Tidioute and Warren Oil Company," will be found in the sketch of Mr. J. L. GRANDIN, and we omit further mention of it here.]

The development of this property followed early in the spring of 1860. The spring freshets came, and upon them, Mr. Hyde started with his stock of lumber for Pittsburgh. A day or two after reaching the Allegany river, he was overtaken by a raftsman from above Tidioute. Mr. HYDE inquired of him, in regard to the oil prospects in that locality, and especially in reference to developments upon " the Tidioute and Warren Oil Company's " lands. "They've got a big well up there," said the raftsman, "and upon that very farm, too !" " Have they?" said Mr. HYDE,—" I own one-tenth of that property !" " Do you?" was the reply.

"Then you'd better jump from that raft and go straight home! You've no business running lumber for a living, if you own an interest in that property. Your fortune is made!"

The information thus given, proved to be reliable, for the first well sank upon this property, produced forty barrels per day. This Mr. HYDE learned after he reached Pittsburgh, with his lumber, whither he went, and where he remained superintending the sale and delivery of his stock, until all was disposed of.

During the winter of 1860, "the Hydetown Oil Company" was organized, Mr. HYDE subscribing for two shares. The Company had leased the McClintock farm, from Brewer, Watson & Co. at an advance of royalty, to the latter Company. The Company —Mr. Hyde being the lessee and developer—commenced operations upon the property early in 1860,—completing his first well to the depth of 280 feet during the summer of that year. A small steam engine was used to drill the well, and the oil they obtained, was "second sand rock" product. When struck, the well began to flow largely—150 barrels—and continued to produce for nearly a year.

Oil, at this date, was a "drug in the markets." One dollar per barrel was paid for it, in the spring and summer of 1860, but later in the season, large quantities were sold at thirty, twenty, ten, and even five cents per barrel, "at the wells!" The McClintock farm development as well as the Tidioute investment, were of course non-paying, and Mr. HYDE realized little or nothing from either, during 1860, or 1861.

About the date of these briefly referred to oil operations, Mr. HYDE turned his attention to individual ventures. He obtained a half acre lease upon the Clapp farm, adjoining the McClintock farm, and during the summer of 1860, drilled two wells upon it. One of these, at the date mentioned, was the deepest well drilled on "the Creek"—if we except one put down by Jonathan Watson, Esq., in 1865, to the depth of 2130 feet, at the base of the hill, where Spring Brook comes into the city of Titusville, a detailed statement of which is given elsewhere—and one other drilled on the Watson flats, in 1864, to the depth of 1,200 feet. One of Mr. HYDE's wells, was put down to the depth of 982 feet, and the other was 675 feet deep. We have no record of these wells, save that given from recollection by Mr. HYDE. No other "sands" were discovered below the "third," however, and this was found at the depth of 456 feet. "Second" and "third sand" oil, was found, but not in paying quantities, and both wells were shortly after abandoned.

The "Hyde Town Well," continued to flow nearly up to its first year's product, through the winter of 1861, and spring of 1862, and better prices were realized for oil in the Pittsburgh market. Mr. HYDE had, in the meantime, engaged quite extensively in the purchase and shipment of the staple, in flat-boats, barges, barrels, &c., to that point. Upon one of his return trips from Pittsburgh, in the spring of 1862, he met Dr. A. G. Egbert, then operating in a limited way, upon the widow Davidson farm, at Petroleum Centre. Dr. E., held a contract for the purchase of this property, and in the course of the interview offered to dispose of one half his interest, upon what he termed "the ground floor." The offer was accepted by Mr. HYDE, and upon the following day, the transaction was completed, and the widow Davidson farm passed into the hands of Hyde & Egbert, and by this name it has been known ever since. The fabulous product of this farm, gave it a world-wide notoriety, and brought untold wealth to its fortunate owners. Mr. HYDE paid Dr. Egbert, $2,625 for one half his interest in this farm, which was in fee, with a reservation of one-twelfth of the oil to the original owner, as royalty.

On and after this purchase by Mr. HYDE, we may safely assume, he began his career as one of the largest producers of "the Creek." He spent much of his time upon the farm, superintending its development, and having a general care of the valuable interest he had acquired. In rapid succession, after developments began in 1862, and in 1863, many large wells were obtained, among them,

"The Jersey Well," "The Maple Shade," and other smaller "flowers," the product of which aggregated, in cash receipts, many thousands of dollars per day. Later, in 1865, "The Coquet Well," and half a dozen others of less product, went far toward doubling and quadrupling his income of previous years. During the years 1864, and 1865, there were TWENTY-THREE FLOWING WELLS upon the Hyde and Egbert farm, which altogether comprised only about forty acres of land, and one-half the product of these "flowers," and nearly as many good pumping wells, went to swell the immense income of the fortunate owners.

Details of the development of this wonderful mine of wealth, we know would be interesting to the general reader, but we have scarcely room for them, or time to devote to their rehearsal. Very many facts connected with the history of the Hyde and Egbert farm property, are given in other chapters of this work, and we leave this portion of Mr. HYDE's personal history, with the single remark, that more than $3,000,000 was gathered into the coffers of the lucky, and we may add, always industrious and enterprising proprietors. Mr. HYDE still holds his original interest in this property, and it brings him in a small monthly revenue.

"The Second National Bank of Titusville," was organized on the 11th of February, 1865, with a capital of $100,000. Mr. HYDE was the principal mover in the enterprise, and became a subscriber to its stock to the amount of $57,000. At its organization he was unanimously elected President, a position he has been annually re-elected to fill ever since.

In January, 1866, the capital of the bank was increased $100,000, without an increase of circulation. In December, 1867, the charter of "The First National Bank of Titusville," was purchased, with the privilege of its circulation. This institution had a capital of $100,000. On the 1st of September, 1871, the bank purchased $100,000 of the capital stock of "The First National Bank of Meadville," and thereupon increased its capital to $300,000.

In all these additions to the capital stock of "The Second National Bank of Titusville," Mr. HYDE has maintained his position as its largest stock subscriber, and he is to-day the owner of about $170,000 of its capital. "The Second National Bank of Titusville," is one of the successful, as it is the most substantial banking institution in the oil region, or indeed in Western Pennsylvania. In addition to the large ownership of stock of the bank named above, Mr. HYDE has on deposit with the Treasurer of the United States, $300,000 in government bonds, his own personal property, left there as additional security to bill holders and creditors of the institution, of any and every character—should it be required.

In 1869, Mr. HYDE became largely interested in real estate investments at Plainfield, New Jersey, and removed thither with his family, in that year, in order to give to the enterprise his undivided attention. Since his residence at Plainfield, he has added extensively to his landed property, and has now under the highest state of cultivation, three or four hundred acres of the very finest lands in the State of New Jersey. Under his skillful hand these have been made to "blossom like the rose." The close proximity of this large body of land to the city of Plainfield, for it lays just outside the city limits, may give the reader some idea of its great value. Upon this property Mr. HYDE has a very fine private residence, with commodious out-buildings, carriage-houses, etc., not lavishly expensive structures, but substantial, and in entire keeping with his good taste, and ample for his wants. It is safe to say Mr. HYDE would refuse $350,000 for this property alone. The city of Plainfield has a population of quite 12,000, is admirably situated, and elaborately laid out. It is a city of very many magnificent private residences, owned and occupied by merchants, manufacturers, and professional men, as well, doing business in the city of New York, Brooklyn, Jersey City, etc. Mr. HYDE's residence, and grounds surrounding it, very elaborately laid out, containing shrubbery, flower gardens, walks, boating pond, fountain, etc., are all within the city limits. But the larger portion of his property, contiguous to his residence, spreads out as far as the eye can reach, and embraces about 400 acres of beautifully located lands, which are mainly used now for agricultural purposes. When, four years since, Mr. HYDE came into possession of this splendid property, the lands were exhausted, and nearly valueless for farming uses. His thorough knowledge of the art of agriculture here found ample scope for practice. The soil has been greatly enriched, and to-day, every acre of it is as productive as it ever was.

The management of this property absorbs the time and attention of the enterprising owner. It is soon to become an addition to the city of Plainfield, and already streets are laid out, and partially graded through its ample boundaries. When it is put into market, the income from it will reach a fabulous sum.

CHARLES HYDE is a man of deeds, and not of words. Of modest deportment and reticent manner, he is, nevertheless, a gentleman of solid worth, and has excellencies of character, that brighten and become more and more prominent and evident as acquaintance and association familiarize one to him. One of his chief characteristics is the thorough knowledge he has of his own business affairs, and the scrupulous exactness with which he devotes his personal attention to them. Educated in the practice of a rigid economy, he maintains the same careful, judicious conduct of his business affairs that have characterized him from his youth up. Waste and extravagance have no apologist in CHARLES HYDE. He lives temperately, walks humbly, and deals justly by all. He pays dollar for dollar of his indebtedness, and pays it promptly and cheerfully. Plain of speech, and plain of manner and dress, he seems at home in the drawing-room, in the counting-room, or in the harvest field. Honor, honesty, and unblemished integrity are his, by a life of practical devotion to the observance of these virtues. He is just in the vigor of his manhood and the years of his usefulness, and to all appearances, has before him a lengthened lease of life, which we doubt not will be marked by good deeds, sure to bring happiness and a green old age.

WILLIAM H. ABBOTT.

TITUSVILLE, PENNA.

WILLIAM HAWKINS ABBOTT, is a New Englander by birth—born in the town of Middlebury, New Haven County, Connecticut, on the 27th day of October, 1814, and is the eldest son of a family of twelve children, six sons and six daughters. His father was an industrious, thrifty Connecticut farmer, and from him, young AB-BOTT, as he grew to boyhood, and merged into manhood, imbibed and practiced those rare virtues of integrity, sobriety and honesty, that have been his shield through life. During his early years he had the advantages of a good common school education, embracing these from three to four months of the year; the remaining months were devoted to the interests of his father in the conduct of the farm. He continued in the discharge of these duties until he was eighteen years old, when, with the approval of his father, he entered the mercantile establishment of Gen. Hemmingway, at Water-town, Litchfield County, Conn., as a salesman and clerk. He remained in this position seven years, or until the spring of 1844, challenging from first to last the confidence of his employer, and the public as well, for industry, unquestioned honesty and excellences of both head and heart.

At the age of twenty-five he caught what was then known as the " Western Fever," and removed to Newton Falls, Trumbull Co., Ohio, where he soon after entered the employ of Bronson & Warren, then largely engaged in the general mercantile business. This position he held for one year, when the firm dissolved by the withdrawal of Mr. Warren, and the substitution of Mr. ABBOTT. The firm of Bronson & Abbott continued another year, and then dissolved, both the partners continuing business separately. A year subsequent, Mr. ABBOTT purchased the stock and business of

Woodburytype A. P. R. P. Co., Phila.

WILLIAM H. ABBOTT.

his former partner, and the real estate of Bronson & Warren, and continued the enterprise from that date, 1848, upon his own account, until 1862.

Of course, during these eighteen years of mercantile life, there were fluctuations in trade, "ups and downs" with business-men, sudden changes in values, bank panics, bank failures, "wild cat" schemes for defrauding the public, many of them bountiful successes, but amidst all these, and above and beyond them all, Mr. ABBOTT maintained an unblemished credit, always paying one hundred cents upon the dollar of his indebtedness.

Few merchants of fifteen to twenty years standing, but have at one time or another encountered the embarrassments and discouragements incident to "hard times," "no trade," "no money," no public or private confidence. All these were the experience of Mr. ABBOTT in the conduct of his large and yearly increasing business, but he surmounted them all, and maintained for himself an unsullied repute for reliability in all his engagements, promptness in all his obligations, and integrity of an unimpeachable character in his intercourse with all.

Early in February, 1860, Mr. ABBOTT resolved to visit the oil region of western Pennsylvania, partly from curiosity, but mainly as a matter of business. He reached Titusville on the 8th of February, 1860, and remained only a portion of two days. He visited the only oil well then in existence—"The COLONEL DRAKE"— which had been producing small quantities of petroleum for some months, and at a glance saw, that here was a field for his enterprise and business tact, and that in all probability untold wealth coursed through the rocks beneath him. With his accustomed promptness, he purchased one half of the one-quarter interest WILLIAM BARNSDALL owned in the James Parker farm, upon which a well was then going down, including the "Crossley well" and lease, together with a like interest in one hundred acres leased by Mr. BARNSDALL, at Shreve Rock, all lying a short distance below Titusville. For these three one-eighths, in three different tracts, he paid $10,000, and immediately set out for his home in Ohio. Two days after reaching Newton Falls, he received information that a fifty-barrel well had been struck on the Parker farm lease!

This well was known far and near as "The Barnsdall Well," and was the second struck—the "Colonel Drake," completed in August, 1859, being the first. This, like the "Drake well," was put down with a "spring pole," and was one hundred and twelve feet deep, and produced about fifty barrels per day, of first sand-rock oil.

Soon after receiving this news, Mr. ABBOTT returned to Titusville, and thence proceeded to New York, with a view to making a market for the product of this second well. Mr. BOON MEADE, one of the owners of the well and lease, accompanied him. Mr. Henry R. Rouse, afterward burned on the Buchanan farm, held an interest in this lease.

Mr. ABBOTT while in New York made the acquaintance of Mr. George M. Mowbray, then as now, a chemist of rare acquirements, and through him, obtained an introduction to the extensive Drug and chemical house of Shefflin, Bros. To this firm he sold 200 barrels of oil at 35 cents per gallon, delivered in New York. This may be said to be the beginning of the oil trade with New York city, which has since grown into such enormous proportions, requiring millions of barrels to supply the demand for both foreign· and domestic consumption.

This oil Mr. ABBOTT subsequently shipped to New York in old oil barrels, turpentine barrels, molasses barrels, whiskey barrels, a few new barrels, and indeed every conceivable kind of barrel or cask that promised to hold its oily contents. The result of this shipment proved the almost utter worthlessness of packages of this character for oil shipments, for upon reaching their destination, there was scarcely a barrel but had leaked and wasted from one-quarter to one-third its original contents. While the owners and shippers made "a good sale," they were "handsomely short" on delivery; and yet, the balances, when closed up, were largely in their favor. One thing, however, had been accomplished—a market for the sale of petroleum oil—and this, to a man of Mr. ABBOTT's enterprise, was an ample return for the vexations and losses attending the first effort to bring this great staple to the notice of commerce and commercial men.

Of the experiments subsequently made by and through Shefflin Bros.. to utilize petroleum and make it serve some other and better purpose than that to which it had theretofore been applied, we will not stop here to detail. Suffice it to say, that the experiments made by these gentlemen demonstrated its entire adaptation to illuminating purposes. Of course, other experiments had proven this fact, prior to this, but we think it fair to assume that the tests and chemical appliances of Shefflin Bros., practically settled the question of refining petroleum oil as an illuminator, and gave to the world light as cheap as daylight. What a change has all this wrought! To-day, refined petroleum illuminates more households and domestic hearths upon this, and the continent of Europe, and wherever civilization extends, than all other modes combined.

To make this first shipment to Shefflin Bros., Mr. ABBOTT and his partners purchased the necessary barrels, in and about Titusville, "teamed" them to the "Barnsdall well," filled, and thence sent them across the country by wagon-loads, to Union, on the ·Atlantic & Great Western Railway—a distance of twenty-two miles. The cost of this mode of transportation varied, depending wholly upon the condition of the roads. When "wheeling" was good, 75 cents per barrel was the ruling price from Titusville to Union. When the roads were "heavy" $1 and $1.25 was paid—bringing the average to about $1 per barrel. The Railway charges to New York were $1.56 per barrel, making the total per barrel, when laid down in New York, not far from $4. Had the packages held their contents safely, this first large shipment must have proved very profitable, netting the owners a clear margin of $2,000 to $2,250, after deducting all expenses. As it was they realized a handsome sum, and were far from being discouraged with their first venture.

Very many of the subsequent shipments of oil to the sea board by Mr. ABBOTT, as well as others—in 1861, and '62—were not a source of profit. Barrels that at first could be purchased for 60 and 70 cents each, were now worth $1.50 and $1.60, and they were very scarce at that price. Mr. ABBOTT bought old barrels, however, at these new prices, sent them by teams to the old "Empire Well," on the Funk Farm—then the largest producing well on the creek, where they were filled at 25 cents per barrel, and returned in like manner to Titusville. The price paid for this teaming was 50 cents per barrel, the round trip. Arrived at Titusville every barrel was re-inspected, hoops tightened, bungs replaced or re-driven, and then reloaded and hauled to Union, at an additional cost of 75 cents per barrel. The general depot for this trans-shipment at Titusville, was upon the grounds now occupied by the residences of Col. Pitcher and Dr. Barr, on Washington street.

We have said these subsequent shipments were often a source of expense to their enterprising projectors. Mr. ABBOTT tells us of two instances ; a large quantity shipped to New York, about this date, when sold and accounted for, involved him in an absolute loss of $1 per barrel, which he paid to close the account, and this after furnishing the oil and barrels gratis!

In the fall of 1860, Mr. ABBOTT, having associated with him Mr. James Parker and Mr. William Barnsdall, commenced the erection of the first Refinery built in the oil regions. Work upon this new enterprise was begun on the 6th of November, 1860, and on the 22d day of January following—a little over two months and a half—the fires were lit, and the refining of oil commenced in Titusville. The cost of this establishment, with subsequent improvements, was $15,000. Mr. ABBOTT continued his connection with the enterprise for nearly three years, with moderate profits, the prices of oil fluctuating so remarkably as to render any anticipation of margins upon the manufacture of refined, extremely doubtful, and oftentimes hazardous. The first lot of crude—a few hundred barrels—run through their new works, cost $10 per barrel, and before this was put upon the market, and realized from, crude oil could be bought at $2 and $2.50 per barrel. Of course the prices of refined did not at all times sympathize with the fickleness of the " crude market," and so a fair profit was ordinarily realized upon the manufactured article.

Mr. Abbott had the burden of the labor to perform in the erection of this refinery, and indeed, its superintendence and management after its completion, and during the three years of his joint ownership. The boilers, and other apparatus and machinery, were purchased at Pittsburgh, and thence shipped "by river" to Franklin and Oil City. Here they were "dumped" upon the shore at low-water, and lay in that condition until the streams were filled from bank to bank, by the fall rains, and when with men and teams they sought to raise them from their watery bed, almost entirely submerged, it was found to be an almost impossible task. Two teams, with a full complement of men, made the journey from Titusville to Franklin, charged with the duty of raising these boilers from the water and transporting them to their destination. They came, they saw, and returned as empty as they came! Subsequently, Mr. Abbott personally superintended the work, and they were brought out, loaded, and in four days from the time of starting were on the ground in Titusville, where they were put to the uses intended. In this manner all, or nearly all the machinery for this new undertaking was brought to Titusville.

We have given this detailed account of the first refining enterprise in the oil region, to enable the general reader to form some idea of the obstacles and embarrassments men of enterprise were forced to contend with in the early years of the discovery of Petroleum. This of Mr. Abbott's was only an index to others, bearing upon the business interests of the region. In 1860–'61, and even into 1862, nearly everything making up the machinery of an oil well—engines, boilers, drilling tools, cables, &c., &c.,—had to be transported by wagon, from the railway stations at Corry, Union, Meadville, &c., twenty and twenty-five miles, and over roads that

at times were absolutely appalling to " man and beast." After oil
began to be produced in large quantities, there was a great dearth
of barrels for transporting it from the wells to the railroad stations,
and thence to market. To meet this demand, in part at least, Mr.
ABBOTT contracted for the manufacture of large quantities of barrels
in Ohio, and in many instances sent them by wagons, eighty to ninety
miles, across the country to Titusville. The manufacture of this
indispensable article was, however, soon after extensively carried
on in Titusville, Mr. ABBOTT furnishing a large amount of capital
for this purpose.

In the fall of 1862, Mr. ABBOTT added another branch to his
already extensive business interests—a depot for the sale of coal.
He laid down in Titusville the first car load brought there, himself being consignor, consignee, and retail dealer. This large
supply, for such it was for that day and the locality, was sold off in
small lots of 50 to 200 pounds, to such as were willing to try the
experiment of its use, and on the whole Mr. ABBOTT regarded the
speculation as " tolerably remunerative." He continued the business thus begun, gradually increasing his capital, and enlarging his
boundaries. In October, 1865, he leased for 30 years a large tract
of coal lands, partially developed, at Greenville, Mercer County,
Pa., and continues to this day to mine and ship great quantities into
the oil region and elsewhere.

Of the fluctuations in the price of crude oil, in 1860-'61-'62, we
may as well speak here, and to say they were remarkable, is hardly
sufficient to give the reader a clear comprehension of this branch
of the subject—they were wonderful—and must have been attended
by losses of thousands and thousands of dollars, which no one
positively profited by. In January and February, 1860, as we
have before stated, Mr. ABBOTT paid $10 a barrel for "crude,"
delivered at their refinery—and holders did not care to sell at
these figures ! By the middle of March and April following, oil
was delivered at the same refinery at $1.25 per barrel ! In Octo-
ber, 1862, Howe & Nyce purchased, barreled, and had stored upon

the first platform, erected at Titusville, on the Oil Creek railway, 500 barrels of crude oil. This oil Mr. ABBOTT purchased at $2.62½, including packages, and with a guarantee that every barrel contained 42 gallons. Shortly after, this lot was put upon the market and sold for $3 and $3.50 per barrel. In January and February following, oil was sold from the same platform by Mr. ABBOTT at $12.50 and $14 per barrel, and before the middle of March following, the same lot—for it had not been moved—was sold at $8 per barrel, and thirty days after, the market price at Titusville was $3 per barrel! These transactions involved large amounts of money. The oil bought and sold was usually in round lots of 1,000 to 5,000, and even 10,000 barrels, and destined to eastern markets. While, therefore, the original producer oftentimes received fabulous prices, the purchasers were generally the victims. At the time these extravagant figures were being realized by producers, a few, one at least, known to the writer of this sketch, was dealing out one hundred barrels per day, at 40 cents per barrel! When the great flowing wells on the Funk farm — "lower McEllhenney"—were struck, the old "Empire," producing 3000 barrels per day, and the "Fountain well," pouring out 400 barrels, Mr. Funk, senior, contracted to deliver to Mr. Chas. A. Dean, of Cleveland, Ohio, one hundred barrels of oil per day *for five years*, at 40 cents per barrel! This contract was actually entered into, and the letter of it lived up to by Mr. Funk through two years of its existence! But the contractor became greedy. He was not satisfied with ordinary barrels of 40 to 42 gallons capacity, but would furnish casks and tierces, and almost hogsheads, holding fifty, sixty, and as high as sixty-five gallons, and demanded to have these filled and counted as barrels! To this extraordinary craftiness Mr. A. P. Funk, then in charge of his father's affairs, mildly objected, and finally flatly refused to submit. The contractor applied to the courts for relief, and the courts in turn declared the contract forfeited!

Mr. ABBOTT became interested in the development of oil and oil lands, early in 1860.   From that time until the present, more than twelve years, this connection has been constant and always prominent, both as a producer, a refiner, and a buyer and seller of oil.   In the spring of 1863, he purchased from S. S. Fertig, Esq., a one-eighth " free" interest in the famous " Noble Well," then producing 2,500 barrels per day, and paid for it the princely sum of $27,500 ! Many of Mr. ABBOTT's friends regarded the amount paid as exorbitant, but it proved to be one of the best oil investments of his life. He realized his money thrice over, and never regarded the speculation as at all hazardous or doubtful.   The owners of the " Noble Well," soon after Mr. ABBOTT purchased his interest, determined to buy out the " Caldwell Well," located a short distance from their own, fearing it would injure the flow of the "Noble Well."   They did purchase it, paying for it the extraordinary sum of $145,000 !   The well was producing, when they came in possession of it, about 400 barrels per day.   Mr. ABBOTT, as did the other owners, paid cheerfully his one-eighth of the sum required to purchase the " Caldwell Well," and deemed it a bargain at that.   The owners of the "Caldwell Well," had they held their property forty-eight hours longer, would have been paid as cheerfully $200,000 !

In June, 1867, Mr. ABBOTT formed a co-partnership in Pipe Line interests, and the general transaction of the oil trade, with Mr. Henry Harley, who had just completed his Pipe Line from Benninghoff Run to Shaffer Farm.   Mr. ABBOTT had, previous to this date, purchased the Pipe Line from Pit Hole to Miller Farm, and was operating it successfully.   The consolidation of these two Pipe Lines was, thenceforward, a substantial success in every way— Mr. ABBOTT's judicious management contributing largely to this result.   From that day to this he has been identified with this enterprise, giving to it his large experience and comprehensive business tact, which have gone far to render it not only remunerative to its stock owners, but a source of incalculable convenience to oil producers, as well as shippers and refiners, and all others interested, near or remote, in this great staple.   Out of this first pipe line enterprise has grown up many others, in various parts of the oil region, from Parker's Landing, St. Petersburg, and the lower river country, up to, and including Titusville and points contiguous to it. The history of this first Pipe line is given so fully in the sketch of Mr. Henry Harley, elsewhere in these pages, that we deem further reference to it, in this connection, as only a repetition. Suffice it to say, " The Pennsylvania Transportation Co." now represents a capital invested of nearly $2,000,000,—it has over 400 miles of pipe, stretching its iron lengths at every point where it may be required, from Titusville, Miller and Shaffer Farms, Petroleum Centre and Gregg's Switch, over the mountains to Shamburg, Red Hot, Pleasantville, Trunkeyville, Fagundas, Triumph, Tidioute, Colorado, and Enterprise, including a delivery at Siverly and Oil City. It has, besides, immense platforms and conveniences for railway delivery at all stations on the Oil Creek and Allegany River Rail Road, where connections are made, and maintains tankage capacity, at various points, of quite 300,000 barrels. This vast interest has been brought to its present magnitude and its conceded remunerative basis by and through the rare ability and wise foresight of Mr. ABBOTT and his partner, Mr. HARLEY.

In the Summer of 1865, "The Titusville and Pit-Hole Plank Road Company" was organized, and its construction soon after entered upon. It was completed in the winter of 1866. Mr. ABBOTT, Colonel F. W. Ames, Col. Oliver Keese, and S. Q. Brown, of Pleasantville, were its projectors, furnishing from sixty to eighty per cent. of the capital required for its completion. It was an important work to the region traversed, and went far toward opening the country to oil development, lying between Titusville, Pleasantville and the then great oildorado of the oil region—Pit Hole. It cost $200,000, and more than two-thirds of this sum was advanced by the gentleman we have named.

Union and Titusville Railroad Company.

" The Oil Creek and Titusville Mining and Transportation Company," was granted a charter in April, 1865, authorizing the construction of a railroad from Titusville to Union. James Sill, E. H. Chase, Jno. W. Douglass, H. C. Rogers, Charles Burnham, Joseph Sill, A. C. Bloomfield, and James D. Smith were appointed commissioners, under the act of incorporation, "to open books, receive subscriptions, and organize a company to construct a railway from and to the points named." The capital stock of the company was fixed at $500,000. In April, 1866, a supplemental act was passed by the legislature authorizing the directors to reduce the capital stock to such amount as they deemed proper. On the 17th of April, 1866, at a meeting of the directors, held at Union, the corporate name of the company was changed to "The Union and Titusville Railroad Company." In June 1867, the Company mortgaged its property, real and personal, to secure $150,000 of bonds to be used in its construction. Subsequently, in 1868, there being default in the payment of interest upon these bonds, the trustees named in the mortgage, by due course, sold the property of the company, its privileges and franchises, at public sale, at the office of Jay Cooke & Co., Philadelphia, to E. Cooper and J. C. Frisbee, and executed to them a deed of the entire property. What amount of money had been expended upon this work to this date, is not fully known. Certain it is that the new organization availed themselves of but a small portion of what had been done. In the condition here indicated, Mr. ABBOTT found the enterprise in 1870, and comprehending its great importance to the city of Titusville, and the oil fields adjacent, resolved upon its immediate building. He associated with himself a few gentlemen of known enterprise, and on the 2d day of July, 1870, a meeting of the stockholders was held at the McHenry House, Meadville, and upon a full consultation, it was determined to go on with the work. The following board of directors was thereupon elected:

W. H. ABBOTT, Jno. Fertig, D. H. Cady, P. H. Stranahan, W. R. Davenport, Henry Harley and J. S. Casement. Mr. ABBOTT was subsequently elected President, and D. T. Casement Secretary and Treasurer.

Subscription books were promptly opened, based upon the pledge that Titusville should take $250,000, and the balance, sufficient to complete the work, was to be subscribed by the Casement Brothers, the contractors. Upon this basis the contract was let to the Casement Brothers, and work was begun in August, 1870. The subscriptions to the capital stock by citizens of Titusville fell far short of Mr. ABBOTT's anticipations, but rather than the enterprise should fail, Messrs. ABBOTT & HARLEY, in addition to their already large pledge of $50,000, subscribed $66,000 more, making in all $116,000, to this important work. All this was the labor of but a few days, for within one month after Mr. ABBOTT had identified himself with the enterprise, the contracts were let and work actually begun. Thenceforward, the construction of this important road was prosecuted with great rapidity, and on the 28th of February (1871,) following, it was opened, and the first train of cars passed over it, bearing freight and passengers.

Mr. ABBOTT's connection with the undertaking, illustrates as completely as anything we could assert, his character as a business man. He entered upon the enterprise with no such word as fail within hailing distance of him. When the public interest flagged, he put the whole weight of his personal and financial character upon it, and it went through to completion. It was the first enterprise of this nature, Mr. ABBOTT had identified himself prominently with, and he resolved it should not fail, even if it involved the last dollar of his means.

The road opened in February, 1871, but its progress was not marked with that measure of success so confidently anticipated. Its rival route, the Oil Creek and A. V. R. Road, competed for both freight and passengers, and many of the heavy freighting houses of Titusville, renewed their contracts with the latter road, at reduced rates. The result is easily foretold—the Union and Titusville road did not receive the traffic it had a right to anticipate, if not absolutely claim. This state of things continued through the spring and summer of '71, and in September, the majority of the stock and bondholders of the road entered into negotiations for a transfer of the entire work to the Oil Creek and Allegany River Rail Road. This consummation Mr. ABBOTT opposed from the start, and continued his opposition, with all the power and resources at his command, to the last. At a meeting of the directors, at which a large majority of the stock and bonds were represented, the proposition was carried over Mr. ABBOTT'S voice and vote. From the commencement he had devoted all his energies and time to this work without "fee or reward." He proposed to continue his services without emolument, if the road was retained by its then owners, and give to it all the ability he possessed to render it a success financially. But all to no purpose. The contract was consummated, and the road passed into the hands of the O. C. & A. V. R. R., under a lease, having reservations protecting the interests of the business men of Titusville. When Mr. ABBOTT connected himself with this enterprise the Oil Creek Road were charging $17 a car, for freight to Corry, and 60 cents per barrel for oil. Immediately thereafter this rate was reduced to $15 per car and 30 cents per barrel, and just before the completion of the road the price per car, was reduced to $10. The same liberal policy as to coal and all other freights was granted by this rival route, the advantages of which are still realized by citizens of Titusville and the country adjacent.

Notwithstanding this plain statement, every word of which can be verified by incontestable proof, there are those who assume to believe that Mr. ABBOTT transferred his interest in this railway undertaking at a considerable profit upon his investment! The writer of this knows whereof he speaks, and asserts without fear of contradiction, that the balance was largely against Mr. ABBOTT, amounting to many thousands of dollars. Even this loss did not shake his confidence in the ultimate success of the road. His proposal to add to his already heavy liability to the interest, coupled with an offer to discharge the duties of chief executive officer of the corporation, gratis, until success was attained, is proof of his devotion to its interests, and positive evidence of his entire confidence in its ultimate value as an investment, both for himself and associates, and to the city of his adoption.

A man of the prominence of WILLIAM H. ABBOTT, identified with almost every public enterprise that has engaged the attention of the men of the oil region during the last ten or twelve years, affords an almost exhaustless field of interesting facts for a sketch, of the character of which this work is in part made up. We have given enough we apprehend of the leading incidents of his history to indicate the true character and real worth of the man. Indeed, it seems to be labor lost to assert that he is a man above reproach, and without spot or blemish, either as a public or private citizen. In all his relations he is exemplary, acting always upon his convictions, based upon a broad and comprehensive view of life, its duties and its responsibilities. His generosity and liberality are known of all men. An instance of this will suffice. A year or two since—at a Sabbath School celebration of St. James' Episcopal Church, of which Mr. A. is a consistent communicant, and has occupied the position of Senior Warden for many years—the necessity for a mission branch of the church, to be located in another part of the city of Titusville, was brought to his attention. The lot had been partly secured. Mr. A., with his accustomed promptness, said—" Secure your lot, and I will build your chapel at my own expense, and upon the plan you have submitted ! " The lot was purchased, and Mr. ABBOTT immediately began and completed the church edifice at a cost of about $4,000. Soon after, Bishop Kerfoot visited Titusville, and the little mission church was dedicated with due solemnity, and is to-day an important auxiliary in the work of educating the young in its immediate locality up to a correct standard of Christian duty and Christian responsibility.

Mr. ABBOTT is a man of quick perceptions and rapidity of thought and action. With him it is "yes" or "no," and this promptly and without apparent reflection—and yet he thinks profoundly. His "yes" has cost him thousands and thousands of dollars, and his "no" has been but a slight source of revenue to him! A good man, ever cultivating and acting from the noblest impulses, he is doubtless often imposed upon; but he never permits the opportunity to escape to give bountifully to charitable objects appealing to him. Honorable in the fullest and broadest sense of the word, he has little charity for those whose practices do not come up to this standard. Proverbially he is a man of enlarged liberality, and gives with an open hand, and with none other than a desire to do good. As a man of business he is prompt, thorough and reliable under all circumstances. As a financier he is far-seeing and rarely mistaken in his convictions. As President of the Citizens' Bank of Titusville, he is estimated at his real worth, for he has given to that institution very much of the success and financial solidity that has marked its history in the nearly three years of its existence. As a citizen he is foremost in all enterprises calculated to add to the growth and prosperity of the city of his home. As a neighbor he is obliging—generosity and kindness characterizing his intercourse with all. In his domestic life he is a model of indulgence and excellence, and in all his worldly intercourse a pattern and example for the young about him, as well as those of maturer years.

Few communities can boast better men than WILLIAM H. ABBOTT. His daily life illustrates his worth and illumines his pathway in his declining years. Universally respected, he bears about him those real elements of an unblemished manhood, sure to be honored and beloved as age and infirmity creep upon him. Let us hope he may be spared many years of health and vigor, and be enabled to fill up the measure of his usefulness, reaching the fate of us all, "like a shock of corn fully ripe,"—ripe in those manly virtues now so bountifully possessed—ripe in the practice of the nobler emotions of the human heart—ripe in the assurance of a well-spent life—ripe in "the Christian's hope of a blessed immortality."

Woodburytype. A. P. R. P. Co., Phila.

ORANGE NOBLE.

ORANGE NOBLE.

ERIE, PA.

ORANGE NOBLE is a native of the State of New York, born in Whitehall, Washington County, on the 27th day of April, 1817. He was the eldest of a family of seven children, five daughters and two sons. His father was a farmer of limited means, but of thrifty and industrious habits. He owned a small farm in that part of the State of New York, and cultivated it with industry and frugality, rearing a large family from its products. ORANGE was the eldest of the children, and at an early age became an important aid to his father in the cultivation of the farm. There were no schools within ten miles of the homestead, and so the advantages of even a common education were denied him, if we except a few months of schooling in each year—after he was ten years old— until he attained the age of fifteen. At this age he conceived the idea of becoming a cattle dealer, and importuned his father from time to time to aid him in his enterprise. The father discouraged the undertaking as best he could, pointing to many of his neighbors who had spent their lives in that business, and who were yet poor! This made little impression upon young ORANGE, who could see only success and profit in the business. Finally the father forbid his mentioning the subject in his presence again! But young NOBLE abated nothing of his determination to get into the traffic as soon as opportunity offered—and it presented itself soon after. Visiting a neighbor a few days subsequent to this command of the father, he was shown a very attractive animal—a two year old heifer—and at once resolved to purchase her if he could make terms. The price was agreed upon, and young NOBLE said he would take her if he could have thirty days' time. "Yes," said the owner, "you can have her, and pay for her when you are able!" The animal was driven home, and, within the time named, doubled, and the two were sold at a liberal profit, and the original obligation promptly discharged. This was his first venture in his new business, but it was sufficient to satisfy him that there was money to be made at it.

During the same and the following summer he purchased, on credit, a considerable number of sheep and cattle, and drove them into pasture at home or near there, and sold them off in the fall at fair advances and profit. The following year he borrowed $600 for ninety days, and at the end of eighty-two days had purchased and sold his stock, paid his loan, and counted his profits at $75! This business of buying and selling cattle and sheep, he continued until he was twenty-one, clearing every year from $150 to $250, not meeting with any serious losses. These profits, whatever they were, he cheerfully handed over to his father, to be used in the support of the family.

The fall before reaching his majority he determined to attend, for a single term, the North Granville Academy, located in his native county, and this resolution he carried out, doing chores during the winter for his board. This may be said to constitute all or nearly all of his educational advantages. He had, however, applied himself with so much industry and zeal to his studies, that when the term closed, he felt he had set out anew in life. He had acquired a general knowledge of the common branches of an education, especially of mathematics, English grammar, &c., and was partially satisfied.

The following summer, having become of age, he re-engaged in the business of buying and selling cattle, and counted his profits at the close of the fall trade at $500; and here, we may add, that he continued in the business of purchasing and marketing cattle for nearly fifteen years after he came to man's estate, and with uniform, though moderate success.

In January, 1841, then in his twenty-third year, he married,

and soon after leased a farm, and began his career as an agricul-
turist, in connection with his cattle traffic. Two years later he
purchased a small farm, agreeing to pay $2,200 for it. The terms
of payment were somewhat remarkable. There was a small sum
paid down, and the balance was to be liquidated in annual install-
ments, running twenty-five years! Mr. NOBLE, however, did not
avail himself of all the liberality extended to him, but gradually
lessened his indebtedness, and before the close of the FIFTH year
had paid the last dollar of his bond, and received his deed.

He continued to live upon this farm—to which he added by purchase 60 acres—until the fall of 1851, when he sold it at an advance from the price originally paid, having determined to remove
into western Pennsylvania. After the sale of his farm, for which
he received about one-half cash in hand, and the sale of his stock,
crops, &c., he was enabled to count his worldly wealth in ready
capital, at $5,000.

During his fourteen years' business operations, he had been a regular depositor and patron of the old " BANK OF WHITEHALL,"
always one of the most substantial banking institutions of the
State of New York. His loans and discounts had not been large,
but they were regular, and he had never permitted a note made by
himself to go to protest. He paid promptly, and the officers of
the bank were always ready to take his notes, even when others of
more reputed wealth were denied. A single fact will serve to illustrate the estimate put upon him by " THE WHITEHALL BANK."
When he came to draw the balance due him from the bank, preparatory to his removal West, the officers paid him $3,500 in a new
issue just then being signed, and every note was made payable to
"ORANGE NOBLE, OR BEARER." This was a distinction few customers of a bank had awarded them, but Mr. NOBLE had it extended to him, and at the same time the gratulations and commendations of the officers of the bank.

In the regular transaction of his business through the country, .
it was his custom to give checks upon the BANK OF WHITEHALL

for his purchases. These checks ordinarily, would not reach the bank until after his own return home. There were occasions, however, when the checks preceded him, and not unfrequently his account was overdrawn, $500 to $1,000, and once or twice $2,000. But the bank paid them as freely as if he had $50,000 on deposit!

In the Spring of 1852, Mr. NOBLE removed with his family to Randolph, Crawford County, Penna. Soon after reaching his destination, he purchased two or three farms, in all four hundred acres of land, and settled down to its cultivation, adding the old business of his earlier years, the buying and selling of cattle. He purchased mainly young stock of the better sort, and raised them for teaming and farming purposes. He furnished better cattle than could be found elsewhere in the region of his residence, and always commanded higher prices, because of their excellence and superiority in all respects.

In the Spring of 1855, he entered into co-partnership in the general mercantile trade, with Hon. GEO. B. DELEMATER, the late State Senator from the Crawford district, then a resident of Townville. He was at this time engaged in the manufacture of " Shooks" for the eastern markets, and had built extensively for the conveniences of his increased and increasing business. In the Spring of 1856, the "shook" manufactory was added to the mercantile enterprise, and the capacity of the former largely increased. At times during the following three, four and five years of this mechanical enterprise they employed from sixty to eighty men, in the manufacture of "shooks," and after the discovery of oil, in making barrels for that trade. The firm continued their mercantile and manufacturing operations up to the fall of 1863, some months after the famous "Noble well" was struck.

The assertion is often made, that when Mr. NOBLE, and his partner Mr. DELEMATER, commenced their oil developments, or rather when they struck the " Noble well," both were poor, and were largely in debt. This is very far from the truth. Their business connection had been successful in every direction. They had by their industry and a careful conduct of their affairs, accumulated a handsome property, and owed no man a dollar, they could not liquidate at sight. When they began their operations in 1859–60, in the oil region, the firm was worth in round numbers $50,000. It is proper to state this fact here, and now, that assertions to the contrary may be set at rest.

In October, 1859, or within a few weeks of the striking of the "Drake well," Mr. NOBLE, in company with Mr. Delemater, and Mr. L. L. Lamb,—visited Titusville with no other purpose than to see for themselves the wonderful phenomena of "pumping oil out of the ground." A neighbor of Mr. NOBLE's, who had visited the well, was rehearsing its remarkable operations—"pumping water and Seneca oil in large quantities,"—and descanting upon the "visitation," rather forcibly. "You tell that story very well; one would think you believed it yourself!" said Mr. NOBLE, after listening to a rehearsal of the facts. "And they are facts," said the informant. "I was there and saw it with my own eyes!" But Mr. NOBLE could hardly credit the statement, and he did not. Subsequently, he talked privately with this man, who had seen the strange phenomena, and was more than half induced to give his statement credence. He resolved to see it for himself, and a few days after the interview referred to, Mr. NOBLE, Mr. Delemater, and Mr. Lamb, set out from Townville, twelve miles distant from Titusville, to visit this "well in the rocks," which was reported to be sending forth "Seneca oil" in fabulous quantities. All these gentlemen were disbelievers in the report that had sent them upon this journey of inspection. Mr. NOBLE looked upon it as an utterly impossible thing; a sort of half fact and half hoax, which only a visit would dispel or confirm. The party reached the "Drake well," late in the day, in October, 1859, and were convinced, as well as confounded. The idea of pumping so valuable a commodity as "Seneca oil," an article of rare medicinal value, from the ground, struck them with amazement. There it was, however, and the fact could not be gainsaid. They remained about the well for some hours, and when finally they set out for home, little else was talked of but this wonderful development.

Before reaching their homes they had practically organized an Oil Company, to which Mr. NOBLE and Mr. Delemater each contributed $3,000, and Mr. Lamb $2,000. Mr. NOBLE was deputed to return without delay to "the Creek," and secure leases, and make arrangements for putting down one or more wells. In pursuance of this hurriedly organized effort, Mr. NOBLE went to Titusville, and secured his first lease, upon the Stackpole farm, below the "Drake well." His next lease was upon the Jones farm, in the vicinity of Miller Farm. Derricks were erected, and in the spring of 1860, a well was drilled with a *spring-pole* upon each of these farms. Both were dry holes! Later in the summer of 1860, he secured a lease of seven acres on the Tarr farm, and in 1861 he held a lease of small dimensions on the Hamilton McClintock farm. The Tarr farm lease was developed in 1861. " The Crescent " well was among the first put down, and its history is a peculiar one. It started off at three to four hundred barrels, and flowed steadily at this rate for nearly ten months, and stopped in an hour, and has never since produced a barrel of oil! Efforts were made in 1871 to resuscitate it, but without avail. It was "rimmed out," and re-tubed, and after a month's pumping, was abandoned as a dry hole. During the life of the " Crescent well," in 1861, oil was selling at ten and fifteen cents per barrel, and while it produced largely, it never paid a dollar of profit to its owners!

Three wells were put down on the Hamilton McClintock lease, in 1861–2, two of which produced ten or twelve barrels each per day, and the third was dry!

History of the " Noble Well."

In the spring of 1860, Mr. NOBLE secured for himself and partners—Messrs. Delemater and Lamb—a lease of sixteen acres on the Farrel farm, half a mile above what is now known as Pioneer. For this lease he paid $600 bonus, " spot cash," and one-fourth of the oil. He bound himself and associates to proceed without delay—interpreted to mean twenty to thirty days—to drill a well 134 feet deep. This done his lease was held to be valid for twenty years, whether oil was found at that depth or not. In pursuance of this contract he set immediately about the work. A derrick was erected, a "spring pole" hung, and the labor of drilling began. It required days and weeks, running into months, to put down a well to this depth with a "spring pole," but the work was finally accomplished—Mr. NOBLE personally superintending the entire labor. It was done by "days' work," hiring men from time to time to replace those who had "worn themselves out" in the exhausting labors required of them.

The well reached the required depth—134 feet—late in the fall of 1860, and no oil, and not even "a show" of the coveted article was visible, or had been found from its commencement. Mr. NOBLE was the master spirit of the enterprise, and his judgment and determination was law to the balance of the owners. He resolved to abandon the well for the present, at least, and develop some one or more of his many other leases. This he did without unnecessary delay, and nothing was done upon the property, in the way of developments, until the spring of 1863. Meantime a "third sand rock" had been found at various points on "the Creek" from which oil was being produced in large quantities.

The contract for drilling this abandoned well of 1860, was let to SAM'L S. FERTIG, of Titusville, since become one of the successful oil producers of the region. He employed for the purpose of sinking the well, a small eight or ten horse-power boiler and engine, and with the opening of spring in 1863—in April, or perhaps earlier—he began the work of putting the well down to the "third sand." As a part of his payments for the labor, Mr. NOBLE assigned to him a ONE-SIXTEENTH *working* interest. The well was drilled rapidly, but after reaching the oil rock, at 452 feet from the surface, Mr. Fertig found one or two "crevices," of ten to twelve inches depth, as he judged by the dropping of the drilling-tool. The last "crevice" of ten or twelve inches depth alarmed him, lest he might get his tools fast, and believing he was deep enough in the "sand rock," and prompted by his fears for the safety of his tools, he "shut down" for consultation and further orders. Mr. NOBLE was absent at this juncture of affairs, and did not learn of Mr. Fertig's action for a day or two. When he returned he found everything at a stand-still, and the contractor quite determined to stop where he was, and test the well!

Some time before the well was down, Mr. NOBLE, who had from the commencement in 1860, and so on to the second renewal of operations, sole control of all matters pertaining to it, determined to tube it, when completed, with the best artesian tubing. To obtain this he visited the various manufactories at Cleveland and Pittsburgh, and at last, where he least expected to find it in the latter city, his eye fell upon a few lengths of the desired article, lying in front of a small establishment, in the rear of the St. Charles Hotel. He was not long in ascertaining its "author and finisher," a large establishment in Philadelphia. He ordered six hundred feet shipped at once, and contracted with the Pittsburgh party to take it on its arrival there, mount each joint with brass thimbles, and at the same time manufacture for him a peculiar discharge pipe, which was to cap the tubing, having four stop cocks to facilitate the conveyance of the oil to any desired point. This done he hastened back to the well to await its arrival.

Time and space will not permit a detail of the delays attending the transit of this tubing to its destination. It was "switched off" at Ravenna, Ohio, and lay there a week or two, and was delayed in all about twenty days at various points. It finally reached Corry, and here Mr. NOBLE obtained permission from Dr. STREETER, a director of the road, to have it transferred to a freight car, and this attached to a passenger train nearly ready to start for Titusville. The $10 slipped into the palm of the "train dispatcher" was an amazing help, in the matter of facilitating this unusual manner of transporting large quantities of freight!

The tubing reached its destination in a day or two after its shipment from Corry, and was in the well and ready for operation by 3 o'clock on the 27th day of May, 1863. The well had been standing open for a week or ten days, the surface water, and water veins below pouring into it in great volume. It had not filled up, but the roar of the "falling waters" was almost deafening. Where the water ran to, as the hole did not fill up, the reader can "guess" as well as we can.

After a few hours delay the well was ready "to start up." Mr. NOBLE gave orders to Mr. FERTIG, to "start her slowly and pump steadily." This done he repaired to an eating-house near by, as he says, "to get a bite of something to eat." He had but partially finished his lunch, when a lad at the door remarked, "That well throws water, bully!" Mr. NOBLE heard the remark, but paid no attention to it, until it was thrice repeated; then turning his eye toward the door he saw the "Noble well" spouting oil and water far above the derrick and trees about it! Mr. NOBLE describes the scene most graphically. It was about 4 o'clock, and the expectation was, that it would require pumping some hours before the golden stream would show itself. But it came within twenty minutes after the pump started. The well puffed and blowed, and roared, and the earth about it fairly trembled with agitation. No one dared to approach it, even within the circuit of the falling spray of oil and water. The little ravine near the derrick soon filled up with the great volume of oil rattling and foaming through the two-and-a-half inch tubing. Something must be done to control the discharge and save the oil. Mr. NOBLE offered $50 each, to any three men who would enter the derrick and attach his ingenious device for conducting the oil into tanks. The men stripped to the buff, and entered the derrick. The spray, oil and water completely hid them from view, and nearly drowned them before they could accomplish their task. At the end of an hour, or a little less, they had made the connection and returned to the outer world. For their service Mr. NOBLE gave them $200.

The flow of the well being now under control, tanks became a vital necessity. One, of seven hundred barrel capacity had been provided, but this was filled in an incredibly short space of time ! Soon after the oil began to flow into the tank, Mr. NOBLE despatched men on horseback, down "the Creek" to advise boatmen, that they could have all the oil they wanted at $2 per barrel. And about the hour the 700-barrel tank was full, boats began to arrive from below, and by midnight enough were abreast of the well to take the oil as fast as it could be conveyed to them. The following day more came, and by noon " oil boats" lay in " the Creek" for twenty rods above and below the well, filling the stream from bank to bank.

This was a temporary relief only. Mr. NOBLE immediately employed all the men he could find to assist in supplying tanks. Fifty men were at work the day following " the great strike;" clearing away, and putting together immense tanks made of wood and having capacity of from eight hundred to twelve hundred barrels. Within fifty feet of the " Noble well," stood an empty tank with a capacity of three thousand barrels. None of the owners could be found, but Mr. NOBLE stretched his lead pipes into it, and it was full in less than twenty-four hours after ! Other tanks, wholly empty, were within a hundred feet of the " Noble well ;" and these were pressed into the service and filled. By the time all these tanks were running over full, the boats loaded and dispatched, Mr. NOBLE had an ample supply of his own, ready for use.

We have been thus particular in our detailed account of this well, because of the almost romance with which it was surrounded, and that seemed to mark its history from its commencement—and this even is not its history complete. The well from the start, flowed between 2500 and 3000 barrels daily, and continued at this standard for many months. During its second year, its product gradually fell off, and in the later months of its second year's life in 1865, it produced four to eight hundred barrels a day. The first month's shipments and sales from its product, were 61,300 barrels,—and there were still left in the tanks 15,000 barrels. The loss from waste, leakage, want of tankage, &c., &c., was large, and altogether it is safe to say that during the first months of its existence, its product was not less than 3000 barrels per day. When it began to flow, oil was selling at the wells at $4 per barrel. To save as much of it as possible, Mr. NOBLE sold it by the boat load, *and with little regard to the size of the boat*, at $2 per barrel. While the product was at its height, oil was sold at $6, $7, $8, and even $10 and $13 per barrel, from their great tanks, bringing in return a volume of greenbacks, which went to enrich the fortunate owners and land proprietors. It has come to be a proverb almost, that "The Noble well" earned more money for its owners, than any well ever struck in the oil regions. And it may be added, that every dollar it earned was scrupulously accounted for, and paid over to the rightful owner.

Mr. NOBLE superintended the work of putting down the well, and determined, when it began to produce in such immense volume, that every interest should have its full share of the profits—and he rigidly adhered to this determination, as long as he had control of it. When it subsequently passed into the hands of a stock company he surrendered his charge, conscious that no interest, however small,—and there were many who held interests, drawing only one barrel in one hundred and seventeen—had suffered a loss of a dollar.

"The Caldwell well" was located a short distance from the "Noble Well," and had been producing from 400 to 500 barrels per day for a month before the "Noble" was struck. After the "Noble well" was down—and pouring out its 3000 barrels daily, Mr. NOBLE became interested in the effect it would be likely to produce upon his near neighbors, of "the Caldwell." He frequently met the superintendent, Mr. Brown, and always inquired if he "discovered any change in the product of their well?" "No," was the reply, for several days. But the change not long after this was apparent. "The Caldwell" was surely lessening, day by day, in its product, and Mr. Brown became solicitous for the interests of his company, owners and employers. He made known his fears to Mr. NOBLE, who had from the first been anticipating this very result. He had several interviews with Mr. Brown, who at first intimated and subsequently fairly demanded one-quarter of the product of the "Noble well" as remuneration for the loss to "The Caldwell." The "Noble" was producing at this time—ten days to two weeks after it was struck—3,000 barrels a day. Mr. Brown would be satisfied with one-quarter of the product, and deemed this a fair equivalent for the loss his well was sustaining! Mr. NOBLE, although not legally bound to pay a single dollar to the owners of the "Caldwell well," yet acknowledged a moral obligation to reimburse them. The Noble well had been located and drilled 134 feet three years before the Caldwell well had been projected, and when it was down, and it was absolutely known that it was drawing the oil from his neighbors, there was no legal obligation resting upon him to make good their loss. Nevertheless, he felt it his duty, to in part at least, make good their deficiency. There was still another fear in regard to the "Caldwell well." The owners threatened to draw their tubing and let the water down into the "Noble well," and thus destroy both, in all probability. To avoid this threatened contingency, as well as to deal justly, Mr. NOBLE cheerfully entered into the negotiations we hereafter detail. Mr. Brown was advised to call together his co-partners, at a subsequent day, with a view to some sort of settlement. This meeting was soon after held, and Mr. NOBLE asked them to consider the whole question at issue, and submit two propositions—one for the sale of their property, and the other, the amount of oil they would be satisfied with, from the Noble well. They first demanded one-quarter of the product of the Noble well! The Caldwell had never produced more than 650 barrels per day, even before the "Noble well" was struck, and now they required seven or eight hundred barrels to make good, not their absolute loss, but a prospective injury to their well. ("The Caldwell" was at this time flowing about 350 barrels.) The proposition to sell their property had also been considered, and they had agreed to offer it to Mr. NOBLE for $150,000.

A second council was held, but they adhered to the first demand —one-quarter of the product of the NOBLE well, but had concluded to take $145,000 for their lease, well and property. "When do you want this money?" inquired Mr. NOBLE. "To-day," said Mr. Brown. "That's a good deal of cash to raise in so short a time," answered Mr. NOBLE, "and I am not able to do it. I will tell you what I'll do. I will buy your property at the sum you name, and will pay you to-day $37,500. Fifteen days from to-day I will pay you $40,000 more. Thirty days from to-day I will pay you $40,000 more, and the balance, $28,500, I will pay on or before the end of forty-five days from to-day." This liberal proposition was finally accepted, and the Caldwell well passed into the control and ownership of the Noble Well Company. It continued to produce two or three hundred barrels a day for a few weeks, but in less than one month it "dried up," and never afterwards produced a barrel of oil!

The negotiations and final purchase of this property were all the work of Mr. NOBLE, and were begun and concluded without the knowledge and consent of his associates, if we except W. H. Abbott, Woods & Wright, J. W. Hammond, Esq., and one or two of the land owners—the Messrs. Farrel—who had purchased small working interests. Mr. NOBLE conceded a *moral* obligation to make good, or to a degree repair the loss of his neighbors, but it is morally certain that no court or jury would ever have awarded the fabulous sum he consented to pay to make them whole. He, however, determined to deal justly and honorably with all, and if he erred in his conclusions the error should be against himself. When afterwards his action was submitted to his associates, it was heartily and unqualifiedly approved.

This large sum of money, $145,000, was afterwards paid before it became due ; and every dollar of it was earned and realized from the product of the Noble Well, and within the forty-five days of its maturity!

It is to be regretted that no reliable data can be found now of the immense sums of money this well earned for its owners. The entire product of the well was, according to the books of shipment and sales, 480,000 barrels. It is safe to say, that $2,800,000 is a fair estimate of its fabulous profits. Of this princely sum, one-quarter was paid to the land interest, owned by James, John and Nelson Farrel, now residing at Titusville, and an only sister, Miss Sadie Farrel, since Mrs. W. B. Sterritt, of Titusville. The remaining three-quarters were fairly and equitably divided according to the interest owned, among the ten or twelve fortunate possessors.

Original Owners of the Noble Well.

Orange Noble and Geo. B. Delemater owned one-half the working interest. One-sixteenth of their interest was in 1863, assigned to S. S. Fertig, who subsequently sold it to W. H. Abbott, of Titusville, for $12,500. L. L. Lamb, W. H. Noble, Salmon Noble, father of Orange Noble, Charles Delemater, Thomas Delemater, G. T. Churchill, James Hall, Rev. L. Reed, L. H. Hall, and Rollin Thompson—these last ten owned, altogether, a little less than a one-quarter interest, which, when divided, gave each one barrel of every 117 barrels the well produced. Even this small fractional interest gave to each from sixteen to twenty-two barrels per day. After the well was down and producing 3000 barrels daily, Mr. W. H. Abbott purchased the one-sixteenth interest, owned by Mr. Fertig, who, while drilling the well, purchased a sixteenth land interest from Jno. Farrel, at a nominal price. This he also sold to Mr. Abbott for $14,500. Jno. W. Hammond, Esq., of Erie, Pa., purchased an interest after the well was struck, as did also Woods & Wright, of Petroleum Centre.

There is one fact connected with the history of " The Noble well " which we venture to give here as entirely new to most of the readers of the present day. Ten days, or perhaps two weeks before the well was down, a gentleman called upon Mr. NOBLE, and after much circuitous conversation, asked, "What will you take for your interest in this well, Mr. NOBLE?" Mr. N. disclaimed any desire to sell. "Will you take $10,000?" said the stranger. "Oh, you don't want to pay that amount of money for an uncertain piece of property," replied Mr. NOBLE. But the stranger was in earnest. He offered $10,000, and subsequently $20,000, and $50,000, and finally $100,000 was tendered for the Noble and Delemater one-half interest in the well! Mr. NOBLE declined all, not because he deemed the well worth more than that sum of money, but because he had determined to see it down and tested before he parted with another fraction of his interest. The wisdom of his resolute action was fully realized in the subsequent history of the well, for he received as his part of the profits from its product, nearly if not quite $800,000!

The subsequent history of the Noble well may be briefly stated. During the spring of 1864, "The Noble & Delemater Oil Company" was organized, the interests of Mr. NOBLE and Mr. Delemater in the well, forming the-basis of its capital stock, which was fixed at $500,000. Of this sum Messrs. NOBLE & Delemater held $200,000 in shares, the par value of which was $10 per share. The well was still producing 500 to 800 barrels daily, and continued to do this for some months after the organization of the stock company. Several monthly dividends were declared, and everything seemed to be going on promisingly. Of course with the organization of the stock company, the control of the well passed from the hands of Mr. NOBLE, to that of the company. During the early months of 1865, the well still flowing 300 to 400 barrels, the Superintendent and President of the Company deemed it necessary to take out the tubing and clean out the well. This was very earnestly opposed by Mr. NOBLE, who seems to have had well-grounded fears in regard to the proceeding. His opposition delayed the determinations of these gentlemen for a few weeks, and meantime Mr. NOBLE sold, as did also Mr. Delemater, every share of their stock at a little more than its par value. Later in the spring of 1865, the tubing was taken out, the well thoroughly cleaned, and again put into operation. But the life of the grand old flower had fled! With the exception of a few barrels of oil pumped for a day or two after "starting up," it came to a dead stand, and was shortly after abandoned as a dry hole!

In April, 1864, Mr. NOBLE removed from Townville to Erie, Pa., where he had already purchased a very handsome residence and grounds. The dwelling is situated in the western portion of the city, upon a slight eminence, overlooking the city, harbor and lake. Mr. NOBLE has expended a large amount of money in furnishing his home, and in embellishing the ample grounds about it. Everything is substantial, luxurious and attractive.

In January, 1865, "The Keystone National Bank" of Erie, Pa., was organized with a capital of $150,000—Mr. NOBLE subscribing $75,000 to its stock. At the first meeting for the election of officers, Mr. NOBLE was elected its President, and has been annually re-elected to this responsible position, since. THE KEYSTONE NATIONAL BANK of Erie, Pa., is one of the best managed and most substantial institutions of its character, in the Commonwealth of Pennsylvania. Among its directors are the ablest men of the growing city of Erie, who give character, thrift and success to financial operations of this magnitude. Before leaving this portion of Mr. NOBLE's history, it is not deemed inappropriate to say, that the capital stock of the Keystone National Bank has been increasd to $280,000, and this within a year after its organization.

In April, 1868, Mr. NOBLE was made the candidate of his party for the responsible office of Mayor, and was elected by a large majority. He was re-elected to serve a second term, and pending an amendment of the City's Charter, he held over one term, and was re-elected the third time without opposition. During his four years' administration many valuable and much needed city improvements were inaugurated, and carried through to completion. Among these we may name Erie's splendid system of water works, costing $800,000. The water is taken from the lake, given an elevation of two hundred feet, and thence distributed to every part of the city, through immense mains and in exhaustless quantities. These water-works are after the pattern of the city of Chicago, and may well be the pride and boast of the good people of the city of Erie.

A thorough and elaborate system of sewerage was also adopted and put into operation during Mr. NOBLE's term of office. To this should be added the adoption of a general street paving policy, inaugurated and carried into successful operation during his administration.

Mr. NOBLE is largely identified with many of the industrial and financial enterprises of the city of Erie. He was made the President of The Alps Insurance Company, upon its organization in 1871, and has been twice re-elected to this position.

Upon its organization in 1869, he was elected and has been re-elected since to the Presidency of the Erie City Passenger Railway Company, and is among its largest stockholders.

He is a director of "The Second National Bank of Erie," and is also a director of "The Erie Dime Savings Bank," an institution of steady growth and solidity. He is also a large stockholder and a director of the Foxburg and St. Petersburg Savings Bank, in Clarion County. In 1866, Mr. NOBLE erected the "Noble Block," a magnificent brick and stone structure, four stories high, and covering nearly half a square at the corner of State and Eighth Streets. This improvement, for it is one of the finest blocks in the city, cost him in the neighborhood of $140,000. He was the projector, builder and one-half owner of the first elevator built in the city of Erie. He is also half owner of "The Bay State Iron Works," one of the successful industries of Erie, in which he has invested $50,000. He is besides the owner of a one-quarter interest in the extensive blast furnace of Rawle, Noble & Co.. in which he has invested $60,000.

In 1872, he added to his other enterprises "The Erie Paper

Mill," for the manufacture of paper from wood and other material. This undertaking, at first an experiment, has now come to be a positive success, and to it Mr. NOBLE has contributed of his ample means to make it so, its entire capital—nearly $70,000.

It will be hardly expected that we should make mention of *all* the enterprises with which Mr. NOBLE has identified himself during his ten years' residence in the city of Erie. It is due to him to say, however, that whatever promised growth and prosperity to the city of his residence has met his warm approval and co-operation.

Mr. NOBLE is a gentleman of quiet deportment and simple tastes and habits. He is not proud, nor can he be said to be a handsome man. There is however a cordiality always about him that renders him attractive and agreeable to all who make his acquaintance. He is a rigidly upright, honest and honorable man, of unimpeachable integrity and real private worth. In business circles he is recognized as a gentleman of liberal views and clear conceptions, enterprising, and generously so, with the ample means at his control. He is a man of even temper, amounting to amiability. He is cool in judgment and candid in the expression of his opinions. There may be, and possibly is, a little self-will in his general character, but this is guided and governed by a matured experience that always controls his action. There is no deception in the man. He is frank in his friendships and as steadfast and true as the needle to the pole. Integrity of character, devotion to his friends, and a readiness to serve them at a personal inconvenience, if need be, make up the personal worth and comprehend the general outline of his character.

Although practically retired from the oil trade, he is still the possessor of large tracts of oil lands located in the vicinity of Tidioute. These he leases upon liberal terms, as opportunity offers. His extensive and extending business cares absorb much of his time and attention. But he is blessed with a vigorous constitution and robust health, and bids fair to live to a ripe old age to enjoy the fruits of his industry and deserved success.

Woodburytype. A. P. R. P. Co., Phila.

Dᴿ. F. B. BREWER.

DR. F. B. BREWER.

WESTFIELD, N. Y.

FRANCIS B. BREWER was born in Keene, N. H., October 20th, 1820. His father, Ebenezer Brewer, Esq., moved in 1822, to McIndoe's Falls, Vermont, and there carried on an extensive mercantile and lumbering business until the year 1840, when he and his associates in business purchased several thousand acres of land on Oil Creek, and continued there their mercantile and lumbering operations.

The subject of this sketch is a graduate of Dartmouth College, class of 1843. He continued his professional studies in the medical department of the same institution, and completed them in the Jefferson Medical School, of Philadelphia. He began the practice of medicine in Barnet, Vermont. Soon after the establishment of the lumbering business on Oil Creek, at Titusville, he had heard with interest the accounts of a natural bituminous oil which was found in that immediate vicinity and along the valley of Oil Creek, and in conversation with persons who had returned from that region he learned that it was highly esteemed as a domestic remedy of great efficacy in several diseases, such as rheumatism, neuralgia and affections of the throat. In the year 1848, or 1849, he procured several gallons of the oil and used it with marked success in his practice. The product became an interesting study, and excited his surprise that an article so easily procured and so intrinsically valuable should be so entirely neglected.

In looking at the history of Petroleum he learned that it was found in many portions of the world, and in the East, was to some extent utilized, and entered into the commercial statistics of the countries which produced it.

In the fall of 1850, Dr. BREWER made a journey to western

Pennsylvania, and while at Titusville, where the firm of Brewer, Watson & Co. were operating largely in lumber, he visited and examined the original oil spring a few rods below what was known as the " Upper Mill," and just south of the line dividing Venango and Crawford counties, and on the spot where the Drake Well was subsequently developed.   After visiting the timber lands of the company, and their several mills, and conversing with Mr. Jonathan Watson, who was the active partner then at Titusville, Dr. BREWER determined to accept the propositions of the firm and become a partner in the business.

Although the development of the Petroleum Springs, was often discussed, nothing was really done in the matter for two or three years.   In the fall of 1852, Dr. BREWER took a quantity of the oil to Hanover, N. H., and submitted it, for chemical analysis, to Prof. O. P. Hubbard, of Dartmouth College.   He was assured by Prof. Hubbard that the product was a very valuable one, but that oil would not be found in quantities sufficient for commercial purposes.   The first proposition Dr. BREWER gladly admitted to be true ; the second, however, he felt sure was made at a venture, and did not accord with his own conclusions, based on a careful exploration of the territory and confirmed by abundant surface indications throughout the entire valley of Oil Creek.   Specimens of the oil were left with Prof. Dixie Crosby, of Hanover, and were there seen by individuals, who, after several years, became interested in the oil business.

In the summer of 1854, A. H. Crosby, Esq., son of Dr. Dixie Crosby, visited Titusville at the request of Dr. Brewer, and as the result of this visit, a proposition was made which was the first real step toward the great end now accomplished, of good and cheap light for all the world.   There was formed in New York City, a Joint-Stock Co., called " The Pennsylvania Rock Oil Company," to the organization and working, of which Dr. Brewer gave much time and attention.   On a subsequent visit to New Haven, he was introduced, to Col. E. L. Drake, who remarked that he had been solicited to take stock in an oil company, and wished Dr. BREWER to give him some information on the subject, which was cheerfully done. The stock was purchased, and Col. Drake some years after, moved to Titusville, and developed the first oil well ever bored or drilled in the rock.

From this time till 1864, Dr. BREWER gave most of his time to the oil business. The firm of Brewer, Watson & Co., owning large tracts of land, and leasing many farms, required the active labors of all their partners.

In 1864, the company sold most of their territory, and Dr. BREWER, who had removed to Westfield, Chautauqua Co., New York, organized the "First National Bank," in that village, of which he has always been the president. He is also proprietor of a large manufacturing establishment, called the "Westfield Lock Works."

During the rebellion, Dr. BREWER received the appointment of Special State Agent, with rank as Major, and spent much time in visiting the soldiers of New York, looking after their welfare, and supplying their wants when sick, and suffering in the camps and hospitals, and on the field throughout the Eastern and Southern Divisions of the army.

In the year 1867, the doctor made the tour of Europe, and during this time, he carefully examined into the progress made in the refining business, and the comparative merits of the several processes used by the European manufacturers of refined oil, as compared with the American methods of accomplishing the same results. The samples exhibited at the Paris exposition, were numerous, but the merit belonged to the American Refiners—as was clearly demonstrated by the purity of the oil, and the brilliancy of the light.

Dr. BREWER, has still some free interests in the oil regions, and occasionally visits Titusville, and the valley of Oil Creek, where he is most cordially welcomed by the old residents of the country as well as by numerous gentlemen, who have known him only through business operations. He enjoys the confidence of the people among whom he now resides, as is evinced by his occupying the position of Chairman of the County Board of Supervisors, and Member of Assembly from the First District of Chautauqua County.

Dr. BREWER is a gentleman of solid worth, and we may add, of solid wealth, as well. A man of superior education and rare literary culture, he is fitted for almost any position he may aspire to. Of robust physique, excellent presence, and attractive manner, he wins, and merits distinction in whatever circle he may move. He is enterprising with his large means, liberal in his views of public affairs, and thoroughly honest and honorable in all his private and public relations. He acts from his convictions always, never doing violence to his own sense of right and wrong. He has due respect for the opinions of others, and is proverbially reserved in the expression of his own. He is eminently a man of the people, and is universally respected for his integrity and purity of character.

Woodburytype. A. P. R. P. Co., Phila.

JOHN FERTIG.

JOHN FERTIG.

TITUSVILLE, PA.

THE subject of the following sketch may be classed among the most active and successful business men of the oil region, distinguished alike for integrity and reliability, and possessing those rare qualifications of head and heart, sure to work out success in almost any undertaking.

JOHN FERTIG, "is native and to the manor born." He comes of humble parentage. His father was one of the early settlers of Venango County, taking up a small farm near what is now known as Gas City, in 1833–4. Here he reared a large family, JOHN, being the third, of twelve children, six sons and six daughters. JOHN was born on the 17th day of March, 1837. His boyhood was spent under the parental roof. As soon as his age would permit, he was sent to the nearest school in the neighborhood, a mile or more from his father's house. The struggles and privations of the early settlers of this part of the Commonwealth of Pennsylvania, were fully realized by all whose lot was cast in the almost trackless wilderness which then, and for years afterward, surrounded them. Without entering into a detailed statement of these, we may say that luxuries of any and every name and nature, were denied them, if we except a peaceful home! They were miles and miles from civilization, and a market for their products. The pretentious settlements, "where stores and shops abound," could only be reached by paths, and roads picked through the forests, and over the rugged mountains, and bridgeless streams intervening. Schools and school-houses, were "few and far between," and these were of limited duration and of very ordinary character, so far as teachers were concerned.

It was under adverse circumstances like these, faintly pictured, that young FERTIG grew almost to manhood—working with his father upon the little farm, that at best, afforded but a bare subsistence to a large family, and attending a district school two or three months in each year, until he was sixteen years old. At this age he "hired out" to a farmer near by, at $8 per month, for one year. His wages were cheerfully handed over to his father to assist in the maintenance of the family, and young FERTIG faithfully worked out his contract.

At the age of seventeen, he gathered together his entire fortune, amounting to about $5 in cash, and with his scanty supply of clothing packed in a primitive valise, he set out for the great lumbering regions of the Susquehanna River, more than one hundred and fifty miles from his home. He performed this journey on foot and alone, reaching his destination late in the fall of 1855. Here he sought and obtained employment as "a sawyer" in one of the extensive lumbering establishments of that section. He remained here through the winter and spring of 1855–6, and when the early floods came, he "rafted" his way homeward.

He was now nineteen years old. He had seen something of the world, and began to feel the need of an education. Thus far in life his opportunities in this direction, had been few, and to him, lamentably so. He resolved to educate himself, and straightway went about the task before him—securing books and whatever he needed to attain this desired end. How well he succeeded, may be understood when we state, that a year later we find him in charge of a district school, in the vicinity of Neilltown, and subsequently at Steam Mills, in the neighborhood of West Hickory. The little red school-house, on the right of the road from Titusville, to Fagundas and Tidioute, half a mile east of Steam Mills, yet remains, and here JOHN FERTIG, taught "reading, writing, cyphering, syntax and grammar," as also the higher branches of a common school course. He taught four years during the winter, four months of each year, his wages averaging $20 per month. This sum aggregated $80 for his winter services, and constituted the entire fund at his command for the expenses of clothing, board and tuition during the remaining eight months of the year, if we except twenty-five or thirty dollars, earned in " haying and harvesting." During this four years of effort, to secure for himself an education, he attended an academic institution, located at Neilltown, Warren County, Pa., which offered very many advantages. Joseph A. Neil, Esq., a lawyer of prominence, now residing at Titusville, was, during some of the terms of this Academy attended by Mr. FERTIG, a teacher in, or a professor of the Institution.

Mr. FERTIG, spent the winter of 1859-60, in charge of " Deerfield District school No. 8," at Steam Mills, and this was the last of his teaching. He received for his services from the District, $18 per month and board. To this amount, Captain A. B. Funk, then a large lumber manufacturer and dealer at that place, and a man of proverbial benevolence and good deeds, added a like amount, from his own purse. With this addition to his ready means, Mr. FERTIG hoped to complete his higher course of studies, and looked forward to this consummation, with earnest solicitude. During the fall and winter of 1859-60, however, the oil developments of Col. DRAKE, had been attracting public attention, and it would be strange indeed, if young men of Mr. FERTIG's temperament and ambition, did not catch the inspiration of the hour. He visited the old "Drake Well," and subsequently, the " Barnsdall Well " in the same vicinity—and as soon, almost as we can write it, the earlier aspirations of his youth, were laid aside, and he determined to have an oil well ! He was without capital, save his winter's salary, but he had health, two hands and a will, and determination ample for the emergency. He obtained from his friend, Capt. A. B. FUNK, a sub-lease of five acres, on the upper McElhenny farm, then wholly undeveloped territory. The farm at this date had been purchased by Capt. Funk, and he was drilling the first hole upon it—the old " Fountain Well."

In this, his first oil well enterprise, Mr. FERTIG had two partners.

Mr. David Beatty, then of West Hickory, and since a successful operator in that locality, and now a wealthy citizen of Warren, Pa., was one of these, and Michael Gorman, then a small farmer near Steam Mills, and since become a wealthy oil-man, and removed to Ohio, was the other. The well was put under way, as soon as possible after the spring floods had subsided. The drilling was done with a spring pole, by contract, and to pay his part of the expense of putting down the well, Mr. FERTIG hired to the contractor at one dollar a day! This engagement, however, lasted only three or four days, for Mr. FERTIG, found he was "kicking" two to five feet per day, for which the contractor received $2 per foot,— one-third of which was chargeable to his own account—and he was only receiving $1 per day, for his services!

Three or four months were used up in drilling this well down to the depth at which oil was found on the Watson flats—175 to 200 feet. At the depth this well was drilled, 200 feet, no signs of oil were visible! A whole summer gone, and the little all Mr. FERTIG could call his own at the commencement, had long since been exhausted!

The well was abandoned and the territory condemned! Mr. FERTIG, who had by his three or four days' services upon his own well, learned enough of the "art" to warrant him in embarking in the business of drilling wells, supplied himself with a set of tools, and contracted to put down two or more wells at Walnut Bend, on the Allegany River, above Oil City. The receipts from these were sufficient to pay up his "assessments" on the "Fertig Well," on the Upper McElhenny. Meantime Capt. Funk, who had been at work upon "The Fountain Well," during the same summer, and who, upon reaching a depth of 200 feet, resolved to go deeper, assuming from "the surface indications" that "oil *must* be there, somewhere this side of China!" After the "spring pole" upon Capt. Funk's well gave out, a horse-power was substituted, and subsequently a small steam engine was employed to complete it to the depth of 500 feet. This was the "Fountain Well," and started off, "flowing" at

300 barrels per day. Soon after this well was struck Mr. FERTIG resumed operations upon his abandoned lease of the year previous—1860—and without detailing the trials and vexations that attended this, as all others of the early developments of that day, we may simply add, that when the well reached the third sand it commenced to "flow" at the rate of 300 barrels per day !

This was in the spring of 1861, and here we leave Mr. FERTIG in the possession and enjoyment of his good fortune, to bring up a sketch of the life of his partner, Jno. W. Hammond, Esq., of Erie, Pa., who shortly after, purchased an interest in this well, then and afterwards known as the "Fertig Well." This done, we shall trace the progress of the firm of Fertig & Hammond through the twelve years of its existence.

Mr. FERTIG is of unassuming manners and attractive address. A thorough business man, he is rarely excited, but is always "pushing" whatever demands his attention. Self-poised and self-possessed, he wins the good opinions of all with whom he comes in contact, by a manly straight-forwardness always indicative of the true gentleman. Scrupulously honorable, there is an air of manliness about him at once noticeable and attractive. Socially he is a man to cultivate. In private life he is above reproach, and in all his intercourse with his fellow-citizens he maintains an unblemished reputation. Always exhibiting a careful regard of the opinions of others, he does not obtrude his own offensively upon any. He is, however, a man of decided convictions, and acts upon them without fear or favor. In business, commercial or mercantile circles, he takes high rank, not more for his uniform reliability and promptness than for his efficiency, candor and acknowledged practical views of whatever engrosses his mind.

He cultivates a liberal estimate of men, and is generous in his dealings with all. Enterprises calculated to add to the growth and enhance the importance of the city of his adoption, have his ready approval with both purse and effort. It cannot be said, that he is prodigal in the use of his ample means, and yet he gives bountifully to charities, whether of a private or public character. He is, above and beyond all, a gentleman of high moral tone, modest in his deportment, frugal in all his business affairs, and temperate in all things.

In the acquisition of his large fortune, he has paid dollar for dollar of his indebtedness, and no man can say he has been wronged by him. His investments have been successes, perhaps beyond the common lot of men. But his possessions are the fruit of his own industry and his rigid attention to his own affairs. This has brought him to be among the largest property owners of the City of Titusville, and has made him one of her prominent, enterprising citizens. In this connection we may add, he is the owner of the "Fertig block," one of the substantial business structures of the city, located at the corners of Spring, Martin and Diamond Streets. He is half owner of the large flouring mill on Franklin Street, in the same city, and has recently built for his own use, a very elegant residence at the "East End," on Main Street. He is besides the possessor of valuable building lots, located, many of them, in the heart of the city, which, as the city expands, will become a mine of wealth to him.

In April, 1873, Mr. FERTIG became the nominee of his party, (Democratic,) for the office of Mayor—a distinction he neither sought nor declined. His opponent was the then Mayor in office, Dr. W. B. ROBERTS, one of the representative and popular men of the Republican party, and the contest was a sharp and decisive one. Mr. FERTIG was successful, his majority being nearly 500— the largest ever given to any candidate whose election had been contested.

Woodburytype. A. P. R. P. Co., Phila.

JOHN W. HAMMOND.

JOHN W. HAMMOND.

ERIE, PA.

Mr. HAMMOND is a native of Carthage, Jefferson County, New York, where he was born on the 6th day of May, 1829. He was the ninth of a family of twelve children—seven sons and five daughters. When he was but seven years old, his father, a civil engineer and surveyor, died, leaving a large family, not in affluent circumstances, to the sole care of his mother. Young HAMMOND at this early age seems to have had a realizing sense of his responsibilities, and the necessity for self-reliance and self-support, for soon after the death of his father he sought and obtained employment as a clerk, in one of the mercantile establishments of his native town, receiving for his services $5 a month, and boarding himself. He was at this time nine years old, and his opportunities for securing an education had been very limited. He had attended school, however, and had learned to read. Beyond this his acquirements were quite limited. As opportunity offered, he employed his leisure hours in learning to write, and in obtaining a knowledge of arithmetic, grammar, etc. He attended school during the winter months until he was seventeen, and continued his clerkship the remainder of the year, his salary having been advanced to $8 per month and board. At the age of seventeen he left his home in Carthage, and with less than five dollars in money in his pocket, and a scanty wardrobe, packed in a hand valise, made by Mr. ALEXANDER MORSE, then a young mechanic of Carthage, but now a resident of Titusville, and an oil producer of prominence, he made his way by stage to Utica, New York, in the spring of 1847. With the letters of recommendation he held, from his former employers and many prominent citizens, he sought and found employment in the wholesale grocery establishment of Mr. Caleb Watkins, one of the largest in the city of Utica. He remained with this house two years, steadily advancing in the confidence of his employer, and assuming more and more of the responsibilities of the increasing business.

During the fall and winter of 1848, and '49, the country from the ocean to the far western rivers, was in a feverish excitement caused by the great gold discoveries in the lately acquired territory of California. Companies were organizing in various parts of the country, destined in the early spring of 1849, for the golden shores of this new Eldorado. Young HAMMOND, then nearly twenty years of age, caught the "golden fever," and without great ado, made his arrangements to "get there" as soon as possible. In March, 1849, he left Utica, and proceeded by way of Buffalo, Pittsburgh, and Cincinnati to St. Louis, where he completed his outfit, and joined one of the many emigrating parties leaving for Fort Independence, the great plains, and the Pacific slopes. The experience of this company, numbering more than one hundred and fifty persons, has its counterpart in the scores and hundreds that followed it, for but few had preceded them—and so we omit mention of any of the numerous incidents or accidents that came and passed, as they made their way across the great sand plains, mountains, rivers and chasms, making up the 3,000 miles they were compelled to traverse to reach their journey's end.

The company reached the mining regions of Nevada City, in the early days of September, having been nearly four months in crossing the plains. They were in good spirits, and in the enjoyment of general good health. They commenced at once their mining operations, "panning" for gold, and with ordinary success. At the end of six weeks, their supplies getting low, young HAMMOND was delegated to proceed to Sacramento for a new stock. He executed his charge faithfully, and returned to the camp in due time. The late fall rains had begun, and winter was at hand. Fearing its rigors, Mr. HAMMOND and his brother, Dr. C. B. HAMMOND,

of Titusville, Pa.,—whom, we should have before said, accompanied him from the outset—resolved to return to Sacramento, to
remain until the opening of spring to prosecute their mining operations. While here they erected a building, and let it for mercantile
purposes—occupying a portion for the conveniences of their own
business, that of the sale and transit of provisions and supplies to
the mining regions, near the head-waters of the Uba river.

In the spring of 1850, they returned to the mining regions of the
American river, their party consisting of but four persons. Soon
after reaching their destination they resolved to change the current
or course of the river, the better to enable them to prosecute their
searches for the precious metal in the bed of the stream. This was
a great undertaking for *four* men; but having settled upon its
necessity, they were not long in determining to accomplish it.
Four months of labor, such as few men can endure, saw the completion of their undertaking, and the old bed of the river, bare
before them. They had worked sixteen hours a day, most of the
time, in two and three feet of water, sleeping at night in swung
hammocks, and subsisting upon food not over-abundantly nutri-
tious—but their work was accomplished, and they looked forward
to a rich harvest of golden sands and nuggets, from the great river
bed now spread out before them. But disappointment lurks
everywhere. After days of toil, "panning out" the mud, and
sand of the old river bed, they found "neither gold nor precious
stones," in paying quantities, but only the gravel, sand, and the
debris of the mountains, which were yearly cleared out by the
floods and replaced by melting ice and snow from their sides and
summits.

Late in August of the same year—1850—the mining enterprise
upon the American River was abandoned, and the party divided,
one going one way, and another another. Young HAMMOND
resolved to go to Sacramento. He owned a mule, had a small
amount of money—about $100—and so he started down the Nevada
Mountains to Sacramento, now acknowledged to be one of the

most beautiful and enterprising cities of the Golden State, and since made the Capital of California. When he reached Sacramento, the cholera was raging with fearful fatality. A day or two spent in examining the situation, convinced him of one great need to the stricken city and its suffering populace. It was ice! Promptly he set about supplying this important article. He purchased an additional mule, bought a heavy wagon, and set out for the snow-capped mountains of the Sierras—80 miles distant. Here he cut great glaciers of ice from the mountain gorges; and loading them upon his wagon—a ton or more—he hastened back to the pestilent city. When he reached Sacramento, nearly half his cargo had melted away under a hot August sun. But he had ten or twelve hundred pounds left; and this he sold in an incredible short time at *one dollar per pound!* Besides his large profits—over $1,000— he had made himself a benefactor, and all classes of people thanked —aye, blessed him for his enterprise and foresight.

This undertaking, and its wonderful results, set the whole town in a furious excitement. It was a bran new business, and large numbers quickly engaged in it. The prices of horses and mules advanced in a very few days, 25 to 30 per cent., and long trains of teams and wagons were daily seen making their way across the valley of the Sacramento River, to the perpetual snow and ice regions of the Sierra Nevada Mountains. Mr. HAMMOND, after closing out his stock of ice, was offered an almost fabulous price for his team and wagon, to which was to be added correct information as to the sources of his ice supply. He accepted the terms of sale, and quit the business $1,500 richer than when he entered the afflicted city. The ice traffic was subsequently overdone, and the commodity was sold before the season was over, at 6 cents per pound both at Sacramento and San Francisco, and points interme-diate along the Sacramento river.

The ice enterprise disposed of, Mr. HAMMOND went largely into the traffic in horses and mules. He leased a tract of land just out-side the city limits, on the great emigrant route of travel, and as they came out of the Nevada Mountains, their teams weary, worn-out and foot-sore, he purchased them at low figures, put them in condition for market, and re-sold them at a handsome profit. He continued in this business two or three months, or as long as it "paid," meanwhile, he made money. Late in October, he re-solved to make his way down the coast to San Francisco, and thence, by sea, to New York and his home.

Arriving at San Francisco without incident or accident, he soon after engaged passage in a merchant sailing vessel, that had been re-fitted and re-painted, and pressed into the passenger business, now grown to enormous proportions, both upon the Pacific and Atlantic sides of the Isthmus.* The vessel, "THE TALMA," sailed upon the advertised day, down the magnificent bay of San Francisco, out through the Golden Gate, into the broad expanse of the Pacific Ocean. For days and days they "beat against the winds," and made little progress. They ran before the wind two or three days at a time, passing Honolulu and the Sandwich Islands, but were making no less "the waste of waters" between them and their destination—the Isthmus. Discontent and bitter complaint were heard upon every hand. They should have made the voyage in ten to twelve days. They had been "bounding about upon the deep," for nearly thirty days! Mutterings of disappointment were becoming more and more audible. To add to their calamity, the ship's provisions were nearly exhausted, and their fresh water sup-ply had given out! Cholera had broken out, and numbers had died. The ship's officers would give no satisfactory replies to the reasonable inquiries of those who approached them for information.

---

* He remained at San Francisco five or six days, waiting for the vessel to sail. Meantime, he was upon the look out for his brother, Dr. C. B. HAMMOND, whom he had left in the mountains two or three months before, and had not since seen. The sequel shows that the Doctor had arrived at San Francisco about the time his brother reached there, and had engaged passage in the steamer for Panama, and that the ves-sel he sailed in lay alongside the merchantman at the same dock, and that they passed and re-passed each other daily in their visits to their respective ships, and did not meet! The Doctor reached home in due time, and John W. had an overland passage through Mexico and up the Mississippi River, to reach the same port, which we detail hereafter.

The passengers counselled together. They resolved upon a desperate expedient for their release from the perils that confronted them
—a seizure of the officers and crew with a view to compel them to
steer for the nearest port! They promptly executed their purpose,
and after securing the captain and his officers, and extorting from
one of their number a promise to accede to the wishes of their cap-
tors, they were released, and the ship headed for the port of Acu-
pulco, on the coast of Mexico. They landed there after a drifting
voyage of 44 days from San Francisco, in the rotten old hulk in
which they had been induced to take passage. The vessel was
there seized by the proper authority, and after a thorough examination, was condemned as unseaworthy, and sold at auction for the
benefit of whom it concerned.

From this point, seventeen of the passengers, among them Mr.
HAMMOND, after securing horses and suitable outfits, guns, pistols,
bowie-knives and ammunition, proceeded across the Mexican do-
main, taking in the City of Mexico in their route, to Vera Cruz.
The country, as ever, was infested with Guerilla bands, ready to rob
and murder any and all who chanced to fall into their hands—especially Americans—towards whom the people of Mexico generally
cherished very little respect or admiration,—the results of the war
their own insolence had brought upon them. The party, however reached Vera Cruz in safety, and found the steamer Alabama
ready to sail for New Orleans. Engaging passages, a few days
later they landed in the Crescent City. Mr. HAMMOND shortly
after left for his home in northern New York, which he reached in
due time, having been absent about two years. Upon counting up
the profits of his trip he found a balance in hand of a little more
than $5,000.

In the spring of 1851 he commenced business upon his own account in the city of New York, mainly dealing, in a wholesale way,
in foreign and domestic fruits, to which he added a general commission business. He continued in this lucrative trade for many
years, establishing for himself an excellent repute as a merchant

and successful tradesman of undoubted credit. He built up an extensive and profitable business, his customers and patrons being located and resident in almost every State in the Union. While enjoying this deserved prosperity, in 1857, he married the grand-daughter of his first employer, at Utica, N. Y., Mr. Caleb Watkins.

In the winter of 1860, and '61, he visited the Oil Regions, more from motives of curiosity than a desire for investment. He had heard much of the wonderful wealth of this region of country, and determined to see it for himself. He had meantime disposed of his business in New York City, and counted his gains and resources at $35,000 to $40,000. The civil war was impending, and he feared the consequences upon trade and commerce generally, and so gave up his lucrative trade, reserving the privilege of re-purchasing, at the end of twelve months. It was at this particular juncture of his own affairs, and the threatened national calamity, that he came into the Oil Region—as we have before said, with no intention either to invest or in any manner engage in this new development of nature's great riches. Once here, however, he could not resist the temptation to interest himself in the then developing wealth of the locality. He secured a lease of Mr. John Watson, on the flats below Titusville, and shortly after contracted for the drilling of four wells.

Returning to New York, he awaited anxiously reports of the progress of his enterprise. He could get no word from his contractors, and becoming impatient, he returned and superintended the work himself. He put down three wells, abandoning the fourth. This was in the spring of 1861. When the wells started up they pumped from twelve to fourteen barrels each, per day, and these were regarded as first-class producers! Oil was selling at $10 per barrel when these wells were struck, and this price was realized for some weeks after. Three months later, however, it had a dragging sale, at $1.50 per barrel! This was in consequence of the great flowing wells of Capt. Funk, and others on the lower

McElhenny Farm, two or three of which were producing largely, and the oil from these was selling at the wells at 25, 50 and 75 cents per barrel !

Mr. HAMMOND was not long in determining upon his future field of operations. His pumping wells on the Watson Flats had ceased to be remunerative. He therefore resolved to go into the " flowing well region." Prior to this time, and while oil was selling at $10 per barrel, some gentlemen at Dunkirk, N. Y., had begun the erection of a refinery, and Mr. HAMMOND secured the contract for supplying it with crude for refining. The terms of this contract were somewhat remarkable, and we deem them of sufficient interest to rehearse in this connection. The chemist employed by this refining enterprise had, from his experiments, made himself and associates believe he could realize 90 per cent. of refined oil from one hundred barrels of crude ! He had demonstrated, as he alleged, that this result was easily and practically attainable. With this scientific (?) assurance before them, the proprietors contracted with Mr. HAMMOND, for one year's supply of crude, which they bound themselves to manufacture into refined oil, giving Mr. HAMMOND nearly *one half*, or 45 barrels of every 100 distilled ! It is needless to add that Mr. HAMMOND had by far the best of the bargain.

In pursuance of this contract, Mr. HAMMOND, who had previously met Mr. John Fertig, then the owner of a large flowing well on the Upper McElhenny Farm, entered into a written agreement with him, to furnish the larger part of the oil contracted to this Dunkirk refinery, at $1.25 per barrel, *at the well!* The contract with the Dunkirk parties was subsequently greatly modified, and finally wholly surrendered by Mr. HAMMOND, after realizing a large profit. And this involved the cancelling of the contract with Mr. Fertig, which resulted in the purchase by Mr. HAMMOND of an interest owned by that gentleman in the old "Fertig Well" and lease on the Upper McElhenny Farm, and the organizing of the firm of Fertig & Hammond, a sketch of which is given herewith.

Mr. HAMMOND, is a gentleman of rare business comprehension,

and superior executive ability. His whole career, thus far, bears us out in this judgment of his character. His life has been a clearly eventful one, but he has been equal to every emergency. He is a studious man, of deep reflection, and rapid in his movements when his mark is set. He comprehends readily, and this done his work is more than half accomplished. In business circles, Mr. HAMMOND is known as an eminently practical and careful man. His investments are made after a thorough and rigid examination as to probable results, and hence it is, that success has, in the main, followed close upon all his financial transactions. In the city of his residence, he is identified with whatever adds to her growth and prosperity, devoting of his ample means, liberal sums to this end. In private life, he is universally esteemed for his social excellences, and for the warmth and steadfastness of his friendships and attachments.

As before remarked, he has a comprehensive mind, and executive ability, rarely found in men of his peculiar mould. Upright, honorable and free from conventionalisms of every character, whereby deception is fostered, he deals squarely and fairly with all men, and matters presented to him. An old proverb says, that " punctuality is the politeness of kings." Mr. HAMMOND illustrates the truth of this province of royalty, by a punctuality in all his engagements, that might be imitated with profit by others. A man of generous impulses, he gives of his abundance to all worthy charities and to the needy poor about him. He is yet in middle life, possesses a rugged constitution and unimpaired health, and bids fair, even with his great activity of brain, to live to a good old age, to enjoy the fruits of a thus far, profitable and well-spent life.

FERTIG & HAMMOND.

THE firm of FERTIG & HAMMOND was formed, or rather terms of co-partnership were entered into, soon after Mr. HAMMOND purchased an interest in the FERTIG well and lease, in the spring of 1861. The terms of this partnership were simple, and yet amply comprehensive for the men who made it. Only " HONOR AND HONESTY, THE ONE TO THE OTHER." Upon this really substantial foundation, they began their operations, purchasing interests, sinking wells, producing and refining oil, &c., &c.

During the years, 1862-3, they added four new wells to the " Fertig Well," on the same lease. These were all excellent producers, and helped to swell the cash balances of the thrifty and industrious owners and operators.

In the fall of 1861, the firm built a refinery at Erie, Pa. The material, stills, &c., for this enterprise, had been ordered for, and shipped direct to, Titusville. Reaching Erie, the then nearest Railway station to the oil fields, it was deemed to be an impossible task to remove it across the country to its destination, and Mr. HAMMOND promptly resolved to put it into operation at Erie. He set about the task before him, and in 21 days from its commencement, the refinery was in running order, and a few days subsequent, a car load of Crude Oil was refined, and shipped to an eastern consignee. The firm subsequently owned and operated another refinery at Erie, transporting their crude oil from their wells on the Fertig lease, on the Upper McElhenny farm.

In the spring of 1864, they put down one well on the widow McClintock farm, which produced for a long time, between 600 and 700 barrels per day. Jno. W. Steele, since so famous as an " oil prince," was at this time the owner of this farm, and was on and after this time, in receipt of the fabulous sums of money so generously paid, and so recklessly wasted. In 1863, the firm purchased the Young farm near Titusville, consisting of 106 acres, paying $2,800 for it. In the oil land speculations of the following year, they sold it for $30,000 ! Two years thereafter, they re-possessed themselves of it, at a Sheriff's sale, for $3,000—about its value for farming purposes !

In 1864, they purchased 22 acres, of Custer and Drake's addi-tion to Titusville. This purchase embraced what is now known as Drake street, both sides, and extending Eastward to the West bounds of lots on Kerr St., and Westerly to Martin St. The tract had been surveyed into city lots, and was soon after offered for sale. The rapid growth of this chief city of the Oil Regions, following the revulsions of 1864-5, doubled and quadrupled the value of these lots, and the profits upon this transaction alone, were very large. The last of this property, was disposed of in 1872. Before closing this part of the history of the operations of this firm, we may state, without fear of contradiction, that at least one-tenth of all the titles to city property in Titusville, bear the signatures in convey-ance, of JOHN FERTIG, and JNO. W. HAMMOND.

In the summer of 1864, they made sales of their interest in the Fertig lease, and five wells, together with the Hammond Well on the widow McClintock Farm, and some others scattered along "The Creek," receiving therefor $220,000 in cash, and a consider-able amount of the stock of " The Hammond Oil Co.," which was soon after organized with a capital of $500,000. Several hand-some dividends were made upon the earnings of this company, and its stock at one time was quoted above par.

At the close of the war, in 1865, the firm having disposed of many of their valuable oil interests, their active operations lapsed, but did not cease altogether. Mr. HAMMOND removed to Utica, N. Y., where he purchased a handsome residence and lands ad-joining, which he afterwards divided into city lots, disposing of sufficient to make his investment a source of considerable profit. Mr. FERTIG removed to Painsville, Ohio, still making Titusville his business head-quarters. Both, however, were occasionally in the region buying, selling and giving attention to their remaining interests and property.

During the memorable years of 1865–6—memorable for the disasters and revulsions that were experienced by all the enterprises of the oil region—Messrs. FERTIG & HAMMOND put down in various localities, or were interested in their drilling, including Church Run, Hyde Town, Pit Hole, Sweitzer Farm, Dawson Centre and elsewhere, *forty-five* wells, without obtaining a single barrel of oil from either! Here was an expenditure of nearly or quite $100,-000, and no return, save the establishment of the fact that the territory they had tested was barren and dry of petroleum deposits.

During the spring and early summer of 1867, they put down the " Maple Shade Well," at Pleasantville, which proved to be one of the best in that rich district. Other successful investments followed this, throughout the years 1867–8–9 and '70, which required the attention of one or both members of the firm, and in 1870, Mr. FERTIG again took up his residence in Titusville. Mr. HAMMOND in 1870, visited Europe in search of health and recreation, and soon after his return home, or in 1870, removed from Utica, N. Y., to Erie, Pa., where he now resides.

In 1869, they purchased the Dutchess Farm, at Parker's Landing. Jno. L. McKinney, of Titusville, was subsequently admitted as an equal partner in this purchase, and shortly after a small portion of it was sold for development. It proved to be superior oil property, and the development opened up an entirely new field of explorations. In the spring of 1870, they put down the first well upon what is now known as the flourishing borough of Foxburgh, and later we find them leasing, purchasing and developing territory in the St. Petersburg District, far beyond the circuit occupied by others, and with uniform success.

" The Keystone National Bank," and " The Erie Dime Savings Bank," at Erie, Pa., were organized in 1868, and Mr. HAMMOND and Mr. FERTIG became interested in both as stockholders, and both were subsequently elected directors of each of these institutions, and are now holding these responsible positions.

In 1871, they established "The Foxburgh Savings Bank," with a capital of $100,000. Mr. HAMMOND was elected President, and Mr. FERTIG, Vice-President—these gentlemen owning a majority of the stock. The St. Petersburg Savings Bank is a branch of the Foxburgh institution, and both are upon a solid basis, and are known and acknowledged to be among the most substantial in the Oil Region.

Of course we are unable to give a complete detail of all the extensive and extended operations of these gentlemen as oil producers, nor can we, in this connection, mention but a moiety of their multitude of financial operations. They are emphatically men of business. They are to-day the owners and lessees of more than four thousand acres of valuable oil tracts, located all along the Allegany and its tributaries, from Parker's Landing to Tidioute and Titusville, which under their skillful control must bring them, in years to come, wealth untold.

We said at the commencement of this brief sketch of the career of this firm, that the basis of their co-partnership was "honor and honesty, the one to the other." When they concluded to go on together, both agreed to this proposition—" *When either deceives or attempts to defraud the other, the partnership ends.*" This was the beginning and the end of stipulations between them. No others have ever been entered into, and to this day—twelve years since the firm had an existence—no WRITTEN terms of co-partnership have been made, and none exist. Their business transactions have from year to year largely increased, involving at times, thousands and thousands of dollars in capital and credit, and yet no man can say the firm of FERTIG & HAMMOND has not promptly met and discharged every obligation it has assumed. And this is also true of each, in his individual operations. This has been, and is the character they have established for themselves, and maintained through a series of years, as marked and remarkable for their changing hues of depression and success, as for the wonderful profit and prosperity brought to their doors. During their twelve years of co-partnership, no word of rebuke, censure, or complaint has passed between them—each in turn approving the operations of the other, where the firm's interest has been involved, or in any manner affected. Of course they have made some losing ventures; but the losses, whatever they may have been, have been mutually borne, and as cheerfully so as they have divided their large gains. They have maintained an unsullied and an unquestioned credit, from the outset of their business connection, by promptly meeting all their engagements, and discharging all their liabilities at maturity. In the conduct of their extensive, and steadily augmenting financial operations, involving hundreds of thousands of dollars, with individuals, with corporations, with banks, with merchants and manufacturers, they have never yet been under protest! A record like this is something to boast of, and is of itself, capital in any community, or for any enterprise, and we question if the annals of many commercial circles, near or remote, afford a parallel to the facts here briefly alluded to.

Woodburytype. A. P. R. P. Co., Phila.

Dr. W. B. ROBERTS.

DR. W. B. ROBERTS.

TITUSVILLE, PA.

WALTER BROOKS ROBERTS was born in Moreau, Saratoga County, New York, on the 15th day of May, 1823. His early years were spent under the paternal roof, his winters at a district school, and his summers in labors upon the farm. At the age of seventeen he accepted a clerkship in a banking office in the city of Albany, N. Y. He held this position but a few months, however, severe illness compelling him to return to his father's house, where he remained until his health was fully restored.

In the summer of 1841, he entered the Academy at Evans' Mills, in Jefferson County, New York, with the determination to qualify himself as a teacher. A few months later we find him in charge of a district school, in the town of Northumberland, in his native county, at a salary of $11 per month. His success in this responsible calling was marked and gratifying, not alone to young ROBERTS, but to all who knew him. The following four winters were devoted to school teaching, and the summer vacations to the study of Mathematics at the Glens' Falls Academy, and of Medicine, with Dr. Sheldon, of Glens' Falls, N. Y. Two years of subsequent study of medicine, and he turned his attention to the Dental profession, acquiring a thorough knowledge of this science in all its delicate and important branches.

During the summer of 1845, with an ample outfit for the practice of his new art, he traveled through New Hampshire, a few months, visiting Meredith, New Hampton, Holderness, &c., winning golden opinions for skill and superior knowledge of his profession. Closing up his business affairs in New Hampshire, he returned to Poughkeepsie, N. Y., to establish himself permanently in the business of Dentistry. Here misfortune overtook him, for he was seized with a violent illness—typhoid fever—and for nearly five months his life hung as if by a hair. Although among strangers, he yet found friends, who, through all the weary days, weeks and months of his helplessness, "stuck to him closer than a brother," emblazoning indelibly, upon his memory, remembrances of their kindness and devotion. Passing the relapses of his serious illness, hemorrhage of the lungs, &c., he finally regained his health and strength sufficient to return to his professional labors, and in 1849, he opened a Dental office in connection with Dr. C. H. ROBERTS, at Poughkeepsie, N. Y. Less than a year's confinement to practice, which grew rapidly upon his hands, brought on an attack of hemorrhage of the lungs, and he promptly determined upon a sea voyage for the benefit of his declining health. In February, 1850, he sailed for the West Indies, spending a month or more upon the Island of Cuba, then a great resort for invalids of a pulmonary character. Returning home by way of New Orleans, Cincinnati and Washington city in the late summer months of 1850, he soon after disposed of his interest in the Poughkeepsie Dental enterprise, and continued the practice of his profession in many of the principal towns of Dutchess County, taking healthful out-door exercise, and employing such other means as were at his disposal, for strengthening and developing a rugged manhood, since vouchsafed to him.

In 1853, Dr. ROBERTS, with a view to entering into mercantile pursuits, visited Nicaragua, Central America, and after a careful survey and examination of the country, its products, &c., decided to engage in the purchase and shipment of deer skins and cattle hides. Returning home, a company was shortly after organized, under the firm name of Churchill, Roberts, Mills & Co., of which he was one of the principal partners and business managers. The partnership formed, agents were dispatched to Grenada, with ample means for the purchase and shipment of the commodities which formed the basis of this new commercial enterprise. In a very short time a large and lucrative business was established and an unbounded credit given the firm. Its drafts, letters of credit, &c., were termed "gilt-edged," and were preferred, and commanded a readier sale in commercial circles than any house in Central America.

After the successful establishment of this Nicaragua enterprise, he again turned his attention to his profession, and in connection with his brother, Col. E. A. L. Roberts, opened an elegant, sumptuously furnished and appointed Dental office in New York City. A year later he purchased the interest of his brother, and located himself in Bond street, in the same city, where he continued his practice until some time in 1868.

The position Dr. ROBERTS held in the Dental profession was always marked and prominent. He received the First Medal awarded by the American Institute, for the best artificial teeth, a distinction which at once placed him in the front rank of men of the Science of Dentistry. He was during this period, Editor and Proprietor of THE NEW YORK DENTAL JOURNAL, a publication devoted to Dental Science and interests, and continued his editorial connection with it about four years, its columns from month to month, containing from his pen, many of the most practical and useful articles upon dentistry ever published. He was one of the leading spirits in the organization and establishment of the New York Dental College, and is at present one of its trustees.

The internal feuds of Central America had grown into a fearful civil war, destroying values, and deranging business of every character. The new firm found it necessary to close their commercial relations with that country, and to this end Dr. ROBERTS revisited Nicaragua. After months of hardship, endured in traversing swamps, mule paths and unbroken jungle fields with hair-breadth escapes from bullets, banditti and yellow fever, he succeeded in reaching the camp of the insurgents, accomplished the object of his mission, and returned home, with the ills incident to that climate fastened upon him, and these clung to him, resisting all medical appliances, for nearly a year.

In the early spring of 1863, Dr. ROBERTS was appointed by Rev. Dr. Bellows, President of the National Sanitary Commission, an agent, to visit Gen. Hunter's Division, then having its head-quarters at Beaufort, South Carolina, to examine into the condition, sanitary and otherwise of that portion of the Union Army. To complete this duty, required a month's time, and upon his return, his report was published in full in THE NEW YORK DENTAL JOURNAL, and widely copied and commended, throughout the Northern States. It contained a detailed account of his visits to the various camps of our soldiers, and was exhaustive in practical suggestions for the amelioration of their condition in all regards.

In the Spring of 1864, he was induced to subscribe to the stock of an oil company, scores of which were at this time " beating about " for patrons. This "investment" proved like many others of its cotemporaries, to be a permanent one. Dr. ROBERTS, upon visiting the Oil Region, soon after this venture, found his large tract of oil territory located miles away from developments, and where none but speculators in oil stock companies would think of looking for oil wells. This investment remains unproductive, and is likely to remain so, "as long as grass grows and water runs." His visit, however, he turned to good account, for he made a thorough examination of the then producing regions of " the Creek," and settled into the conviction that wealth untold coursed through the rocks beneath, and that with capital, business tact and mechanical skill it could be brought forth. On his return to New York City, he sought to enlist his brother, E. A. L. Roberts, in his plans for developing, by tendering him the superintendency of some wells he had resolved to put down. Col. R. declined the position, but suggested, that he had an enterprise of far more value, and if the Doctor would furnish the necessary capital to test his torpedo, he would assign him a half interest. A theoretical examination convinced Dr. R. that not only was the torpedo of vital importance to the uses intended, but he saw a mine of wealth in it, and promptly accepted the proposition of his brother, Col. Roberts, and entered heartily into his plans for testing its practicability.

In January, 1865, Col. ROBERTS went to Titusville, having previously made six torpedoes, to test their power and efficiency. Meantime, application for a patent had been made. The experiments were, in all respects, successful; and not only the utility, but the necessity for blasting oil wells to increase their production, was established. This accomplished, the Torpedo became a fixed fact, and a rich harvest awaited the enterprising discoverer. But this was not to be had without an effort. Others sought to supersede the ROBERTS' discovery, and filch from them the rewards of their genius and enterprise. Half a dozen applications were filed in the Patent Office at Washington for torpedoes for like uses, and the claims of the contestants were two years in traversing the various departments of the patent boards, and finally taken to the United States District Court, before Judge Carter, where the priority of invention was awarded to ROBERTS, as had been the result in all preceding examinations, to reach that point. Dr. ROBERTS had the management of all these litigations, acting as attorney for the patentee, and at the end of two years found himself "master of the situation." The patent was issued to his brother, and upon counting the cost of the whole effort, it was found to be less than is ordinarily paid now to an attorney as "a retainer" in an important suit.

In the spring of 1865, he organized THE ROBERTS PETROLEUM TORPEDO COMPANY. In 1866, he was elected its Secretary, and in 1867 its President, which responsible position he now holds.

In the fall of 1866, he was elected to the Common Council of the City of New York, for two years, and served his constituents faithfully and well. Although a pronounced Republican, and a candidate in a strong Democratic district, he was elected by a handsome majority. In a political minority in the Board of Councilmen, he yet very soon after becoming a member of it, assumed a prominent position among his colleagues, and before the public.

In the Council of 1867–8, he was the recognized leader of the Republican minority, and in the winter of 1868, was the candidate of his party for President of that body, the vote standing 13 Democrats to 11 Republicans.

In the summer of 1867, he visited Europe in quest of health and recreation. He visited various portions of France, Switzerland, Austria, Prussia, Holland, Belgium, England, Ireland, &c. In 1868, he removed to Titusville, surrendering his lucrative practice in New York, that he might give his undivided attention to his growing interests in the oil region. The torpedo infringements were assuming proportions, and Dr. ROBERTS reluctantly entered upon the defence of his rights and franchises in that direction, and after months and months of labor, and the expenditure of a handsome fortune, success crowned his efforts, the courts sustaining the Torpedo patent at every stage of the proceedings.

During his residence in Titusville he has, with a liberality rarely exhibited, expended thousands of dollars in the erection of substantial and elegant business blocks, which have gone far to make the city of his residence the pride of its people and the emporium of the oil regions.

In March, 1872, he became the candidate of his fellow-citizens for Mayor ; and though opposed by one of the strongest men of the opposition, after a hotly contested canvass, he was elected by a large majority.

When the "South Improvement Company" threatened the prosperity and very life of the oil-producing interest, Dr. ROBERTS was the first to raise his voice against it, and penned the first call for a public meeting. In all the subsequent struggles with this towering monopoly, he was foremost and persistent in his efforts to strangle it, even before it had a being. At the first meeting of producers, held at the Opera House, in Titusville, to devise measures to thwart the schemes of this company, Dr. ROBERTS was a leading spirit, and by his wise counsel and determined voice, did much to crush the monster monopoly. To his energy and characteristic enterprise is largely due the present flattering prospects for the early completion of the Buffalo & Titusville Railroad. When this subject was presented to him, he did not hesitate to embrace the earliest opportunity to give it his powerful aid by becoming a subscriber to its capital stock in the princely sum of $50,000. When subsequently an organization was perfected, he was unanimously chosen President of the Corporation.

On the 1st of January, 1872, he, in company with his brother, E. A. L. Roberts, and John Potter, Esq., of Meadville, L. B. Silliman, of Titusville, organized a banking firm at Titusville, under the name of ROBERTS & Co., BANKERS, Mr. JNO. PORTER, Cashier. It is among the most substantial monied institutions of the oil regions, and indeed of the State of Pennsylvania—with an ample capital, an unlimited credit, unquestioned public confidence and a business ability of the highest order, this, like all the enterprises Dr. ROBERTS puts his mind and efforts upon, is to be a bountiful success.

The ancestors of this branch of the ROBERTS family, were distinguished both in the diplomacy and in the sterner realities of war. The great-grandfather on the maternal side, Andre Everard Van Braam, was the second embassador of the Dutch East India Company to the Court of Pekin, China, and in this capacity perfected the treaty with the Chinese government, that enabled the Hollanders to hold and control the trade of that peculiar people so many years, to the exclusion of all other nations. He was also the publisher of one of the first books in the English and French languages, detailing the habits, customs and peculiarities of that wonderful people.

The great-grandfather on the paternal side was a distinguished officer in the Revolutionary War. He was a native of Wales, England, and at one period an officer in the British Army. When the mother country resolved to subdue her rebellious colonists in America, Col. Owen Roberts was a citizen of Charleston, South Carolina, and patriotically espoused the cause of his adopted country. He was, however, tendered his commission in His Majesty's service, which he promptly and indignantly declined, defiantly returning as his answer, his assurances of devotion to the land of his adoption, and an avowal of his determination to "stand by her fortunes, come weal or come woe." When hostilities began, he was commissioned a Colonel of the 4th South Carolina Artillery, and was subsequently killed at the battle of Stono, while gallantly leading his command in an effort to prevent the landing of British troops at that point. Mortally wounded by a cannon ball through one of his lower limbs, he was carried from the field and placed under the shade of a tree and out of the range of the battle, still raging. His son, Richard Brooks Roberts, grandfather of the subject of this sketch, learning of the terrible disaster to his father, hastened to his side. (See Alexander Garden's Anecdotes of the Revolution.) His father, observing the emotions of his son, said : "Take this sword, which has never been tarnished by dishonor, and never sheathe it while the liberties of your country are in danger. Accept my last blessing, and return to your duty." A short time after he breathed his last, upon the spot where his comrades had placed him. His son, Richard B. Roberts, was a youth, scarcely eighteen years of age, holding a captain's commission in his father's regiment. He faithfully and patriotically lived up to his father's dying injunctions, remaining in the service of his country until the close of the revolutionary struggle, and after, and was commissioned a Major in the regular army by General Washington. He died at the early age of 37, leaving three sons, the eldest of whom was Lucius Quintius Cincinnatus Roberts, father of Dr. W. B., and Col. E. A. L. Roberts. This name was given him in honor of the Cincinnati Society, of which he was a distinguished member, and to the privileges of which, his eldest son attained, upon his father's death.

Dr. ROBERTS is a man of marked characteristics, mentally as well as physically. In person he is about six feet in height, with a well knit, powerful frame, capable of enduring any physical effort he may undertake. Rarely, and we may add difficult to arouse, he seems always to be of an even temper, and absolutely free from mental excitement. He is, however, a profound thinker, and never discharges a subject that at any time engrosses his attention, without fully comprehending it, in all its points and bearings. Having done this he is riveted, so to speak, to his convictions and conclusions, and will defend both with consummate skill and determination. When thoroughly awakened upon any matter of personal or public concern, he does not hesitate to give the whole weight of his personal and intellectual power to the issue before him, and seldom with doubtful results.

In his private intercourse he is sociable and companionable, drawing men to him as if by magic, and retaining their friendship and confidence ever after. In business circles he is known for his reliability, and for the sacredness which he attaches to his promises and pledges. Notwithstanding the vast wealth he represents, he is a plain liver, unostentatious in his intercourse with his fellow-citizens, and frugal and temperate in all his daily walk and life. In all public enterprises promising advantage and prosperity to the city of his residence, he is foremost with voice and means to secure the coveted prize. He is generous and confiding toward personal friends, kind and benevolent to the poor, giving with a willing and bountiful hand to relieve those who have claims upon the charity of their more fortunate fellows. Dr. ROBERTS is not the man to shirk any responsibility that may fairly be put upon him, and when he sees his way clear—and he never moves till this is reached—he advances with resolute step. In business circles he is recognized for reliability and undoubted sagacity. In council he is cool, but never timid, and generally carries those who consult with him over to his own convictions.

He is in the prime of life, and in the height of his prosperity and usefulness. If life and health are spared him he is destined to fill an important place in the public esteem, confidence and service.

SAMUEL Q. BROWN.*

PLEASANTVILLE, PA.

No other modern enterprise, so plainly as the development of Petroleum, has demonstrated the fact, that endowed as he is, with the highest faculties bestowed upon the creations of an Almighty God, man still needs opportunity, before, with all his reason, with all his cultivation, with all his wonderful capabilities, he can achieve that success which will leave the impress of his career upon the world, the state, or the community—in any degree upon the generation in which he has lived, or the generations that follow.

Not overlooking the importance of preparation, or the value of experience, this is so palpably the truth, that none but a very successful man would ever deny it; and even a very successful man would only deny it in so far, as to claim that himself, had created the opportunity.

Men who succeed are prone to believe in themselves; men who fail are sure to believe in Fate. The wisest probably believe in neither. But it is very natural that a successful man should seek to enhance the merit of his success by overrating the difficulties surmounted; and it is equally natural for the unsuccessful man to excuse his failure by attributing it to a power beyond his control.

The many instances in which the plain sons of Venango County, not unlettered always, but certainly uneducated in the management of any but the scantiest finances, seized the opportunity of "that tide in the affairs of men which, taken at its flood, leads on to fortune," would seem to encourage a belief in the theory of "mute, inglorious Miltons;" for who can doubt that without the opportunity afforded by the development of Petroleum, they must have

* No Photograph furnished.

lived and died, not useless perhaps, but certainly obscure; whereas taking advantage of opportunity, many have developed into the most extensive, intelligent and successful financiers the commercial history of the country can boast. Prominent among these stands the man whose history it is the purpose of this imperfect sketch to record.

SAMUEL Q. BROWN was born at Pleasantville, Venango Co. Pa., on the 19th of September, 1835. His parents had formerly lived in the City of New York, where his father was at one time engaged in commercial pursuits. But being a man of sensitive religious convictions, and believing that a family of children would be less exposed to temptation, and could be more easily reared in the fear of the Lord, in the country, he determined to remove thither, and, together with a number of other families of some means, employed an agent to come to what was then (1833,) the "West," and, after careful inspection, select a location for the little colony.

The agent looked upon Crawford County as offering, in the greatest degree, the advantages required; but, instead of locating the farms together, he selected them several miles apart, and fixed upon a spot for Mr. BROWN in the howling wilderness several miles above Titusville.

Leaving his wife to come on with the household effects and a stock of merchandise—for it was his intention to unite with his farming, the business of a small country store—by way of the Erie Canal, then the great thoroughfare to the West, Mr. BROWN hastened across the country, via Harrisburg, to prepare in advance, the home for her reception.

When he came to view the land, he was so disappointed and disgusted with the selection of the agent that he went at once to Pleasantville, where there were already several thrifty settlers, and established a home.

When his wife joined him, they found the prospect so unpromising, that they determined to return to New York, as soon as their stock of merchandize could be disposed of, and for more than half

a year, left the most of their household goods packed, prepared to
return at the earliest possible moment. But to sell the merchandize,
he had to let the goods go on credit, and before he could make
collections the family became so attached to the place, and the
hearty hospitality which always characterizes new settlements, that
they were willing to remain, and two years after their arrival the
subject of this memoir was born.

He obtained a good common-school education, and was even
prepared for a classical course in Allegany College, at Meadville,
then in a very prosperous condition, and having some four hundred
students; but he only completed the freshman year, for feeling ill
from exposure, he lingered in a critical condition for several years,
which prevented the renewal of his studies, and after final recovery,
which was not till he had nearly attained his majority, he gave up
the notion of a collegiate education, and after a preparatory commercial course at Duff's college, in Pittsburgh, he joined his father
in the management of the store in Pleasantville.

For the place and time, his educational advantages were exceptional. Both parents had taken great interest in the cultivation of
his mind, and his natural aptitude for study was evinced by the
remarkable fact, that at the age of thirteen he entered college.

There was very little of incident in his life up to the year 1859;
nothing at all, but the careful management of a very small country
store, to fit him for that brilliant financial career, which afterward
distinguished him among the remarkably brilliant and active set of
men, brought out in that eventful period of history.

A few weeks after the discovery of petroleum, in 1859, Mr.
BROWN, in connection with Messrs. Mitchell and Rouse, obtained
control of the Buchanan farm at Rouseville, which soon, under their
management, became one of the best producing farms on the Creek.
In the year 1864, after having thoroughly developed this farm by
the sinking of nearly a hundred wells by lessees, and producing
great quantities of oil, Mr. BROWN put the property into the " Buchanan Farm Oil Company," with a capital stock of four million
dollars, realizing by the transaction himself one million dollars.

He at once established a broker's office in Philadelphia, and the following year one in New York, and dealt extensively and successfully in oil stocks (after which people were at that time crazy ;) travelling by night from one city to the other, and devoting alternate days to the business of each office.

Among other companies organized by Mr. BROWN, was "The Titus Oil Company." He also put the Rynd Farm into a stock company, in connection with several other owners of the property.

Among the most useful acts of his career was obtaining the charter of the Farmers' Railroad along Oil Creek, which, owing to the opposition of Senator Scott, he was unable for several years to accomplish. The original charter was granted for a horse-power railroad, and after it had been extended to the employment of steam-power, Mr. BROWN disposed of the charter to Messrs. Bissell, Bishop and others, who constructed the Oil Creek Railroad.

In 1862, the first pipe line was chartered by him in connection with others, but its purpose was only to conduct the oil from the wells along the Creek to Oil City, and thus obviate the expensive and disastrous system of pond-freshet conveyance by which great quantities of oil were lost.

The joints employed, however, were such a poor affair that on account of leakage the enterprise was abandoned, and Mr. BROWN afterwards sold out to a gentleman who carried the matter to perfection and covered the region with a net-work of pipes to conduct the oil to the principal shipping and storing points.

In 1866, Mr. BROWN became partner in a large mercantile house in New York, continuing his connection with the wholesale and retail branch at Pleasantville, for the Oil Region.

A few months later, he married Miss Lamb, of his native village, an estimable woman whom he had known from childhood. Shortly afterwards, he was taken down with hemorrhage of the lungs, which prostrated him for the rest of the summer, and kept him for weeks at the very point of death.

But, though a delicate person in appearance, and though all attention to a naturally fragile physique, had been overlooked in the absorbing enterprises of the last few years of his life, the spark of vitality that still remained, proved sufficient to recuperate him, and in the winter of 1867–68, he travelled in the Southern States, by which he was so benefited that, when the excitement following the discovery of oil at Pleasantville commenced, he was able to return, and take an active part in the development of his own territory, of which he held several hundred acres in the vicinity, that proved to be among the best in that field.

By this happy turn of affairs, the large fortune he had already acquired, was greatly increased, and Mr. BROWN stands now among the richest men in the oil region.

His religious convictions are very decided, and though a "successful man," his success has never lessened his reliance on the directing care of Providence. Though a Presbyterian himself, he still has encouraged the establishment of every other denomination, in his village, by liberal donations, and the Christian hospitality that so beautifully distinguishes his own charming home, is by no means restricted to so narrow a sphere. He is still in the prime of life—only thirty-seven—and the busy life through which he has already passed, will probably be crowned by many years of even more extended usefulness.

Mr. BROWN opened a banking establishment at Pleasantville, in company with Mr. Mitchell which, on a dissolution, was carried on under the firm of S. Q. Brown & ———; this firm was during the present year superceded by a stock company, called the " Pleasantville Banking Co.," with the subject of our sketch as President. Mr. Brown is the senior partner of the extensive mercantile house of Brown Bros., of Pleasantville.

Woodburytype. A. P. R. P. Co., Phila.

J. L. GRANDIN.

J. L. GRANDIN.

TIDIOUTE, PA.

IN preparing the biographical sketches which accompany and form so important a part of this work, it has been the aim of the author to select for special reference gentlemen prominently identified, first and last, with the production of Petroleum in Western Pennsylvania, and known as successful operators, or those who stand in the front rank, in one or another branch of the development of this wonderful phenomena and gift of nature. The majority of these, have been individual cases, whose various and varied enterprises have been briefly detailed, the better to enable the reader to comprehend the magnitude of the subject we are elaborating, and thus to realize, in some degree at least, the vast amount of labor and capital involved in its successful conduct. From the earliest days of the petroleum discoveries in western Pennsylvania, " the GRANDINS " at Tidioute have been advanced and advancing producers and operators, largely successful and always reliable and representative men of the Oil Region.

SAMUEL GRANDIN, the father of the three brothers GRANDIN, was a pioneer in Western Pennsylvania, having removed to Allegany Township (near Pleasantville), Venango County, from New Jersey, in 1822. His early ambition was to own lands. The old Grandin farm, within a mile of Pleasantville, which Mr. G. took up in 1822, and subsequently cleared, and successfully cultivated for many years, affords the best proof we can adduce of his industry and ability to overcome embarrassments and discouragements inseparably connected with early settlements in a new country, and that country almost a wilderness. During nearly half a century's residence in Venango and Warren Counties, Mr.

GRANDIN has maintained a high character for integrity and scrupulous honesty, and in the conduct of his large business interests, has commanded the confidence and retained the esteem of all who know him. In earlier years, his enterprise led him into heavy lumber operations; and these continued and increased with his mercantile, mechanical and lumbering operations, and always gave him a prominence and prestige, as success crowded upon him, few men in Western Pennsylvania have attained. Mr. GRANDIN, Sr., now over seventy-three years of age, is yet hale and vigorous. Some twelve years since he withdrew from active pursuits, not perhaps because of his age, but that his sons, three in number, whom he had reared and educated with care, might assume the greater portion of the responsibilities he had spent the best part of his own life in gathering upon himself.

This brief reference to the father of the GRANDIN brothers is made, because it seemed to be eminently due to him, as also to the sons, who succeeded to his responsibilities and large business cares, quadrupling them all, and who, with so much fidelity and tact, have maintained the credit and unsullied repute he established for himself through more than thirty years of active life.

JOHN LIVINGSTON GRANDIN was born in Venango County, Penn., near the village of Pleasantville, on the 20th day of December, 1836. He is the eldest of three sons living, an elder brother having been accidentally drowned, at Tidioute, twenty odd years ago. In 1839, the family removed from Pleasantville to Tidioute, where Mr. GRANDIN, Sen., had acquired by purchase, large tracts of timber lands, and here they have continued to reside for a third of a century. The subject of this sketch, here grew into boyhood, having the advantages of a Common School education. Later he was sent to the Academy at Warren, Pa., and subsequently to the Jamestown, N. Y., Union Institute, and at the age of eighteen, he entered Allegany College, at Meadville, Pa., and remained here, pursuing mathematics, and the higher branches of a classical education. When he connected himself with Allegany College, his determination was to take to the law as a profession. A few months' observation, however, among his fellow-students and class-mates convinced him that his success in life did not depend upon his becoming a lawyer. Of the class graduating, the year young GRANDIN left College, very many had fixed upon the profession of the law as a life calling. He estimated that not more than one-third of these *could* succeed, and contrasting his own chances of success with the small fraction of his class-mates, he determined to abandon his earlier resolves, and seek success in other and less crowded fields.

He left college in 1857, and returned to Tidioute, and engaged in business with his father, then largely interested in mercantile and lumbering enterprises. He readily acquired a full knowledge of his father's business, and later, assumed almost entire control of it, and was generally recognized as its responsible head. This position he has continued to occupy to this day, and has always been regarded as the master spirit of very many of the great financial, commercial and business operations of the successors of SAMUEL GRANDIN.

When in 1859, the oil developments of Col. Drake, upon the Watson Flats, became known, J. L. GRANDIN was the first to in-augurate measures for its production at Tidioute. He had known of the existence of an oil spring on Gordon Run, one of the tribu-taries of the Allegany river, at that point, from which oil had in years past, been gathered in small quantities. Within a day or two after the Drake well was struck, and while Col. Robinson, then a resident of Titusville, was rehearsing the particulars of the wonder-ful success which had attended Col. Drake's efforts, to a listening group in his father's store, young GRANDIN saddled his horse, and within an hour, was pressing negotiations for the purchase of thirty acres of the Campbell farm, upon which the oil spring was located. The terms of sale were agreed upon—30 acres at $10 per acre—a payment made, and Mr. GRANDIN returned to Tidioute.

Later, upon the same day, he visited the spring, had it thoroughly cleaned out, and immediately the oil began to rise to the surface, in small globules, and before leaving it, a pint bottle of the green fluid was obtained.

On the morning following his purchase and explorations about the spring, Mr. GRANDIN visited Mr. H. H. DENNIS, an old favorite of the Grandin family, and a man of superior mechanical ingenuity, still residing at Tidioute, then living at Dennis' Mills, near what is now known as New London. He hurriedly disclosed to him his plans, the purchase of the oil spring, and its flattering "surface indications." The Drake well and its remarkable product,—eight barrels per day, and selling at the well for 75 cents per gallon!— this and much more was discussed, and finally Mr. G. made known the object of his visit. "I think," said Mr. GRANDIN, "we can find oil in paying quantities on Gordon Run. The old oil spring gives strong proofs of its abundance there—and I have determined to put down a well, right in the centre of the spring! I have come to see if you can provide the tools and put down the well?" "Well," said Mr. Dennis, after a moment's hesitation, "I think, by jolly, I can do it, if anybody can!" A bargain was struck at once, and a contract entered into between the parties, and the following day a derrick, consisting of four twenty-feet scantling, was erected—a spring pole procured, and everything necessary for a commencement of the work of drilling, put at the disposal of Mr. Dennis as rapidly as possible. The spring hole was excavated to the rock bottom, the drilling tool "swung," and all accomplished that could be, before the setting of the sun on the THIRD day after Mr. GRANDIN's purchase!

As this was the FIRST well started at Tidioute, and probably the very next commenced after Col. Drake's well was completed, in the oil region, we have deemed its history of sufficient importance and interest, to give a detailed statement of its inception, manner of drilling, and the incidents attending its progress and ultimate failure.

Mr. Dennis, though a mechanic of remarkable skill and rare genius, had never seen an oil well, nor indeed had he ever turned his attention to rock drilling, in any regard—even to rock blasting, but he at once comprehended the undertaking, and set about supplying himself with the necessary implements for the work before him. He had seen the old style "churn drill," used for blasting, —and procuring a bar of inch and a quarter iron, three feet in length, he soon fashioned it to his needs. One end was flattened to form a cutting bit, two and a half inches in breadth, this being the diameter of the hole to be drilled. In the upper end of this iron bar or bit, he made a socket into which, as the work proceeded, he put an inch bar of round iron, tapered to fit the socket, and fastened by means of a key, and this riveted, and made perfectly straight and solidly fast. These continued additions, constituted his drilling tools, drilling jars, auger stem, &c., &c. When it became necessary, as it did ordinarily twice or thrice a day, to remove the drill, or bit, to sharpen or repair it, the rivet or key had to be cut off, and the drill removed, and thus every succeeding bit—and he afterwards made several—was operated.

In this manner, and with this rude outfit, the first well was drilled at Tidioute, and as before remarked—the SECOND well in this region was begun. Of course the process of drilling was slow, and had to be done with great care. The first break in the drill-point—half an inch or more being taken off—was deemed fatal to the enterprise. Mr. Dennis, however, tried the blunting of the point of his drilling tool, and pounding away for a day or two, finally drove the "offender" into the walls of the well. Thus, one after another of the innumerable obstacles encountered, were overcome, and the well drilled down to the depth of 134 feet! It was begun in the month of August, 1859, and spite of embarrassments and hindrances, more easily imagined than described, was completed to the depth stated, in the last days of October of the same year, or a little more than eight weeks from its commencement. During the progress of drilling, no "show" of oil or gas had been visible, but the depth was deemed ample for oil purposes. The Drake Well, was sixty-nine feet six inches deep, and this was double the depth of this first test well, and was regarded as sufficiently deep for all practical purposes.

Mr. GRANDIN, immediately ordered from a Pittsburgh manufactory, copper tubing for the well, giving its dimensions, depth, &c., as also a pump of sufficient capacity and power to draw the fluid from its great depth. The manufacturers replied, that they made no tubing of the size required, and informed Mr. GRANDIN, that his well-hole was too small by nearly one half! That it must be *four inches* in diameter, in order that tubing, and a pump of sufficient power be used to make it practicable and successful!

Here, indeed, was time, money and effort expended for naught. The well had been put down upon the best information attainable, and to appearances was valueless because of its size! Mr. GRANDIN, accompanied by Mr. Dennis, had visited the Drake Well, while drilling their own, purposely to obtain information to enable them to proceed correctly, and to learn how to remove broken bits and rock cuttings from their well, and such other facts as would aid them in accomplishing the end in view. When they reached the Drake Well, all was boarded up tight, and the entrances barred, bolted and locked ! Col. Drake himself was absent, and the wonderful " wonder," was in charge of a German fellow-citizen, who denied all access to the inner courts of the derrick, and refused utterly to give any information upon the subject! Of course Messrs. GRANDIN and DENNIS, returned to their own enterprise, as wise as they went to Col. Drake's well, and the following day Mr. Dennis called upon Mr. GRANDIN, and detailed to him his plan for enlarging the dimensions of the hole. He wanted a bar of iron two inches in diameter, and from six to eight feet in length. But where to get such a ponderous piece of metal, was the next suggestion. No hardware establishment in all that section of the country, kept iron bars of this character, and it was questionable if it could be had short of Pittsburgh. During the same day, however, Mr. Dennis' eye fell upon "just what he wanted !" It was a discarded axle, used upon a tram-railway, running out of Gordon Run, used to transport lumber to the Allegany river! "It was just the thing needed !" It was six feet in length, two inches in diameter, made of wrought iron, and would weigh nearly one hundred pounds. This was quickly transformed into the desired shape, a block of steel run through the bar, welded and riveted, four inches from one end, and the steel on either side of the bar, was flattened to a cutting edge, two inches in breadth. To the other end he attached an inch and an eighth cable, and fastened this to his spring pole, and thus began the work of enlarging the well hole. For a sand pump, he used while drilling the two and a half inch well, three feet of an inch and a half copper pipe, cut from a boiler water pipe, used at a neighboring saw-mill, fastening a leather valve at or near the lower end, so as to securely hold whatever entered it. This sand pump he continued to use success-fully, until the enlargement was completed!

Of course it was the labor of days and weeks, running into months, to "rim out" this hole, with their home-made tool, but it was ultimately accomplished. During the fall and winter of 1859, and '60, Mr. Dennis toiled on, some days, "rimming out" six inches and others as many as two feet per day. As he proceeded he discovered the necessity for "jars" in his drilling apparatus, and so constructed "a pair" to meet the demand. He made of inch and a quarter bar iron, two links, similar to the links of a log chain—two feet in length—and attached these to his "car-axle" drill and cable, and this constituted his drilling jars ! Mr. Dennis admits, cleverly, that the principle he adopted for "drill jars," has been *enlarged* as well as improved !

But the saddest part of the story of this second well, drilled for oil in the Western Pennsylvania oil fields, and which we have given with such particular detail, and for a purpose that the reader will not fail to detect, remains to be told. When the "rimming out" to the bottom of the first well had been completed, and the "car-axle " drill, had been run down for the last time, " just to make sure that the work was finished," it in some unaccountable and inexplicable manner became fastened to the rocks below, and " never saw day-light again ! " Days and days were spent in the vain effort to release it, but all to no purpose. Mr. Dennis constructed a rude torpedo out of the remaining portion of his copper boiler feed pipe, and charged this with blasting powder. After some experiments with a fuse, as to the time required to reach the top of his drilling tools, he made one or two efforts to explode it, and finally accomplished his purpose. The explosion was sensibly felt upon the surface, for Mr. Dennis says " the ground trembled like an earthquake under his feet ! " The explosion was effected, doubtless, nearer the surface than he intended or anticipated. But the effect of this " first torpedo exploded in an oil well," was a failure to release the embedded tools. Other expedients were resorted to, but all to no purpose. Each succeeding effort only served to put it beyond the mechanical skill of Mr. Dennis to recover the lost implements, and they were finally abandoned. What a curiosity that rude drilling tool would be now, and what a contrast would be discernible in them and those in use at the present day !

We may as well state here that while other wells followed this first one upon Gordon Run—one as late as 1865, no oil has ever been brought to the surface in that locality. The oil spring, which first induced Mr. GRANDIN to purchase it and the land surrounding it, is still in existence, and the gravel two feet from the surface is thoroughly impregnated with heavy petroleum. It is supposed the oil is forced up through the rocks from Dennis Run, half a mile distant, and finds an exit in this spring.

Before the completion of this well, or perhaps we should say before its abandonment, during the winter and spring of 1860, Mr. GRANDIN secured leases upon the river flats, on lands belonging to " The Tidioute and Warren Oil Company." These leases were sub-let to other parties, and in the early spring and summer of that year, two wells were put down, one of which produced eight barrels and the other forty barrels daily. Mr. GRANDIN retained a " free interest " in the property leased, and was thus made a party to the profits without great outlay. The wells were shallow, not being in any case, we believe, during the development of 1860, and '61, more than 125 to 150 feet in depth.

During the summer and fall of 1860, developments had extended across and down the river from Tidioute a mile or more. Mr. GRANDIN leased a number of farms in the direction indicated, put down a good many wells himself, and sub-leased to other parties upon various terms, but the developments were without marked success. Little or no oil was discovered, and the consequence was an innumerable number of " dry holes " rewarded the industry and enterprise of the explorers. Efforts to discover the producing oil-rock were continued by Mr. GRANDIN, with unabated zeal through 1860, '61, '62 and '63, and with moderate success. During these early years of the petroleum excitement, and while known as a prominent producer, he became largely interested in the purchase, sale and shipment of oil to the sea-board.

In 1863, Mr. GRANDIN, partly as a speculative operation, but mainly to introduce refined petroleum to western dealers and consumers, shipped 130 barrels of oil, refined, at Irvineton, Pa., upon shares, to Chicago, and went thither himself to attend to its sale. He found much difficulty in interesting dealers in its traffic. Their prejudices, the results of a total lack of knowledge of this new illuminating agent, would admit of no arguments, or proofs, or suggestions. They " had a coal oil that gave ample light," and they would not touch the new material. Mr. GRANDIN did, however, finally succeed in interesting one or two dealers in its sale—but the speculation was not a flattering one, financially. But it served to introduce the new " illuminating fluid " to the people of the west, and later, opened up one of the best markets on the continent for the sale of refined oil.

The first oil-producing well put down at Tidioute was struck in the fall of 1860. It was located upon the river-flat below Tidioute, and was drilled by Messrs. KING & FERRIS, then both residents of Titusville. The land upon which it was situated belonged to W. W. WALLACE, of Pittsburgh, and in 1860, was purchased by "THE TIDIOUTE AND WARREN OIL COMPANY." This was, probably, the first oil company organized at Tidioute, if not in the Oil Region, and its proprietorship is worthy of preservation in these pages. There were ten shares of the stock at $1,000 each, and the following gentlemen were its original owners:

Samuel Grandin, . . .	Tidioute.
Brewer, Watson & Co., . .	Titusville.
Charles Hyde, ...	Hydetown.
Robert Brown, ...	Milltown.
W. T. Neill, ...	Neilltown.
E. T. F. Vallentine, . . .	Warren.
L. L. Lowry, ...	"
C. B. Curtis, ...	"
L. D. Wetmore	"

The Company purchased the "undivided one-half" of the Wallace farm property, which consisted of about 500 acres of land, and leased the other half at "an eighth royalty." The investment proved to be a very profitable one, the shareholders receiving large sums as dividends. In 1865, Mr. GRANDIN became interested by purchase, in this valuable property, and was soon after elected Treasurer of the Company, and one of the Executive Committee. He had been identified and indirectly connected with the purchase and development of this property from the first, representing his father's interest, but became a purchaser of a portion of its stock himself, as we have stated. He holds the position of Treasurer and Manager at this writing, directing its affairs almost exclusively, the remaining members of the Committee being residents of Pittsburgh and Warren, and rarely at Tidioute. As evidence of the success with which Mr. G. has developed this property, we may add, that during his eight years' supervision of it, he has paid over to its stockholders more than $1,200,000, in dividends!

This is quite $150,000 per annum as profits upon an investment of only $10,000 made in 1860!

In 1866, Mr. GRANDIN became interested with Pierce & Neyhart in the purchase, sale and development of oil lands, building of iron tankage, and the general traffic incident to the oil trade. Later the firm added the shipment of oil to the sea-board, and elsewhere to their extensive operations. Few firms or individuals in the oil region have been as large buyers and shippers, as Pierce & Neyhart, and Mr. Adnah Neyhart, who succeeded Pierce & Neyhart. Buying upon a falling market as readily as upon an advancing one, their ample capital permitted the purchase of large quantities of oil which they were enabled to hold at various points, and ship as required. Mr. GRANDIN retired from this firm in 1868, after two years of successful operation, and Messrs. Pierce & Neyhart succeeded to the shipping traffic.

In 1866, the first Pipe Line was put into successful operation upon Dennis Run. It was owned by J. L. GRANDIN, Pierce & Neyhart, Fisher Bros., and C. W. Ellis. It extended from the river, through Dennis Run to the "New York and Dennis Run Oil Company's" lands, below Triumph. This line was in 1867, sold to the "New York and Dennis Run Oil Co.," and passed into their control at that date.

Another Pipe Line was laid in 1867, and by the same parties, with M. G. Cushing and James Parshall, added to their number. The pipe used for this second enterprise was three inch, the first being but two inch capacity. In 1870, the Fagundas developments startled the oil region from end to end. Prior to these extraordinary productions, or in the fall and winter of 1869, J. L. GRANDIN, with his brothers E. B. and W. J., and the brothers Neyhart, purchased the undivided one-half interest in the David Beaty farm, on Hickory Creek, and adjoining the Fagundas farm on the west, for $91,000. The following year, 1870, Messrs. J. L. & E. B. GRANDIN, A. Neyhart and David Beaty purchased one-half of the Hiram Scott farm at Fagundas, paying for it, with developments within a stone's throw of its boundaries, $5,000.

Shortly after this transaction, the same parties, in conjunction with Fisher Bros., purchased *five-sixths* of the Fagundas Farm, John Fagundas retaining a one-sixth interest. For this the parties paid $100,000! Subsequently Mr. A. Neyhart purchased one-half of John Fagundas' one-sixth interest, for which he paid $25,000!

During the same summer, and after developments began to prove the great value of property at Fagundas, the GRANDINS and Neyhart, and David Beaty, leased *ten* acres of the afterwards widely-known Wilkins Farm—five of which they re-leased to James McNair. Upon the remaining five acres they sank three wells, one of which, "The McQuade Well," produced from the start, 400 barrels a day! This was by far the largest well obtained upon the Fagundas belt, and brought to its fortunate owners a large amount of solid profit. Besides these returns, here briefly referred to, these gentlemen possessed valuable interests at various points in this new field, all which, or nearly all, added to the volume of wealth literally pouring into their coffers. At this writing, May, 1873, they are owners of nearly 100 producing wells, scattered over the Fagundas, Scott, Wilkins, Beaty, and contiguous Farms, the aggregate production of which is not less than 500 barrels per day.

In closing this brief mention of the Fagundas operations, we may add, by way of a summary statement, that in the purchase and development of the property above referred to, more than $600,000 have been expended upon the Fagundas and Beaty Farms alone! Of course the "right side" of the ledger will exhibit a "stretch of numerals" largely in excess of even these fabulous expenditures.

During the summer of 1868, J. L. & E. B. GRANDIN, Fisher Bros., and Adnah Neyhart, purchased the remaining interests in the Tidioute Pipe Line—these gentlemen being from the first its largest stockholders—and in 1869, with David Beaty and Jahu Hunter added to their number, began and completed a Pipe Line from Fagundas to Trunkeyville, a station on the O. C. and A. R. R. R., a mile and a half from the Fagundas oil field.

These lines of pipe were successfully operated through 1869–
'70 and '71, at which date "The Pennsylvania Transportation
Company," the most extensive Pipe Line incorporation in the oil
region, became interested in them. Messrs. Grandins & Neyhart are now equal owners with this Company in the Pipe-lines
from Fagundas to Trunkeyville, and from Fagundas to Tidioute,
and from Fagundas to Titusville. They also own, in connection
with the same Company, one-half of the Pipe-line running
from Triumph to Tidioute, which, with its extensions, reaches
to all the producing territory lying between Colorado and
Tidioute.

The storage or Iron tankage belonging to Grandins & Neyhart, and David Beaty, who is also part owner with these gentlemen in some portions of their tank investments, is larger than
that of any private firm or individual in the oil regions. At Parker's Landing, J. L. & E. B. GRANDIN, and Messrs. Neyhart and
David Beaty, have a tankage capacity of 36,000 barrels. The
tankage at Tidioute, owned by J. L. & E. B. Grandin, and Adnah
Neyhart, amounts to 75,000 barrels. Upon the Beaty Farm, the
same parties, with Mr. Beaty, have tankage for 18,000 barrels,
and 11,000 barrels of tank capacity is owned by the same parties
at Miller Farm, on Oil Creek. These gentlemen are now erecting at
Titusville for the conveniences of their own private business, a
tankage capacity of 18,000 barrels. Here are nearly 160,000 barrels of storage capacity owned and controlled by the enterprise of
these gentlemen, and it is but truth to say they have "no room to
spare." Their production and purchase of oil, month in and month
out, test the capacity of all the storage they have at their disposal.
As with their ample means, Grandins & Neyhart are enabled to conduct this branch of their business upon a thoroughly successful
basis, and at the same time keep their large capital actively employed.
They are large buyers at all times, but specially so upon a depressed
market. It is no uncommon fact for Grandins & Neyhart to pur-

chase and carry over to a higher market 150,000 barrels of oil.*

In December, 1865, Wadsworth, Baum & Co., opened a private banking·office at Tidioute. In February following, (1866,) J. L. GRANDIN associated himself with Dr. Baum, one of the original partners of Wadsworth, Baum & Co., and purchased the interests of the remaining partners, and thenceforward the enterprise was known as " Grandin & Baum's Bank." In February, 1870, after four years of financial success, this banking firm was succeeded by J. L. & W. J. Grandin, and these brothers have since conducted the banking business solely upon their own account and responsibility, under the firm name of Grandin Brothers. The institution is wholly a private enterprise, owned and managed exclusively by these two gentlemen, both of whom, it is needless to add, command the confidence of the public, and take rank among the leading and substantial bankers of the oil region and Western Pennsylvania.— The amount of capital required to conduct their extensive banking transactions is not named, nor does it seem necessary that it should be. Whenever required, their resources will supply every demand made upon them. Their deposits are very large, and their line of discounts are probably equal with any other banking institution in the oil region.

JOHN LIVINGSTON GRANDIN is a gentleman of large business calibre and experience. In commercial as in financial circles he is regarded as amply comprehensive and reliable beyond doubt or question. He is self-possessed under whatever burden his large business affairs and engagements impose, and never seems disturbed or deprived of his equanimity, either by reverses or successes. He is prompt, energetic, and thorough, grappling large transactions, involving thousands of dollars of outlay, as he would the ordinary operations of every-day life. He is not what the world

* Since the above was written, an addition of 75,000 barrels tankage capacity has been added—a portion of it in the Parker's District—a 20,000 barrel tank on the Jamestown road, below Oil City, and another at Titusville of like dimensions.

terms a bold operator, for that expression oftentimes involves suspicion of recklessness. But he is a clear-headed, far-seeing, sound financial thinker, and accepts and adopts his own convictions and theories, and acts upon them without special regard to the impressions of others. He has an active temperament, with a well balanced brain and a thoroughly cultured mind. He thinks and moves not alone for himself, but for others, and is therefore among that class of commercial and financial head-workers, whose opinion and judgment upon all important subjects of this character will always be sought after, and heeded.

In private life he is a man of rare personal worth and high-toned moral excellencies. Dignified even to reserve, he is nevertheless cordial and companionable toward all with whom he comes into contact. Limited, perhaps to a degree, in his closer friendships, he is yet generously frank and zealously devoted to those so fortunate as to win his confidence and merit.his esteem. In business circles he is irreproachably honorable and scrupulously upright. His integrity is absolutely unassailable, for he has ever regarded his word as his bond, and his bond is always quotable at par.

Such men as JOHN LIVINGSTON GRANDIN, and the firms with which his name is so prominently identified, and has been during the past eight or ten years, give character and stability to the commercial and monetary affairs of the Oil Region,—and in closing this brief resume of his own, and his associates' connection with this important branch of our national wealth and resources, we may be permitted the hope, that both he and they, may yet celebrate other triumphs and still greater successes than those already carried to their credit, which have given him and them, such flattering commendation, and public and private renown. Such men deserve success, and the popular voice is ever an approving one when they attain it.

ADNAH NEYHART.

TIDIOUTE, PENNA.

A POLISHED and, we may add, a piquant, American author has said, that "few virtues are more popular, more fascinating, and unfortunately, more rare, than pluck!" Not "pluck" of that blind, spasmodic, impulsive character usually misnamed under this head, but of a steady, quiet, invincible and persistent quality founded on neither ignorance nor miscalculation, spurred on neither by emulation, nor conceit, following out, through clearly foreseen and fully comprehended dangers, a well defined and thoroughly good purpose. Men thus endowed have elements of greatness about them that sooner or later will crop out, and in the end demand and obtain success in whatever field they may choose to labor. "Pluck," or courage under whatever disaster, is a virtue beneficent solely through its own intrinsic quality. In whatever cause displayed, it is ever a noble and an ennobling trait of human character. No eminent man, in whatever station he be found—in the learned professions, in mechanics or arts, in financial circles, or in the commercial world, has attained distinction without it. Wherever it is discovered, it is applauded, for there are few who do not sympathize with and pay it homage.

There are now and always have been, since its discovery, "plucky" men, identified with the development of petroleum oil in western Pennsylvania—men of unconquerable will, invincible determination, and unabated zeal and industry—men, who under whatever adversity or disaster, have resolved, come what may, to attain success. Among this class of men, deserving recognition and a place in the memory of the reader of these pages, we place the subject of the following sketch, ADNAH NEYHART of Tidioute.

ADNAH NEYHART.

Mr. NEYHART, is a native of Tompkins County, New York, born in the town of Lansing, a few miles from its chief business centre, Ithica, on the 20th of December, 1836. ADNAH, is the eldest of three children, two sons and a daughter. His father was a carpenter and builder, in moderate circumstances, but he early determined to give his children all the educational advantages at his disposal. ADNAH was kept at the district school, from six to eight months of each year, until he was twelve or fourteen years of age, when he was sent to reside with Rev. C. H. A. Bulkley, a Dutch Reformed clergyman, resident in an adjoining town. Here he remained four years, pursuing the higher branches of study, in the public school of the vicinity. At the end of this time he went to reside with Mr. B. G. Ferris, where he remained a year, pursuing his studies—and still, another year spent in like manner, in the family of Mr. D. T. Wood, both of his native county.

About this time—he had not yet attained his majority—he was offered a situation in the large mercantile establishment of J. W. & J. Quigg, at Ithica, and accepted it, entering their employment in 1853. The house did an extensive business in general merchandizing, and the purchase of produce. Here young NEYHART obtained a pretty thorough knowledge of the trade he was engaged in—indeed all his early business experience was obtained while in their employ. Leaving this firm in 1857, he determined to go into trade upon his own account. His cash capital amounted to $200. He secured a partner with about the same amount of means, and with him proceeded to New York—this just after the great panic of that year, 1857, and purchased a large stock of groceries—UPON CREDIT. We may add here, that it was the letters of introduction MR. NEYHART presented, and not his financial ability that enabled him to obtain all the goods "on time," that he desired. This business engagement was maintained for about one year. It proved to be a sure, but by far, too slow a mode of making money to Mr. NEYHART, who had at this early day imbibed and cultivated a speculative turn, in business affairs, and could hardly endure the retail trade of a "corner grocery store" for the gratification of his ambition. It was not the desire to become rich, that prompted him to give up this mercantile enterprise, but rather in the hope, that some more important business engagement, with larger transactions, and more of them, would be presented to him. This retail grocery enterprise we should say here, was a considerable success to the firm, but in the sale and transfer to his partner, Mr. NEYHART was, in the end, the loser of nearly all he had invested. This misfortune, for the time at least changed his business determinations, for he soon after entered the large dry goods establishment of Mr. S. H. Winton, in the same city, as a general salesman. He at once won the esteem and confidence of his employer, who was not long in discovering in his new employee, an especial adaptation to business, as well as his clear, cool judgment, in all matters pertaining to his mercantile trade. Mr. NEYHART remained in this establishment, growing in the esteem of his employer, and the public, for about a year, when Mr. Winton disposed of his business to other parties, with whom Mr. NEYHART remained still another year.

This was early in the fall of 1861. With the accumulations of his two years engagement with Mr. Winton, and his successors, he determined to try his fortunes as a speculator in one of the staples of that portion of south-western New York. Butter was at this time almost unsalable. Its quotable market value was very low. But Mr. NEYHART saw "money in it," and promptly took the risks of its purchase in large quantities. His capital was limited, but the banks gave him generous accommodations, and very soon his profits began to be tangible and substantial.

Later in the conduct of this enterprise, when the staple he dealt in became more abundant, and his capital inadequate to the demands made upon it, Mr. S. H. WINTON, his old employer, with ample means, joined him in his operations. To their butter purchases, which were largely increased under the new firm, was added the buying and shipping of dried apples, now come to be an important article in the list of Army supplies. This business association was continued, and with much profit, until the fall of 1864, just as the rebellion began to wane and give signs of failure. Values were becoming fitful and varying, the finances of the country fluctuated and were unsettled, and commercial transactions had little of permanence, and were fraught with disaster and loss—indeed all business ventures, the country over, presented an uninviting aspect. Messrs. Winton & Neyhart determined to surrender their heretofore profitable enterprise, which they did before the fall business commenced, and the partnership affairs were gradually wound up, and the business discontinued.

While this was being accomplished, Mr. NEYHART, partly from motives of curiosity, and partly as a speculative out-look, visited the oil regions, arriving at Oil City early in the fall of 1864. During his stay of two weeks, he made thorough examinations of the oil-producing localities, noting the details of operating wells, leases, etc., etc. He became fairly interested in this great industry, and resolved to give it his immediate attention and effort. Upon his return to Ithica, he, with other gentlemen of his acquaintance, formed an association or partnership, with a cash capital of $25,000, upon which to commence operations in the new Oil-dorado—Mr. NEYHART receiving and accepting the appointment of Supervising Agent. This completed, Mr. N. returned to Oil City, and thereupon began his operations. His first investment covered a few leases on Cherry Run, above Rouseville, then one of the largest producing points on "the Creek." Leases in this locality commanded a large "bonus," and invariably "one-half the oil," but the few Mr. NEYHART secured, could have been disposed of soon after he obtained them at a fabulous advance, but the association, whose agent he was, decided to hold and operate them in their own behalf, and to this end the capital of the Company was doubled, and the purchase and sale of oil lands added to the original determination. Everything promised well, and large returns were anticipated by

oil men throughout the region—indeed prosperity was upon every-
hand. In 1865, memorable for its devastating floods, its beggarly
prices of oil, and the consequent prostration of this great industry,
leaving in place of thrift and marvellous gains, destruction, ruin,
and bankruptcy, Mr. NEYHART's enterprises met the fate of
others. The leased lands and oil-well properties of the association,
whose agent he was, proved to be, in the main, valueless, and to
add to the general disaster that met him at every turn, the Company was in debt, nearly, "if not quite $30,000!" The home
office declined—perhaps *neglected*, is the better word—to respond
to Mr. NEYHART's appeals for renewed assistance, to enable him
to liquidate some pressing demands, and after a few days of ineffectual effort in this direction, he decided to assume the entire
responsibility himself! Calling upon his creditors, one after
another, he sought their best terms of adjustment, at the same time
assuring them, if they "pushed things," he would be unable to
discharge even a fraction of their just claims—but if allowed to
manage his embarrassed affairs in his own way, every dollar of the
indebtedness would be scrupulously paid. Whatever Mr. NEY-
HART required was conceded by the creditors, and the sequel shows
how thoroughly and faithfully he accomplished his ends. Every
penny of the Association's liabilities were subsequently paid by
him, amounting in all, principal and interest, to more than $30,000!

While these burdens were resting so heavily upon Mr. NEY-
HART, and he was straining every nerve to relieve himself from
them by payment, he made the acquaintance of Mr. Joshua Pierce
of Philadelphia—and together they joined their *experiences*—money
they had none! This partnership began in 1866. The first operations under this new connection were successful—one a lease and
well upon "The Tidioute and Warren Oil Company's" lands at
Tidioute, and another upon the "New York and Allegany Oil
Company's" lands, on Dennis Run. This may be said to be the
commencement of Mr. NEYHART's career as a successful oil producer and operator, and thenceforward new investments and addi-

tional oil enterprises were entered upon. Later in the summer of 1866, the firm purchased a small tract of oil land at Triumph, which proved to be productive, but, unfortunately for the owners, about the date of the completion of their wells upon this property, the price of oil had so far declined as to render its production unremunerative. During the season of 1866, oil was sold at the wells as low as $1.12½ to $1.30. Messrs. Pierce & Neyhart's daily product was quite large, and deciding not to sell at the losing prices offered, they began the erection of great iron tanks—among the first put up at Tidioute—in which to store their own product, discontinuing at the same time their developing enterprises, regarding that part of their business, as almost, if not quite a failure.

The firm determined early in 1867, to engage in the shipment of oil to Philadelphia, Baltimore, and other points, and to this end Mr. Pierce returned to the former city, and after a careful examination of the whole ground and a thorough study of the details of the undertaking, resolved to enter into it—Mr. Pierce to remain at Philadelphia to receive and dispose of the oil, and Mr. NEYHART to remain in the oil region purchasing and having charge of shipments from all points. This traffic was continued, steadily increasing in magnitude and importance, and requiring constantly augmenting capital, until the fall of 1869. A few months subsequent to the establishment of this shipping enterprise, additional capital being requisite to its successful conduct, Mr. J. L. Grandin became a silent partner with Messrs. Pierce & Neyhart, contributing sufficient of his ample means, to warrant its enterprising projectors against possible loss. In the fall of 1869, in consequence of the continual, oftentimes petty, and not unfrequently, insurmountable obstacles at the disposition of chartered transportation companies, which were constantly thrown in their way, Messrs. PIERCE & NEYHART resolved to surrender this portion of their enterprises, and this they did as promptly as possible, and the partnership was dissolved.

Before leaving this point in our sketch of Mr. NEYHART, some-

thing is due to his partner, Mr. Pierce, and justly so. He is still a
resident of Philadelphia, though wholly withdrawn from business
connections with petroleum in any of its varied branches. But
while he was connected with it, and in whatever relation, he maintained a high character for business excellencies and unsullied personal worth. He retired a few years ago with a competency, acquired by industry and the faithful discharge of every obligation.
A man of sterling integrity and rare personal attractions, he is a
man among a thousand to be admired and commended for his success.

Mr. NEYHART did not long remain idle. He immediately turned
his attention exclusively to developing new oil fields, and during
1869, '70, and '71, became, with Grandin Brothers, and David
Beaty, one of the largest producers in the Hickory, Fagundas and
Tidioute districts. Early in 1869, he began negotiations for the
purchase of a one-half interest in the Beaty farm, at Hickory, adjoining the Fagundas farm on the west. Developments in this
vicinity had unmistakably indicated this and contiguous farms as
oil territory, and Mr. N., as before remarked, commenced negotiations for the purchase of all he could obtain of the Beaty farm.
Terms were finally agreed upon, and the transfer promptly made,
and Messrs. J. L., E. B. and W. J. Grandin, with Mr. NEYHART
and his brother Alpheus, became one-half owners *in fee*, of the
David Beatty farm. When this transaction was finally consummated,
all the parties were at home, the brothers J. L. and E. B. Grandin
having returned from their California tour, but Mr. NEYHART, with
the assent and concurrence of his associates, conducted the negotiations generally, and closed the transaction. This may be said to
be true of the subsequent negotiations and purchase of five-sixths
of the Fagundas farm, together with the lease upon the Wilkins
farm.

For the one-half interest in the Beaty farm $91,000 was paid.
For five-sixths of the Fagundas farm, purchased soon after the
Beaty farm transaction, $100,000 was paid. When in 1869, to '70,

developments indicated the great value of the Fagundas farm
property, Mr. NEYHART purchased one-half of the one-sixth interest
reserved by Jno. Fagundas at the first sale, paying $25,000 for it.
This latter purchase, he made upon his own account, and holding
it for only a few months, disposed of it to Pittsburgh parties for
$50,000.

Having been largely instrumental in securing these very valuable oil properties, in conjunction with Mr. E. B. Grandin, he assumed the greater part of the responsibility of developing and
rendering them, as they most assuredly have been, a source of vast
wealth to the owners. Later in the history of this prolific oil-producing locality, Mr. NEYHART gave over the care and control of the
property into the hands of Mr. E. B. Grandin, having previously
determined to re-engage in the purchase and shipment of oil to the
sea board.

In June, 1871, Mr. NEYHART became again largely interested in
oil shipments, principally to New York City.

Throughout the season of 1871, purchases and shipments steadily increased, involving a large amount of capital—at times requiring three, four and five hundred thousand dollars. During
the closing months of 1871, purchases and shipments aggregated
90,000 barrels per month. The average of shipments per month
from all points in the region are about 450,000 barrels. Mr. NEY-
HART, may therefore at this date, be regarded as the buyer and
shipper to the sea-board, of nearly, if not quite one-quarter of the
product of the entire oil region of Western Pennsylvania. His own
ample capital, with that of his partnership connection in other oil
enterprises, promptly at his command when required, afford him facilities in this direction enjoyed or within reach of few engaged in
like transactions.

For the year 1872, negotiations were in progress for the transportation by rail of 100,000 barrels or more of crude and refined
oil, per month. The threatened "South Improvement Company"
monopoly, which alarmed the region during the last months of that

year, to a considerable extent paralyzed this contemplated enterprise, but Mr. NEYHART continued his purchases and shipments,
amounting in the aggregate to more than 100,000 barrels per
month, and requiring the employment of quite half a million of
dollars, and the handling of 1,500,000 barrels of crude and refined
oil per annum.   In the present depressed state of the oil market,
one-half the amount of capital required in 1871, and '72, is ample
for the successful conduct of this important commercial enterprise.
With oil at $4 and $5 per barrel—when shall we see this prosperity again?—$400,000 to $600,000, and even a larger sum, is
often required to meet the demands of this purchasing and shipping
traffic.

In the fall of 1871, from over-taxed energies of both body and
mind, Mr. NEYHART's health failed him, and since that date,
while he has had a general oversight and control of his great business concerns, he has yet practically withdrawn from its immediate
direction and supervision.   In the early months of 1872, he established a commission house in the city of New York for the sale of
refined oil, placing Mr. John D. Archibold, formerly of Titusville,
in full charge.   His crude oil sales, in the same city, are made
through the usual brokerage channels, Mr. Henry C. Ohlen holding the position of principal broker in the crude oil department.

This entire shipping enterprise is under the sole responsibility
and guidance of Mr. NEYHART, and is wholly independent of other
partnership interests held by him with the Grandin Brothers.   He
has an equal interest with these gentlemen in their pipe-line investments, as well as their great iron tankage capacity, scattered over
the region, from Parker's Landing to Titusville and Tidioute.
He is also an equal owner with these gentlemen in nearly all their
extensive oil land properties, located at various points throughout
the oil fields of Western Pennsylvania.

The firm of Grandins & Neyhart was organized in 1868.   Since
that date Mr. NEYHART has partaken largely of its marvellous
successes, and been prominently identified with all its more impor-

tant transactions. We need not remark upon the universally conceded financial solidity of this association of capitalists and business men, so thoroughly recognized throughout the oil region, and wherever they may be known. Suffice it to say, their contracts are greenbacks to any amount they may name, and their word is their bond to be kept to the letter. In this firm, ADNAH NEYHART stands without blot or blemish, the soul of honor, and the representative of that class of American business men, distinguished alike for superior attainments in commercial circles, and for the possession of that unconquerable will and determination to succeed facetiously termed "pluck," and which, sooner or later, hews its own way to power, and in ninety-nine cases out of a hundred, achieves the success it so richly deserves.

In person, Mr. NEYHART is tall, well-proportioned, and in perfect health has a commanding presence, and at all times a dignified manner. In his intercourse with all, he manifests a reserve which may often be mistaken for austerity or a domineering spirit. Nothing is further from his composition. He is a man of thought and reflective inclinations, yet possessing rare social excellencies, of plain, unassuming manners, and simple, unaffected tastes. These prominent characteristics are liable to be attributed to other than their true sources, and thus the true gold of his manhood be misunderstood and misapprehended. In matters of business, Mr. NEYHART says just what he means, and in as few words as will convey his real intent. There is no guile or deceptive traits in his character. Integrity and personal worth are his in an abundant degree. He is sternly honest and rigidly upright in all his life, both private and public. As a business man, he ranks among the leading, successful operators of the Oil Region, and is recognized wherever he is known as a comprehensive, clear-sighted financier, a cool, well-poised man of business, capable, mentally, of working out to successful results any mercantile, commercial or financial problem that may engage his mind or command his energies.

For a year or more, Mr. NEYHART has been compelled to withdraw from the immediate control of his large business affairs, because of impaired health. In the fall of 1872, he sought, by travel and release from business cares, to re-invigorate his over-taxed body and brain. He spent the winter of 1872–3, in traversing the great plains lying west of the Missouri river, to the Rocky Mountains, and into the pure air of Colorado. Thence he went through New Mexico and Texas to Florida and the southern states, returning to his home at Tidioute in the early part of June last, much improved, and with strong hopes of an ultimate restoration to his old-time vigor and health.

Woodburytype. A. P. R. P. Co., Phila.

E. B. GRANDIN.

E. B. GRANDIN.

TIDIOUTE, PENNA.

SUCCESS in life, is said to be a passport to popular favor. This is especially true, where success has been attained through individual effort, and without the aid or assistance of influential friends or wealthy relatives. Men do become rich and great, at one and the same stroke of fortune—at least this seems to be the public estimate in innumerable instances—and we would fain believe the public voice is not always at fault. But the man, who, through years of practical industry, and zealous attention to his business affairs, acquires a competency, and at the same time builds up and rigidly maintains for himself, a character for integrity and unsullied honesty, not only wins, but deserves the plaudits and commendations of his fellows. Among this class of leading men in the Oil Region of Western Pennsylvania, we place the subject of the following sketch.

ELIJAH BISHOP GRANDIN, is a native of Tidioute, Warren Co., Pa., where he was born on the 23d day of November, 1840. He is the youngest son of SAMUEL GRANDIN, Esq., of Tidioute, and brother of J. Livingston Grandin, a sketch of whom immediately precedes this.

As soon as his age would permit, young GRANDIN was sent to school, and kept there, summer and winter, without much interruption, until he was fifteen years old. At this age, with the assent and approval of his father and family, he left his home, and entered the mercantile establishment of S. J. Goodrich, at Warren, Pa., as a general clerk. This was in the early spring of 1856. A year's labor with Mr. Goodrich, gave him a tolerable knowledge of the business, and at the same time won for him the approbation of his

employer, for industry and strict integrity, and the good opinions
of all who knew him.   At the end of a year's engagement with
Mr. Goodrich, that gentleman disposed of his establishment, and
retired from business, and young GRANDIN returned to his home
at Tidioute, and during the winter of 1856, and '57, he attended
school.   In the spring of 1857, he again sought employment as a
clerk, and obtained such a position, including that of book-keeper,
in the mercantile establishment of Mr. Charles Hyde, at Hydetown, Pa., then one of the largest lumber manufacturers and merchants in that part of Western Pennsylvania.   A temporary change
in Mr. Hyde's mercantile operations, occurred a few months later,
and young GRANDIN was permitted to relinquish his position a
short time, and he returned to Tidioute.   A respite of three or
four months, and he again entered the employ of Mr. Hyde, now
assuming the position of confidential clerk, cashier and accountant.
From the first, he had won the confidence and esteem of Mr.
Hyde, who in turn, committed to his care and control, the most
important business interests of his large establishment, lumbering
as well as mercantile.   This confidential relationship continued for
many years, Mr. GRANDIN very generally superintending his
steadily increasing business affairs, and financial operations—selling
goods from the store, buying timber for his mills, and ordinarily
having a secondary charge of his large manufacturing and mercantile concerns—holding at the same time the closer relationship of
cashier and confidential clerk.

It was during the later years of this business engagement, that
the oil excitements, and developments of 1859, '60, '61, and '62,
were inaugurated. . Young GRANDIN, still held his responsible position, and when in 1860, and '61, Mr. Hyde began his career as
an oil producer and oil land operator, and gave his personal attention to these interests, it was part of young GRANDIN's duty to
look after and have an accountant's care of these enhancing responsibilities.   The better to accomplish this increased labor, a
division of his time and labors became necessary, alternating

between Hydetown and "the Creek." When Mr. Hyde began the drilling of wells, upon "The Hydetown Oil Company's" lands below Rouseville, young GRANDIN became a limited subscriber to its capital stock, and thenceforward, had a personal interest in the enterprise. This, with another small venture in development elsewhere upon the same farm, comprised his first oil investments, and they were successful. The interest he had purchased in "The Hydetown Oil Company," he sold before the first well was down, for $900—nearly all profit—receiving in payment therefor, three horses valued at $500, and $400 in cash.

At this date Mr. GRANDIN was known along "the Creek," and among the crowds of oil operators, daily increasing, as the agent and confidential adviser of Mr. Hyde, and was constantly approached by interested parties having oil lands, leases, etc., to dispose of, and thus the principal, who was regarded as among the substantial capitalists, and monied men of the region, was frequently appealed to through his agent (Mr. GRANDIN,) to purchase or lease this or that interest, first as matter of profit, though mainly we dare say, to assist his less fortunately provided neighbors and co-operators, in their oil enterprises. Very many good investments were thus made, a few of which Mr. GRANDIN became limitedly interested in. Some of these subsequently added to his gains, but the larger portion of the property held by him, and his associates, has never been developed.

In the early months of 1862, Mr. Hyde purchased from Dr. A. G. Egbert a one-half interest in the Widow Davidson farm, at Petroleum Centre. His oil land investments and operations now absorbed nearly all his time, and required his undivided attention. Mr. GRANDIN had been, and was at this date, his faithful, invaluable "right hand man," so to speak, and in order to show his appreciation of his services, and at the same time lessen his own cares and responsibilities, he offered Mr. GRANDIN an interest in the mercantile establishment at Hydetown, which was subsequently accepted upon the generous terms proposed. The trade and traffic of this mercantile enterprise had doubled, and quadrupled under Mr. GRANDIN'S supervision, and at the date of this co-partnership (1862,) greater facilities were required and additional room demanded for the transaction of their constantly increasing trade. A new and more commodious structure was erected to accommodate the business of the new firm, and their stock of goods and wares materially added to. Besides this latter responsibility, Mr. GRANDIN still had the care and supervision of Mr. Hyde's daily augmenting oil interests and investments—among the most important of which was the Widow Davidson—or as it is now known, the Hyde and Egbert farm purchase.

At this early day in the history of petroleum, there was little system or order in the conduct of the great oil well interests. The oil was produced and tanked—sales made, the money received, and the amount divided among the fortunate owners of interests, without delay. This, in conjunction with Dr. A. G. Egbert, was Mr. GRANDIN'S portion of the labor—see to the product of each well, sell and deliver the oil on hand, and receive pay for it—and as soon as practicable thereafter, pay over the proceeds to the rightful owners.

The mercantile firm of Hyde, Grandin & Co., (William C. Hyde, now Vice-President of the Second National Bank of Titusville, being also a member of the firm,) was dissolved at the expiration of one year's successful operations, Mr. GRANDIN purchasing the interests of the remaining partners, and thenceforward until 1865, he conducted the enterprise upon his own account. Under his proprietorship and individual control, and we may add because of his rare business tact, prosperity, deserved and positive, was vouchsafed to him.

In 1865, one-half this mercantile enterprise was re-sold to Mr. Charles Hyde, and the firm of Hyde & Grandin, continued the business until 1869, or '70, when the stock was transferred to Mr. R. D. Fletcher, of Titusville, and thereafter the establishment at Hydetown ceased to exist.

During Mr. GRANDIN's sole ownership and direction of the Hydetown mercantile enterprise, from 1862, to 1865, he became quite extensively interested in and identified with the oil developments of "the Creek." In the spring of 1863, he purchased a one-eighth interest in "The Keystone lease," located upon the Hyde & Egbert farm, and during the same summer two wells were drilled on this property—one of which produced from the start 50 to 60 barrels per day. The interest he held in this lease was disposed of before the second well was down, for $9,000—this sum being nearly all "clear profit." Prior to this transaction Mr. G. had purchased interests and secured leases upon territory at or near the junction of Pine Creek with Oil Creek, below Titusville, and these also, from prompt sales, proved to be sources of considerable profit.

This may be said to be true of very many of his subsequent ventures. He acquired interests in leases at various points along "the Creek, (and yet undeveloped,) but he made it a point to dispose of his property thus held whenever he could do so at a fair advance. This policy, uniformly practiced in all his early operations in oil property, enables him to assert that he suffered no loss upon any lease or on well interests held by him. Soon after Messrs. Hyde & Egbert obtained possession of the Davidson farm, they gave leases to parties applying for them, usually at one-half royalty. This continued only a short time, however, for the property began to develop and produce largely—and leases were finally declined altogether, the owners preferring to operate it themselves. In the spring of 1864, because of personal considerations, doubtless, Mr. GRANDIN, A. C. Kepler, William C. Hyde and Titus Ridgeway obtained a small lease upon the farm, a little removed from developments already made. This lease comprised about one acre of land, and was located upon the western boundaries of the Hyde & Egbert farm, and a few rods from Oil Creek. The lessees contracted to give the land owners a royalty of three-quarters of the oil! The usual royalty at this date, at all points

on " the Creek," was one-half. It is safe to add that no lease before
or since made, gave so large a royalty to the land owners.

The well was located by Mr. Kepler, and by him named " The
Coquet," and soon after the work of drilling began. One-half the
" working interest" in this well was owned jointly by Mr. GRAN-
DIN and Mr. Kepler, and the other half of a like interest, was held
by W. C. Hyde and Mr. Ridgeway. Each of these·" half working interests " drew one-eighth of the oil, and the other six-eighths
went to the land proprietors! A few days before the well was
" struck," or completed, Messrs. Hyde & Ridgeway sold their half
of the "working interest," for $10,000! This interest, after
changing hands once or twice, finally became the property of Dr.
M. C. Egbert, a brother of the original purchaser of the farm, who
had meantime become the possessor by purchase of a one-sixth interest *in fee*, of the entire property. At the date of the sale by
Messrs. Hyde and Ridgeway, GRANDIN and Kepler were offered
the same sum for their interest, and declined it. The well was
subsequently completed, and was pumped ten or twelve days, producing daily two or three hundred barrels. The flow of gas inter-
fered materially with its operations, and it was determined to draw
the sucker rods. This done, the well commenced to flow, and for
the first few days its product was variously estimated at 1,000 to
1200 barrels. It finally settled down to 800 barrels, and at this
rate it produced for many months. Immediately after " the Coquet " began to produce in such bountiful quantities, Messrs.
GRANDIN & KEPLER were offered large sums of money for their
one-half working interest. The price they finally put upon it, was
$150,000! Three months after, and while the well was gushing
forth its eight hundred barrel stream, they disposed of it to Mr.
Frank Allen, of New York City, the consideration being $145,-
000, or $75,500 for each one-sixteenth of the product of the well!
During Messrs. GRANDIN & KEPLER's ownership of the interest,
the prices of oil ranged at a high figure, and the first 10,000 barrels sold from the " Coquet," brought $9.00 per barrel, or $90,000
in round greenback numbers!

We have given this detailed history of "The Coquet Well," and Mr. GRANDIN's connection with it, first because of the romantic interest attaching to it, and its marvellous product, as also the great wealth it brought to its owners;—and second, because it may be regarded as the basis upon which Mr. GRANDIN laid the foundations of his later fortunes and present wealth. From this single investment, which at the commencement involved an outlay of $2,000 to $3,000, he realized from the sale of his portion of the product of the well, during his three or four months ownership of it, not less than $10,000. Add to this the sum received for the interest when sold—$72,500, and we have a total of $82,500, all but $2,500 of which may be regarded as profit! While fortune was thus showering its gifts upon him, he was the possessor of many other paying ventures, nearly all of which brought him large and remunerative gains. At the date of these transactions—1864, and '65, Mr. GRANDIN was scarcely twenty-four years of age, and yet by his energy, careful business ventures and successful operations in oil lands, leases and oil well interests, he had accumulated a handsome fortune, all without special aid or assistance from relatives or friends. His father, having ample means, upon several occasions offered to aid him in his enterprises, but he determined to stand or fall upon his own efforts, and accomplished his aims most successfully.

After the sale of the "Coquet Well" interest, Mr. GRANDIN resolved to dispose of his oil properties and interests, of whatever name and nature, as rapidly as possible, and did so, as far as he was able to.

He still held and gave such attention to his mercantile interests at Hydetown, as they required at his hands. The business had steadily increased, and his profits as a sequence, yearly grew apace. This co-partnership, as before remarked, terminated in 1870, when the stock was sold to other parties, and the establishment closed. With the exception of this mercantile enterprise, Mr. GRANDIN had little else to engage his attention for a year or more, save the remaining unsold interests of his earlier operations, and later investments made at West Pit Hole, and elsewhere, in other producing localities.

In 1867, Mr. GRANDIN again became a permanent resident of Tidioute, his ample means husbanded for immediate use. His first operations here, some months later, indicate his well-defined plans for the future. He commenced by purchasing a one-quarter interest in what was then and is now known as the "Valley Tank." Later during the same season, he largely increased his tankage investments, and as occasion offered, bought oil upon a "low market," and carried it over for better prices. There has not been a year since 1862, or '63, that the difference in prices of this great staple has not varied from one to three dollars per barrel, during each year. Not unfrequently prices have fluctuated to even wider points; but the general average, we believe, has been about one dollar per barrel, in favor of the buyer and holder for better rates. Mr. G. steadily enlarged this branch of his petroleum operations, from year to year, subsequently merging his business interests and investments in this direction, with those of his brother, J. L. Grandin, and his brother-in-law, Adnah Neyhart, who are now co-partners in portions of the extensive tankage owned by Grandins' and Neyhart, at Tidioute and vicinity, at Parker's Landing and Brady's Bend, at Miller Farm, on Oil Creek, and at Titusville. At times during every year since 1868, it has been no uncommon occurrence for Mr. GRANDIN and the firm of which he is a member, to have in store more than a hundred thousand barrels of oil, purchased upon a depressed market—and this over and above their own large production, which averaged in 1869, '70, quite fifteen hundred barrels daily, and is at this date fully five hundred barrels per day.

A single statistical statement, and we leave this branch of our subject. The tankage capacity owned by the Messrs. Grandins (J. L. & E. B.) and Mr. Neyhart is in the neighborhood of 75,000 barrels, of which E. B. GRANDIN, is one-third owner. In addition to this, the brothers Grandin, Adnah Neyhart and David Beaty own a tankage capacity, located at Parker's Landing and elsewhere in the region, of 85,000 barrels, of which E. B. GRANDIN is one-quarter owner.*

In 1867, Mr. GRANDIN purchased one-half the OIL INTEREST in the Royal Scott farm at Fagundas. This was a year or more prior to developments in that locality. This property he held until the "Venture Well" was struck in 1868, and then disposed of three-fourths of his purchase to Fisher Bros. and others, reserving a one-quarter ownership and interest to himself. This property proved to be abundantly productive, and the interest reserved in it at the commencement of developments, is still held by Mr. GRANDIN, and is a source of considerable revenue to him. While operations were in progress upon this farm in 1869–'70, Mr. G. obtained a lease upon it, and put down several wells upon his own account, all of which proved to be largely remunerative.

This imperfect sketch of Mr. GRANDIN's early and later operations as an oil producer, with an enlarged and increasing traffic in lands, leases, product, etc., has been necessarily void of detail. To follow up and particularize even a fraction of his very many ventures and investments, would require time and space we could hardly devote to it. It is, however, but "the vindication of the truth of history" to say, that from the beginning of his oil enterprises in 1860, (long before reaching his majority,) he has been prominently identified with this great national benefaction, and is to-day regarded as one of its leading influential men. The firm of Grandins & Neyhart, of which Mr. E. B. GRANDIN is a member, has a representative character, co-extensive with the history of petroleum itself. The enterprises of these gentlemen, the steady and liberal employment of their large capital in seeking new fields for development, place them in the front rank of oil producers in

---

* Mr. GRANDIN is equal owner in the Beaty farm, and holds a like interest in the Fagundas farm with his brothers, J. L. and W. J. Grandin, and Mr. A. Neyhart. There are other oil interests held by the firm, unnecessary to mention here, in which Mr. G. is an equal owner with the remaining partners.

Western Pennsylvania. Nearly all their developing operations of later years have been inaugurated by lease or purchase, in the name of one or other of the partners, and subsequently by a sale to each of the remaining partners. Thus have their interests harmonized, and all things have worked together for good. There are and have been, of course, individual ownerships of valuable leases and oil lands held by one or all the partners, at various dates during their associated operations. This is true of some recent purchases of Mr. E. B. GRANDIN. He holds by late investments, interests he deems to be valuable, and has determined to develop them himself. But very many of the larger and more important and lucrative undertakings of the firm of Grandins & Neyhart, have in later years thus had their inception—one of the parties has purchased or leased tracts of land for oil development, and the firm has immediately assumed the transaction and executed the contract to the letter. This mode and manner of conducting and augmenting their extended and constantly extending business operations, has given them a prominence as producers, buyers and sellers of petroleum, second to no firm or association of capitalists in the Pennsylvania Oil region.

Mr. GRANDIN is the owner, independent of his associates, of considerable oil property of undoubted value, which it is his determination to operate and develop in the future as his own enterprise. He is besides part owner in the extensive oil pipe lines of the firm of Grandins & Neyhart, a detailed statement of which will be found in the sketch of Mr. J. L. Grandin, in the preceding pages.

We have thus imperfectly sketched the career of ELIJAH BISHOP GRANDIN, who began life as a merchant's clerk, and who with the early discoveries of petroleum, and while yet a minor, had earned for himself an enviable position as a successful oil producer and operator. All this he accomplished by his own personal efforts. There are those who may charge this flattering record, and its attendant successes, to " good luck." Very much of it may be ascribed to this, but the major part of it, is due to superior business sagacity and clear-headed financial discernment. That he has deserved his success, none will dispute or question, and deserving and attaining ought always, as in this instance, to go hand in hand.

In person, Mr. GRANDIN is below the medium height and size, but he is a man of compact build, of nervous-lymphatic temperament, of active, well developed brain, and substantial physical power. Much of this is due to his temperate, unexceptionable habits of life. Quick in perception, his conclusions are arrived at without circumlocution, and his movements are rapid, and his aims high. He evidently enjoys the making of money, more than he does its hoarding after it is acquired. Yet he has a just estimate of its value, and disposes of it wisely and well. In his private relations as well as in his associations with neighbors and friends, he is genial and full of good-nature, with enough of the milk of human kindness to prompt him to aid the deserving, encourage the unfortunate, and lighten the burdens of the needy. In all respects, he may be said to be a generous man, liberal, without ostentation with his ample means, in every good work. Instances of his generosity bountifully dealt out to young tradesmen, and deserving industrious mechanics of his native town, and elsewhere in the oil region, are not wanting in proof of this trait of his character, many of which are rehearsed by the recipients of his favors, with gratitude and grateful remembrances.

As a business man, he ranks among the best, as he is unquestionably among the most successful, in the oil region. He is active, comprehensive, energetic, and always reliable. Honorable in the fullest sense of the word, he requires neither seals nor bonds to hold him to his engagements. His word once given, if involving even the expenditure of thousands of dollars on his part, is as faithfully adhered to and executed as if bound by forfeitures of double the sum of the original transaction. He is, in short, a thoroughly earnest and honest man, one whom the people of all grades of

society regard as above the trickery and the sharpness of mere
"money making," honoring at all times, and illustrating in every-
day life, not only the outward appearance and manners, but the genuine traits and kindly feelings of a true gentleman. He is yet a young
man—scarcely thirty-three years of age—in the full vigor of manhood and usefulness.   His past experiences and triumphs—for
such his career has been—are but indexes to other successes and
more important achievements.   To such a future we confidently
commit him, assured that he will not alone maintain his unblemished repute for integrity, probity, and high personal and commercial honor, now so happily united to rare business talent, and
private worth, but that he will add to his renown in these regards,
as the years roll by, and age, with its attendant and increasing responsibilities, creeps upon him.

Woodburytype. A. P. R. P. Co., Phila.

S. D. KARNS.

STEPHEN DUNCAN KARNS.

MR. KARNS is a native of the State of Pennsylvania, born in the county of Allegheny, some twenty miles above the city of Pittsburgh, on the Allegheny River, in what is known as the Salt Well Region. His ancestors were of Irish extraction, and emigrated to this country early in the last century. They came to Pittsburgh, and shortly after took up a large tract of land on the banks of the Allegheny River, and began to clear it up for agricultural purposes. Years after their settlement there, salt was discovered, and the lands in the vicinity were pretty generally given over to this new development— the father of the subject of this sketch entering largely into the enterprise. It was here S. DUNCAN KARNS was born, on the 21st day of September, 1843. He was the eldest of three sons. As soon as he attained a suitable age he was sent to school, and kept there steadily until he reached the age of fourteen years. He then entered the Turtle Creek Academy, located near Pittsburgh, and remained two years in that institution. Subsequently he spent some months as a student in the "Iron City Commercial College," at Pittsburgh, and when in his sixteenth year, he graduated at "Duff's Commercial College," in the same city. This briefly given record constituted all his educational advantages.

In the spring of 1859, then in his sixteenth year, he resolved to see the "western country," and set out with a party bound for Colorado and the Rocky Mountain silver region. This trip, which seems to have been undertaken more as an adventure than for profit, occupied six months of his time, but he returned to his home, with broader views of life and its responsibilities, and a better comprehension of the extent and almost limitless western boundaries of his native land.

Soon after his return from his western trip, or early in 1860–61, he was sent by his father to superintend his interests in West Virginia, where he had leased for salt, and other mineral deposits, the subsequently widely known Rathbone Farm, consisting of 800 acres of land. Upon this farm a salt well had been drilled years before, but it did not prove to be especially valuable or productive. Young KARNS was dispatched to this farm, and well, with instructions to test the old salt development for oil. The well was cleaned out, tubed, and the pump set in motion. For a few days the prospect was anything but promising. A little oil was from time to time visible upon the large volume of salt water the well afforded, but when the water had been exhausted the oil began to come in goodly quantity. The well for months after produced 40 barrels of petroleum oil per day! This may be said to have been the commencement of oil developments in West Virginia, and to Mr. Karns, Sen., and to the subject of this sketch, may be awarded the credit of its development. This was in the fall, and winter, and spring of 1860 and '61.

During the winter and spring months of 1861, civil war was threatening and impending at various points in the nation. South Carolina was the first to precipitate it, and inaugurate the strife which filled the land with armed men, and the tramp of great armies upon either side. When Sumter had been fired upon, young KARNS, then scarcely seventeen years old, hastened away from West Virginia, then the home of outspoken treason and unchecked rebellion, back to his father's house, and soon after entered the service of his country, as a private in Company C, 9th Pennsylvania Reserve Volunteers. These were three years' volunteers. The Regiment was promptly marched to the front, and early in 1862, participated in the battle of Drainsville, and in the following spring, as part and parcel of the "Army of the Potomac," under McCLELLAN, was in the Peninsula Campaign, and the advance upon Richmond. The Regiment had its full share of the seven days' retreat, and the battles of that terrible campaign, from Mechanicsville to Malvern Hill, and Harrison's Landing. Not long after these disasters, young KARNS, who had distinguished himself for bravery and gallant conduct in these fearful battles for the Union, was mustered out for promotion, by order of the Secretary of War, and upon recommendation of the Colonel of his Regiment, he was soon after commissioned a Second Lieutenant in Company I, 123d Regiment, Pennsylvania Volunteers, and subsequently was engaged in the battle of Fredericksburg, where he was wounded slightly. In this battle, the Captain of his company was seriously wounded, and the First Lieutenant killed; of the rank and file of the command, forty-three of eighty men were killed and wounded. Lieut. KARNS was soon after promoted to a First Lieutenancy, and the Captain remaining away, in consequence of his wounds, Lieut. KARNS was in command, and led his Company in the battles of Antietam and Chancellorville, and retained this position until his term of service ended, when he was mustered out, and honorably discharged. He remained at Washington for a short time, in expectation of further promotion, but it came too slow for him, and in the spring of 1864, he returned to his home, and assumed control of his father's business, continuing in the discharge of his duties for two years.

In the spring of 1866, then in his twenty-third year, he determined to visit the oil region, with a view to embarking in the business as a producer. He went to Parker's Landing, and shortly after secured his first lease upon the Fullerton Parker farm, lying on the river above and within the present limits of the borough of Parker's Landing. The lease consisted of one acre of land, for which he paid a bonus of $1,000, and "one-quarter of the oil." The well he put down himself, taking his regular "tower," and superintending the work generally. Wells in the immediate vicinity found rock-bottom for driving pipe, at about fifteen feet. He first excavated twenty-eight feet. It was almost an impossi-

bility to find men who would risk themselves in the labor of excavation, the earth being of a sort of quick-sand, or loam and rock
together, which constantly caved in, rendering it extremely hazardous to operate. Indeed, once when the excavation had been nearly
completed, Mr. KARNS, who did most of the work himself, left for
dinner, and during his absence the banks caved in, and the hole
filled to within a few feet of the surface. The " Conductor hole"
was again cleaned out by Mr. K., and the pipe driven fourteen feet
in addition to the twenty-eight feet of excavation. All this labor
accomplished, ordinarily, in a few days, required six weeks of
valuable time before the drilling began !

The well was completed in September following. The oil rock
was found at the same depth as that in neighboring wells; but as
a matter of experiment, and to settle the question of other and lower
sand rocks, Mr. KARNS put it down to the depth of 1,065 feet—
250 feet below any well before or since drilled in the immediate
vicinity of Parker's Landing No additional sand rock was found,
and the well was thereupon " tubed and tested." After pumping
three days, it commenced to produce less than a barrel of oil per
day. The pumping continued on through the fall and winter,
gradually increasing, and at the end of three months' active operations the product was about three barrels per day. In the spring
of 1867, the oil from the well was sold at $2.40 per barrel, leaving
the plucky proprietor largely indebted for expenses of the fall and
winter's operations. Far from discouragement, and with no idea
of abandoning his enterprise, as he was earnestly advised to do
upon all hands, he redoubled his efforts, and from April, 1867,
up to late in the summer of 1868, he " steadily clung to his task,"
the well meantime, gradually increasing its product up to eight, ten
and twelve barrels per day. Oil during these twelve to sixteen
months was sold at more remunerative prices, and the return from
the well was a source of revenue and profit. Later, the product lessened, and in the fall of 1869, it was a good day's work to pump
one barrel from her. At this juncture it was torpedoed, and the

product immediately increased to twenty-seven barrels per day! continuing at this standard for five or six months, when it began to fall off, and steadily went the wrong way, until it reached four barrels and a half, which is the product of the well at this writing —March, 1873.

During the life of this first well, named "Karns' Well, No. 1," it has realized to its enterprising owners, by judicious management and care, over $30,000 in profits!

We have been particular in our detailed account of this well, because it illustrates, better than any language we can employ, the determined character of the gentleman whose sketch we are endeavoring to write. Not one man in one hundred, would have regarded this well as worth more than a month's effort. Mr. KARNS, however, persuaded himself that the well could be made to pay by steady pumping, and the correctness of his impressions was fully realized in the subsequent history of its product and large profit.

In the summer of 1868, Mr. KARNS leased from the land owners, an abandoned well, belonging to, or put down by " The Miles Oil Co., of N. Y." some time in 1866. The first operators supposed they had reached the "third sand," or oil rock, and had given it up as a "dry hole." Mr. KARNS, full in the faith that the rock was still below, leased the property from the original land owners, and put the drill through the third sand, and from the start it produced twenty-four barrels per day! This well had been abandoned more than eighteen months. It was located south of, or below Parker's Landing, nearly a mile from other developments. When Mr. K. avowed his determination to resuscitate it, his neighbors and cotemporaries fairly jeered at him. Every means was resorted to, to dissuade him from his purpose, but without avail. Regardless of the universally expressed opinion that it was dry territory, he began his enterprise, cleaned out the old well, put in the drill, ran it down and through the oil rock, and had a "good show." Those who had watched the undertaking began to experience a change of opinion! When subsequently the well was tubed and tested, and proved to be one of the best of the locality, they were generous enough to applaud Mr. KARNS for his perseverance. This well continued to produce in paying quantities up to the spring of 1871, but the development had a still more important significance. It settled the question of oil deposits below Parker's Landing, and went far toward encouraging the development of the extensive oil fields soon after undertaken at Lawrenceburg, and Bear Creek, and later at Petrolia, Fairview, Millerstown, Karns City, Greece City, &c.

In the spring of 1869, Mr. KARNS secured four additional leases, of limited boundaries, upon the Fullerton Parker Farm, at Parker's Landing, and drilled a single well upon each, during that summer. All these wells were successful, averaging for many months at least 20 barrels each.

During the same season he leased the Farran Farm, lying south of Lawrenceburg, and containing fifty acres. For this lease he paid a bonus of $2,800, and " one-eighth of the oil." He put down during the summer of 1869, two wells upon this property, one of which produced eight barrels per day, and the other one barrel every twenty-four hours ! The larger portion of this farm he subsequently leased to other parties at a quarter royalty—giving him a one-eighth free interest. He has upon this farm four paying wells of his own, and has one-eighth royalty in six or eight others. This farm was for a long time very productive, paying to all interested about $9,000 per month.

Developments upon the Farran Farm and upon Bear Creek, still beyond, and which were prosecuted during the same year by others, had demonstrated to Mr. KARNS the existence of deposits still in advance of those already discovered. A thorough believer in the " Belt Theory," he caused two lines to be run—south and west,— one of which terminated upon the Esquire Campbell farm, and the other upon the Stonehouse farm, two miles or more in advance of any operations. One of these lines he denominated the " Middle Belt," and the other the " Western Belt." Having secured his lines, he promptly set about the work of obtaining leases, He at once leased thirty acres of the Esquire Campbell farm, upon his own account, and in company with Mr. C. P. Badger of Parker's Landing, obtained a lease of the entire Stonehouse farm, consisting of 240 acres. Both these tracts were leased at an "Eighth" royalty.

During the following spring and summer, 1870, he put down one well on the Campbell farm, which from the start produced thirty barrels per day. Another well, drilled during the same season upon the Stonehouse farm, started off at fifty barrels per day! Soon after the Campbell farm well was struck, he sold the lease and property to Keese, Perry & Co., of Titusville, for $20,000.

A one-third interest in the Stonehouse farm lease, owned jointly by Mr. KARNS and Mr. Badger, was, after the first well was down and producing fifty barrels per day, sold to Rev. J. D. Norton, of Brooklyn, N. Y., at $30,000. This was in the fall of 1870. In 1871, Mr. KARNS purchased the one-third interest owned by Mr. Badger, which gave him two-thirds of this very productive property—for such it has proved to be. In March, 1873, there were six producing wells upon this farm, belonging to the original lessees, netting *two hundred and thirty barrels per day!* Upon the same farm are seven other wells, upon leases at a "quarter-royalty"—one-eighth free to Messrs. Karns and Norton—all of which are producing in abundance. The two wells first mentioned in this connection, were two good Pennsylvania miles in advance of other developments, when they were struck, and remained "at the front," until the summer of 1871.

In the spring of 1871, Mr. KARNS in connection with Gibson & Brown, of Parker's Landing, purchased the Fronsinger Farm, lying directly south of the Esquire Campbell Farm, containing 80 acres. The cost of this farm was $22,000. The purchasers leased it at "a quarter-royalty," and it has been pretty thoroughly developed, and was producing in March and April last, over 400 barrels per day!

These large interests occupied Mr. KARNS' time and attention during 1870–'71, and into 1872, so entirely that he had little inclination to advance with the advancing developments then and since looming up with such wonderful results, still beyond his own field of operations. But he did not lose sight of the new Oildorado! In the spring of 1872, the Cooper Bros. had commenced a well on the McClymonds Farm—two miles in advance of other operations. At the date of Mr. KARNS' visit to this well in May, 1872, the drill had penetrated about 1,390 feet without finding the third sand, and the Cooper Bros. were well nigh disheartened. They believed their well would prove a failure, and were half-inclined to abandon it. Mr. KARNS, who had watched the progress of the work, after a careful survey of the country, and the nearest developments, settled into the conviction that the oil-bearing rock was yet below the drill. He generously offered Cooper Bros. $8,500 for one-half their well and lease, and they accepted! The condition of their enterprise, it is safe to say, made them easy victims to Mr. KARNS' liberal offer. Two days after this purchase the well was down 1,400 feet, and flowing through the casing at the rate of 100 barrels per day. The product of this well for the first ten months of its life, foots up over 20,000 barrels, and there are little signs of decrease.

Upon the conclusion of negotiations for this property, and before the well was struck, Mr. KARNS purchased one-quarter, *in fee*, of the McClymonds Farm—consisting of 200 acres—paying for it $25,000. Four months later this farm was producing 1,000 barrels of oil daily, with only a partial development!

At the same time Mr. K. purchased a lease of the Riddle Farm, lying south of the McClymonds tract—containing 200 acres—for which he paid the fabulous sum of $40,000—and one-eighth of the oil! During the summer of 1872, he put down three wells upon this farm—one of which started off at 200 barrels, another at 100, and the last, 80 barrels per day!

Three other wells upon this farm—two leased at one-quarter,

and one at three-eighths royalty, brings its daily production up to 750 barrels. Eight leases at one-quarter, and one at three-eighths royalty, are now in process of development, and when completed, will doubtless double the product of this farm, for it is pronounced by experienced operators to be the best oil territory yet developed in the Parker's Landing District. Very many leases have been disposed of upon this farm, and in every instance save *one*, a bonus of $1,000 has been paid for every *five* acres—in addition to a royalty of one-quarter of the oil!

Mr. Karns did not stop here. At the date of the above transactions, in conjunction with M. S. Adams, of Lawrenceburg, a lease was taken of the John B. Campbell farm, adjoining the property before mentioned, and containing 130 acres—at an eighth royalty. This farm has since been leased at a quarter royalty, and is being rapidly developed, and proves to be very productive. At this writing March, 1873, there are SEVEN wells upon the farm, producing in the aggregate, 800 barrels a day! Among these seven, are the famous "Salisbury," and "Grace" wells, each averaging 200 barrels per day!

Upon the heels of all these briefly mentioned transactions, Mr. K. purchased the Forcht farm, containing 100 acres, and adjoining the Story farm. This property he is developing himself, and within a few rods "The Thompson Well," which has produced for months past, 150 barrels per day.

Besides these, soon after the "Jameson Well," near what is now known as Greece City, was struck. Mr. KARNS, purchased two half acre plots in the immediate vicinity of this development, and put down a single well upon each, one of which started off at 120 barrels, and the other at 100 barrels. These have been producing without diminution for some months, and promise to hold out for months to come. Three months' product from these two wells is 10,000 barrels.

Karns City, is located upon a branch of Bear Creek, in Fairview Township, and is eligibly situated upon the Riddle &

McClymonds farms—lands owned and leased by Mr. KARNS, and contained in March, 1873, a population of quite 2,000 souls, and this a growth of less than six months.   Before the discovery of oil in June, 1872, the spot now covered with a thrifty, enterprising oil town, was devoted to agricultural purposes.

Upon the McClymonds farm was the old homestead, and a dilapidated saw-mill, and all else was the quiet of a far removed farming region.   To-day the busy hum of industry is heard upon every hand.   Mercantile establishments have multiplied.   Hotels have sprung up as if by magic, and a city of real pretensions, with all the conveniences, if not the luxuries, of older settlements, are at command.   Banks, Insurance offices, telegraph stations, Pipe Lines, and whatever makes up a " live town," are visible, and may be enjoyed by all who visit Karns City.

Upon the site of Karns City, building leases have been disposed of by Mr. K., the income from which, is already $2,500 to $3,000 annually.

This necessarily brief and rapid resume, of Mr. KARNS' oil operations, is sufficient to warrant the assertion that he is among the most prominent, if indeed, he be not the leading oil producer of the entire region of Western Pennsylvania.   He is the possessor, by lease, or by purchase, of large tracts of the best oil producing lands yet developed in the new territory, lately discovered in Butler County, and this is as yet, only partially tested.   How much of wealth in petroleum, still courses beneath the surface of his possessions, time must determine.   It is safe to anticipate for him, however, enough in this world's wealth to satisfy all ordinary demands.

Mr. KARNS, has not confined his enterprise to the production of petroleum oil alone.   His ample means have been used in various ways for the benefit of his neighbors and fellow-citizens.   In the fall of 1868, he laid the first pipe line from Parker's Landing to the railway, on the opposite side of the river.   This was for the convenience of only FOUR wells, then producing at that point—

two of which he owned. Though not an expensive enterprise, for the line was less than half a mile in length, yet he constructed it, entirely at his own expense. Months after its completion and successful operation, and when the production began to increase sufficiently to warrant it, Mr. Fullerton Parker, joined Mr. K., and the facilities were largely increased. During the summer of 1869, Parker, Thompson & Co., opened another line as a competitor to the KARNS & PARKER line. After a year or more of rivalry and competition, the two were united under the name of " The Union Pipe Line," and the conveniences of the line again, largely added to. In January, 1872, Mr. K., sold his interest in The Union Pipe Line," for $25,000.

In the fall of 1871, "The Exchange Bank of Parker's Landing," was organized, with a capital of $140,000, of which Mr. KARNS is one-quarter owner. At the first meeting of the directors he was, with entire unanimity, elected its President, and was re-elected at the succeeding annual meeting in 1872. The bank is organized under the law of the commonwealth of Pennsylvania, each stockholder being individually liable. " The Exchange Bank " has been in existence less than two years, and it already takes rank among the most substantial and successful banking institutions of the Oil Region. Among its stockholders and Board of Directors are many of the solid men of the Parker oil field—men who give character, credit and confidence to whatever bears their endorsement. The following gentlemen make up the present Board of Directors :

S. D. KARNS, Fullerton Parker, Wm. C. Mobley, R. B. Allen, James Fowler, Peter Hutchison, Wm. McKelvy, Jacob H. Walters, Capt. J. T. Stockdale.

" The Parker's Landing Bridge Company " was organized in the spring of 1872, with a capital of $100,000. To this important local enterprise and great public need, Mr. KARNS generously subscribed $55,000. Upon the organization of the Company he was elected President of the Board of Directors, a position

he fills with rare ability, and to the complete satisfaction of stockholders, directors, and the public, so much convenienced by its construction. The bridge is a wrought iron structure, having four spans
of 200 feet each, and is one of the finest specimens of mechanical
solidity, strength and beauty to be found in the State of Pennsylvania. Its cost was a little more than $100,000. It was manufac-
tured and erected by "The Canton Iron Bridge Company," of
Canton, Ohio, and is an imperishable monument to the skill and
superiority of American mechanics. To appearances, it is as solid
and immovable as the everlasting hills overlooking it, from both
sides of the river.

In 1872, Mr. K. purchased a farm of 100 acres, 25 miles
above Pittsburgh, situate on the banks of the Allegany River, and
upon both sides of the Allegany Valley Railroad. Upon this farm
he is erecting an elegant residence for his own use, which, completed,
will cost him $35,000. The farm is located in the bend of the river,
with a beautiful slope descending to the water, and commands a
view of the river for miles above and below. The residence when
ready for occupancy will be the finest country-seat west of the Allegany Mountains.

Mr. KARNS is a man of slight build, of nervous temperament,
and of prepossessing appearance. Easy in his manners, he is always self-poised and self-possessed. In his intercourse with all he
is frank, manly and cordial. He could not disguise his generous
nature by any effort he might make. Just what he is may be
learned by any and all who come within the circle of his acquaintance and friendship. As a business man he is without blemish.
His word is his bond, and either is good for any amount he may
name. He is a man of inflexible integrity and acknowledged personal
worth. Generosity, in its largest sense, is his ruling characteristic.
Those who know him best, admire him most. He is enterprising,
bold in adventure, and rapid in the execution of whatever he may
undertake. Conscious of his own power, he is restive under re-
straint, and would sooner carry any enterprise alone at whatever

cost, than be hampered by the hesitation of the timid, or be held in check by the doubting. In all things he is an emphatic man. It is "yes," or "no," with him, and as promptly as the most impatient could wish. In no regard does he lack boldness of characacter, and he has originality and discrimination largely developed. He is a man sure to be remarked and felt in any assemblage of men. To be comprehensive, we may say of him without fear of offence or contradiction, that with a goodly element of pride, he is a man of firmness amounting to combativeness, and is at all times dignified and determined. To these characteristics may be added a tenacity of purpose rarely excelled, and " a will of his own," of which he is seldom bereft or deprived.

These have been the foundations of his great success in life, and as they are prominent elements in the character of the man, it is easy to prophesy a successful future for him. Yet in the prime of life, blessed with good health, a vigorous constitution, and a mind disciplined to a clear comprehension of all enterprises commanding his attention, it is surely his province to leave upon the great industry now engaging his large means, a name and fame at once honorable and enduring.

C. D. ANGELL.

FORESTVILLE, N. Y.

CYRUS D. ANGELL is a native of the town of Hanover, Chautauqua County, New York, where he was born on the 24th day of April, 1826. Until he attained the age of sixteen, he attended the district and select school of his native town. He had, however, applied himself with so much industry to his studies that when, at the age of seventeen, he entered the academy at Fredonia, N. Y., he was so far advanced in the rudiments of an education, that he maintained his place among the leading classes of that most excellent institution, and from first to last sustained an enviable character for studiousness and application, as well as rectitude in all his intercourse with his superiors and fellow-students. Leaving the Fredonia Academy, after two years of attendance, he entered the Genesee Wesleyan Seminary, at Lima, N. Y., when, upon attaining his majority, he found himself qualified to " go out into the world and work his own way to success."

Returning to his home in Chautauqua County, he was soon after selected by his fellow-citizens for the important position of School Commissioner. This office he held until 1856, giving to its administration, abilities rarely possessed by officials of this character, and yet indispensable to their success in the delicate and responsible duties imposed upon them. Young ANGELL comprehended this fully, and in the discharge of his arduous task had but one ambition—the elevating of the standard of common schools in his native county. In this he was abundantly successful. Many of the reforms he inaugurated are still continued, and the county of Chautauqua may be said to possess to-day the very best system of common schools to be found in western New York.

Woodburytype. A. P. R. P. Co., Phila.

C. D. ANGELL.

In the fall of 1856, Mr. ANGELL entered largely into mercantile pursuits, at Forestville, Chautauqua Co., N. Y. Here, as in all other business connections, he maintained a high character for integrity, moral worth and unsullied fame. As "a country merchant" his credit was unquestioned at home and abroad. His neighbors and fellow-citizens knew him for his high character and personal worth, and his creditors, in whatever part of the country, estimated him at his true value for probity, honor and inflexible honesty. As an instance of his real character, we beg to mention that in 1856, through the treachery of friends, he lost heavily in some business transactions, forcing him to compromise with his creditors. Subsequently he recovered his financial standing, and paid every dollar of his liabilities, principal and interest, accepting from no one of his creditors an abatement of one jot or tittle of their just demands.

In the spring of 1867, Mr. ANGELL, almost empty-handed, came into the oil regions, first effecting a loan of $1,000 from a personal friend. This sum he invested in the purchase of an interest in the Central Petroleum Company's property at Petroleum Centre. The venture proved to be a considerable success, and with his profits he began his career as a producer.

Prior to his disasters of 1866, in connection with capitalists of Buffalo, among whom were Wm. G. Fargo, Esq., S. O. Barnum, Esq., and some twenty others, he had purchased Belle Island, a small tract of land in the Allegany River, twenty-five miles below Oil City, and soon after the "Belle Island Petroleum Company" was organized. His interest in this property, with all else he had, went into the hands of his creditors at the time of his failure. In 1867, he repossessed himself of it, as before stated, by paying every dollar of his indebtedness, and at once took measures for its development. He became the lessee of the Company's property, and in the early part of 1867, drilled three wells upon the Island. These proved to be very productive, all three averaging about 100 barrels each, per day. This product continued for two

or three years, and during this period Mr. ANGELL paid to the
Company as net profit upon their investment of $200,000, 358 per
cent!

While engaged in these developments, Mr. ANGELL gave much
thought to, and critical examinations of a comparatively new theory
in regard to the existence of petroleum in "belts" or deposits.
Of course this *theory* had been mooted by operators who had preceded Mr. ANGELL, but no effort had ever been made to demonstrate its correctness or falsity. From careful observation, however, he came to reject the idea that oil wells could be located by
chance. Following up these convictions, he settled into the belief
that petroleum deposits could be found in "belts," or courses, and
to this theory he confined his examinations and experiments, and
finally, by absolute tests, established its truth beyond question or
doubt. In his researches he determined upon the course of two
"belts," running in a north-easterly and south-westerly direction.
The one from Scrub Grass, on the Allegany River, to Petroleum
Centre, on Oil Creek—and the other from the St. Petersburg district, through Parker's Landing to Bear Creek and Butler County.
There are those who have that implicit confidence in this latter
"belt," now known as the "ANGELL BELT," thus far, extending
by actual developments more than thirty miles in length, and from
three to five miles in width, that they regard it as not altogether
improbable that it may be followed into the oil producing regions
of western Tennessee, a distance of hundreds of miles! The survey of these two lines was intrusted to engineers, Mr. A. usually
accompanying them in person—and after months of thorough, unremitting labor, he completed his examinations and resolved upon his
future. During the surveys of the first line, his convictions became
so deep-seated as to the correctness of his theory, that he purchased
nearly all the lands lying along his lines of survey, and at other
points, within the boundaries of his "belt," from Foster Station to
Scrub Grass, a distance of about five miles. The result of all this
toil, expense and solicitude was most flattering, for, of the great

number of wells put down upon these "belts," at least 95 per cent. have been successful, and many of them of large production, yielding abundant profit to their enterprising owners.

As early as 1868–9, he began seriously his investigations of his belt theory. At this time he had few advocates and fewer believers. He sought by argument and persuasion to interest scores of operators in the development of his property, but he could find none who had "the faith of a grain of mustard seed" in his "visionary notions," as some termed them. Subsequently, however, Messrs. Prentice & Whitney, upon the upper, or "Foster belt," and Messrs. B. B. Campbell, James M. and John A. Lambing, upon the lower, or "Parker's belt," adopted the theory of Mr. ANGELL, and entered heartily into a practical demonstration of their soundness and value. While the great mass of oil producers "on the river" and elsewhere throughout the producing regions scouted the belief that a continuous line of oil producing rock existed anywhere within the oil circuit of western Pennsylvania, these gentlemen quietly proceeded with their investigations and developments, and, as the sequel shows, their efforts were crowned with an abundant financial as well as scientific success.

There could be no "chance" or "luck" about these practical developments upon Mr. ANGELL's "belt theory." They were indeed substantial demonstrations of a sound system of reasoning on the part of their enterprising projector, and future operators will owe it to Mr. A., that he has established beyond contradiction or doubt, that petroleum oil lies in courses of more or less length and breadth, and that with proper efforts to discover these belts, the business of oil producing will be reduced to a basis involving fewer risks than in a crop of wheat or a stock of merchandise. This subject of oil in "belts" or courses, has been so exhaustively discussed and so elaborately presented to the public by the able editor of the Oil City *Derrick*, C. E. BISHOP, Esq., in a communication to the New York *Tribune*, that we deem proper to re-print it, and have obtained permission of Mr. Bishop to make such use of it:

To the Editor of the New York Tribune:

No history of success or failure in the search for oil can be utterly disinterested; and if the sagacious reader should infer from the following description of facts that somebody has oil-lands for sale, he must also admit that, even in the face of such a hypothesis, the facts are of too general value to be confined to an interested few. The history of oil developments in this country has been a chronicle of reckless risks and blind speculation. No systematic, intelligent efforts have been made to detect the sources of petroleum; hence the balance in dollars and cents is against Petroleum in its account with Trade. Science has contributed little, almost nothing, to the practical ends of the business. It is a matter of surprise that these wonderful manifestations of nature's workings have received so little attention from scientists, compared with their merits and their needs. Still more a matter of surprise is it that Trade has so neglected their investigation; while it has systematized the oil-traffic above ground wonderfully. The financial departments of the business are also perfectly organized. An oil-broker can actually handle and transfer a thousand barrels of oil as easily as he could a thousand dollars in money, depositing and checking against the former precisely as the latter. At Oil City, Titusville, and other centers there have been organized Boards of Trade, or Petroleum Exchanges, which keep their members and the trade at large well informed of all that concerns the business *above ground.* Yet none of these organizations have made any effort to collect information, analyze experiences, and make even general deductions which would .offer the driller the net results of all experiments, and thus reduce the tremendous chances against his success. If every driller had kept such a record as he might, and concerted intelligent efforts had been made to collate these records, the lines of subterranean wealth could be traced on the surface with a very considerable degree of accuracy. The result would be the saving of millions of dollars annually wasted in blind probing for oil deposits, and the reduction of the business to as safe a basis as in any pursuit. Strange that nothing like a Bureau of Oil Mining exists to this day!

Such investigation as has been made has been empirical in character. The science of oil development can hardly be said to exist, though recent indications show that it is gestating vigorously. The business of oil development owes what progress it has made to the gain-inspired efforts of practical, unscientific men. The country cannot present a body of men better fitted for practical achievement than the business men of Oildom. Next to the wonderful natural phenomena of this region, the observing spectator will be impressed with the unflaging enterprise of its men. They highly personify that quickness of perception and fertility of resources which Yankees sum up in the word "cuteness." The average ".oil man" is a practical engineer, a consummate financier, and a scientist by intuition and experiment. It is this class of men who have made all the discoveries and originated all the theories of any value yet announced as to the origin, location, and probable future of oil deposits. And they are the ones who will have to construct the new science of oil mining. That such a science is possible—nay, that its truth is near at hand—I propose to show by detailing the experiments of an operator.

Preliminary to the narration, a few well-established facts in oil phenomena should perhaps be laid before the general reader. (1.) Operators and scientists are pretty well agreed on the conclusion that oil exists both in reservoirs or basins of considerable area, and in belts or channels of considerable extent. Whether the two forms of territory have any connection; the distinctions between them; the continuity of the belts; these are unsettled problems. The facts I am about to relate may tend toward their solution. The basins were first discovered, and, it is probable, have been in a large measure exhausted. It was they that supplied the leviathan spouting wells of former days (1860–'65). Such lakes were found at Pit Hole, Tarr Farm, and other points. (2.) Present developments are on extended lines that constitute belts of oil territory; they indicate currents, or stagnant channels, or elongated basins of oil. The ablest thought of the petroleum world is now turned toward discovering the location, direction, and extent of these belts. (3.) Because the first oil was found on the margin of Oil Creek, it was supposed the subterranean oil-courses corresponded in some degree with the external water-courses. Because of this impression, traces of which still exist in the minds of oil men, and because the streams offered the best, sometimes the only means of transit in this rough country, the majority of drillers have chosen locations on streams. Nearly the entire water front of this section is perforated with wells. It is safe to say that not one in twenty of these wells, to-day, by production, vindicates the theory on which their locations were selected. (4.) The oil-bearing rock is the third stratum of sand-rock in order of downward progress. The sand-rock is a conglomerate of flinty pebbles and sandstone. The pebbles in the third, or oil-bearing sand-rock, are of varying degrees of hardness and whiteness, these variations always indicating unerringly the richness of the rock in oil product. The first triturations of the drill brought up from the "third sand" are eagerly scanned by the operator as an index of his success. A handful of "third sand" is the horoscope of the new well. Old drillers can read the value of a well in these specimens. A good geologist might, I presume, collect specimens which drillers and oil companies are now-a-days more in the habit of preserving than formerly, together with the records and information drillers could give, and, by their aid, trace oil lines with some degree of certainty.

This is just what one man, though not a scientist, did accomplish. This brings us to our story, to comprehend which the reader will need to bear in mind the general facts above given. About three years ago Mr. Cyrus D. Angell of Forestville, N. Y., embarked on the petroleum sea. He enjoyed a large degree of *luck*—for it was that more than any thing else that gave his first ventures success. He soon commenced studying the problem that agitates all shrewd "greasers," viz., the location of the oil belt. He began to collect data bearing on this problem, and to study them by day and night. He conceived and assumed as real this postulate: Exactly corresponding geological data, or two or more points of oil development, are proof of a continuous belt between those points. Or, conversely, an oil belt will manifest the same characteristics everywhere. Taking a series of facts regarding paying wells in one locality, he imagined that, if he could find exactly the same series of facts manifest in another paying locality, he would have two bearings on the same oil belt; and at any place between these two points he could be certain of the same practical results as had followed drilling at the extremes, viz., oil in paying quantities.

The data for which he must find duplicates were these: (1.) The depth from the surface to the different sand rocks. (2.) Distance between the upper surfaces of the different sand-rocks. (3.) Thickness of the sand-rocks. (4.) Quantity and quality of the oil indications (called "shows,") found in the *second* sand. (5.)Color and gravity of the oil produced by the third sand. (6.) Texture and temper of the third sand. I have named these indications in the order of their importance as a basis of general opinion as to territory —of course he would look first and most anxiously for those last named, but as these are all fixed scientific data, it is probable that, for purposes of comparing different wells, all are equally important and decisive. It will be noticed that none of these points of inquiry relate to the more superficial and demonstrative manifestations; they are mainly geological, permanent facts.

Mr. Angell started with the data pertaining to his own wells on Belle Island—a little strip of land in the Allegany River just below Scrubgrass Station, on the Allegany Valley Railway, and about 25 miles, by the river, below Oil City. On this island, and also on the river bank above (north of) it, was a small but rich oil field; two thick clusters of wells here had for some years been producing. Making himself master of the whole subterranean history of this development, as told by the log-books of the drillers, the specimens of sand and other rocks, and his own observation while making wells on Belle Island, he turned his steps to other developed territory in search of duplicate "picture in the rocks." Proceeding up the river, the most producing territory he found was at Foster Station, nine miles, by river from Belle Island. Here, on a little area of about 25 acres, rich wells have been for some years pumping, and probably millions of dollars' worth of oil have been taken out. At Foster he heard a driller's tale that was an echo to the one he had learned at Scrubgrass. The depth of drill at the first sand, the texture and color of the third sand, the color and gravity of the oil, and the other data, were remarkably identical at these two points, five miles apart by air line. More than this, Mr. Angell was struck with the fact that this similarity of order, composition and thickness, was not confined to the sand-rocks, but marked also the intervening and overlying strata. The wells at both points were in all respects fac-similes each of the other.

As another evidence, Mr. A. computed the depth of drill at the two points to the upper surface of the third sand. Measuring at the surface of the river, and deducting at Foster for the natural descent of the river between the two points, he discovered that the upper surface of the third sand is on a dead water level at the two points. On his postulate that similarity of rocks indicated identity of belt, he now had two known quantities of his problem. Its solution was still a delicate operation. He knew the belt must be a narrow one, because at the points where the river crossed it (Scrubgrass and Foster) the territory had been by actual development shown to be small; the derricks at these points are huddled as closely together as the necessary operations will admit. On so narrow a belt it would be easy to lose the trail; an error of one degree in the compass would lead off the belt in a few rods of progress.

Mr. A. now employed a civil engineer, and for many months was engaged in running numerous lines between Scrubgrass and Foster. Without detailing all the laborious process and study by which he fixed the line, suffice it to say that he finally settled on a line which passed midway between the *extremes* of development at the two points (Scrubgrass and Foster). In the course of his research, Mr. A. developed another collateral theory, which materially aided him, and if it holds good on all belts is very important. That is: As you digress from the central line of a belt, either way, the third sand grows thinner and the supply of oil less, both finally running out when the limits of the belt laterally have been reached; the nearer the central line the better the territory. This shelving of the rock is from beneath, the upper surface of the third sand being level. This theory was demonstrated by the record of the wells on the belt. He found that running his line midway between the extremes of the development it traversed the derricks of the best wells struck. He was further confirmed in this idea by the fact that his central line between Scrubgrass and Foster passed at one side the celebrated Burning Well, on a bend of the river. This well, several years ago, struck an immense gas vein; the tools stuck in the well, the gas took fire, and for seven years was not extinguished. The owners abandoned it. Its location on the *edge* of what he supposed to be the best width, its failure to produce oil in paying qualities was encouraging.

Having now found the axis of the belt, it was easy to determine its *direction*—the most important point of all. The compass indicated the belt to lie north-east. To confirm all his tests, and "make assurance doubly sure," he now projected his central line on north 16 degrees east *beyond* Foster and across the mountains until he again crossed the tortuous Allegany, four miles from Foster. There was a development on Porter Island, it being the next development above Foster; all "dry holes" between these and Foster. The line left this island several rods at one side, and it is poor paying territory. This fact, and the exact similarity of its data with those collected below were good confirmation of the two theories mentioned. Continuing his line over the mountains five miles further, he again debouched on the river at Reno (two miles below Oil City). Here, to his surprise and satisfaction, his lines crossed the best wells on that territory, and the "testimony of the rocks" was identical with that of Scrubgrass, Foster, and Porter Island. Following his now sure line, he found himself walking through the heart of the rich oil fields of Charley Run, Wood Farm, and Petroleum Center. Thus he had traced his belt for *twenty miles* across the country, crossing the Allegany five times, traversing every paying development on that river, and landing in the centre of the greatest oil basin in the world. He was content, and kept his own counsel.

Mr. ANGELL had now spent a year and a-half of time and several thousand dollars in developing his theory. He could see the strip of oil rock 1,000 feet beneath him as satisfactorily to himself as if all the overlaying strata had been stripped off and the third sand laid bare to his fleshly optics. He proceeded at leisure to gather the fruits of his labor and genius—for it was scarcely less than genius that guided him. I say "at his leisure," for no other operator suspected his great discovery. After considerable shrewd negotiation he managed to secure from every property owner on the belt between Scrubgrass and Foster, a lease on liberal terms to them as to royalty, &c., but including an option clause for the purchase of the land, in fee, at any time within a year at a stated price. He made similar contracts for several hundred acres on the belt north of Foster, also. Before long he, of course, took his option, and in a short time owned all the territory between Scrubgrass and Foster. This was all he could pay for and manage, and the rest of the leases were suffered to lapse. He had now invested $60,000 on faith in his new theory.

The first well was bored one and-a-half miles back from the river, and from all previous development. It found the third-sand rock and oil at the depth of 1,110 feet, and yielded 71 barrels, and is still producing at that rate. The rest of the grand achievement is soon told. Mr. ANGELL, in connection with his partner, Mr. Prentice (a veteran "greaser.") has on his territory at present 16 producing wells, ranging in yield from 12 to 400 barrels. The last strike, made three days since, is "a spouter." It threw oil 150 feet high, took fire, and burned up everything destructible around it. When brought under control, it recorded 400 barrels per day. There has been but a single failure to strike oil in paying quantities by them, and that was an experimental well to test the width of the belt. Every acre of land could be leased by ANGELL & PRENTICE to other parties for $1,000 to $5,000 bonus and one-half royalty, the usual royalty being from a sixteenth to a quarter. They, however, grant no leases; they run no risks in sinking wells. They are now putting down 16 more wells for themselves, with a certainty of results which they would not give any one a dollar to insure. Mr. ANGELL can now set his foot down on the spot that will yield oil "to a dead certainty." Nay, he can tell within a few barrels what a well will do in any spot on the belt, proportioned to its nearness to the central line. Looking across this city of derricks, they "line" almost like a row of shade-trees, and mark to the spectator's eye the limits of the belt with wonderful distinctness, though not any more clearly than Mr. ANGELL saw them by his clairvoyant perception before a blow had been struck, or a "Sampson post" set on this spot.

The belt is less than thirty rods wide; the profitable part of it not half that width. It can be seen, therefore, how easy it is, or was, for a man to miss it while prospecting all oildom for a place to "set in"—or, in petroleum phrase, "wild-catting." His chances were about as 15 square rods are to the whole area of oil country.

Of course, the triumph of this new development created a furore. This soon spent itself. The fever had nothing to work on. The discoverer owned

all the territory, and none was "for sale or to let." Considerable fishing for
the belt further north (between Foster and Reno) was done. One of these
"wild-cat" parties blundered upon a blaze upon a tree, and, assuming that
it was ANGELL'S center line, took a lease and sunk a well. They were upon
one of his experimental lines and not on the center; hence, they found only
a thin third-sand and a small yield—three barrels. They abandoned this
well and moved 40 rods further *south*, which direction, being at an acute
angle with ANGELL'S line (S. 16° W.), brought them within 12 feet of the
center of his belt. There, on the 28th of March, they struck a 100 barrel
well, which has as yet shown no falling off from that figure. This is called
the Milton Well, and is two miles from Reno. A new oil town has since
grown up at this point. With a satirical humor for which the nomenclature
of Oildom is noted, it is called "Driftwood"—it is 400 feet above high
water on the Allegany! The traffic growing up in consequence of the development of this portion of the belt open to operators has caused the erection of a new station on the Allegany Valley Railway, called Prentice,
after Mr. ANGELL'S partner. It is five miles from Oil City.

It would seem that Mr. ANGELL'S successful research tends to establish
the following principles in oil development: (1.) That oil lies in belts of
more or less unbroken continuity; the latter feature having to be established
by tests in each case, and constituting one of the unavoidable contingencies
of belt development. (2.) That the general direction of these belts is N. E.
and S. W. (3.) That different belts may and probably will differ in characteristics from each other; but the same belt is uniform in its leading
physical features throughout. (4.) That these belts do not deviate from a
direct line; at least it is not safe to count on deflections or crooks in making
investments. (5.) That the under surface of the third-sand rock slopes upward each way from its center, and therefore the value of a well will depend
upon its nearness to the central line of the belt. (6.) That the upper surface
of the third sand rock is absolutely level throughout the belt. Mountains
are superadded to it. Therefore there can be no current of oil in the belt,
as some have supposed might be the case. (7.) That superficial water-
courses have no relation to the oil courses or belts. Therefore territory on a
river is worth no more than that on a mountain top, other things being
equal, except that the drill starts a little nearer the third-sand. (8.) That
it is possible to detect an oil belt and stake it out on the surface with a degree of accuracy sufficient for safe business investments. (9.) That when
proper efforts shall be made to discover these belts, the whole business of oil
production (so called) will be reduced to a basis in which there will be fewer
risks than in a crop of wheat, a sea voyage or a stock of goods. (Some of
these principles are well sustained by the experience of operators generally.)

Mr. ANGELL'S success must result—indeed it has already resulted—in
turning the thoughts of practical men to the study of the theory of oil belts.
This may end in the establishment of the whole business on just such a sure
basis as I have indicated. This would make a saving of millions of dollars
annually to the producing interest, and inure to the direct benefit of millions
of consumers of kerosene. It is certain that with the risks and expenses

hitherto attached to oil production, prices cannot remain where they are. Therefore, all who use petroleum oil are concerned to have the risks and expenses reduced. In case the results named flow from Mr. ANGELL's research, he will, while securing a princely fortune to himself, have become a benefactor to his fellows. I believe, as does every practical man who has learned of Mr. ANGELL's achievements, that it and the theory on which it was based are destined to mark an epoch in the history of petroleum. It is certain, whatever the results, that both the nature and manner of his dis- covery of the great oil belt stamp him as a remarkable man—at least as re- markable among oil men, and that means much to any one who knows the class. * * * * * * * * * * * * *

Already the minds of considerate oil men are reaching out to solve the question whether the oil manifestations of this continent have a common origin ; whether there is not a connecting belt running from West Virginia to Canada.

JAMESTOWN, N. Y., May 26th, 1871.

Of Mr. ANGELL's record as a producer, very little has been said in this brief sketch of his connection with the oil region of western Pennsylvania. That he is among the leading and prominent ope- rators of the region, we need not affirm. His enterprise and his unflagging industry are known of all men. His developments at Scrub Grass and on the Foster and Scrub Grass belts, won for him a notoriety he may well be proud of. His late operations in the lower oil fields at Fairview, Greece City, and upon the Moore and Hepler farms, territory he leased in the early months of the pre- sent year, fix his status as one of the successful as well as one of the most indefatigable operators in that section. He is now—midsum- mer '73—in the daily receipt of five hundred to one thousand bar- rels of oil per day, and this product, it is fair to assume, will be largely increased as his developments proceed.

In his native county, Mr. ANGELL is held in high esteem by all classes of his fellow-citizens, not alone for the possession, but for the daily practice of those manly virtues which are the crowning glory of all good men. In no sense an office-seeker, he has not escaped the observation, nor been freed from the importunities of his political friends to permit the use of his name for responsible trusts. In the canvass of 1871, he was named by a large body of his fellow-citizens for the office of State Senator from his senatorial district, partly as a compromise candidate, between the rival factions, but generally because he was regarded, upon all hands, as the most unexceptionable man for the distinguished position. He was, however, not nominated, "the factions" warring to the bitter end—but we doubt not we reflect the sentiments of a very large majority of the partizans of each of these belligerents, when we say that the PEOPLE and not Mr. ANGELL, were immeasurably the losers by his absence from the Senate of the State of New York.

The writer of this brief sketch of CYRUS D. ANGELL has known him personally for many years, and in common with a wide circle of friends, and an extended public acquaintance, recognizes him as a gentleman of pure motives, and clear, conscientious impulses. As a man of business he is sagacious, energetic, and reliable. If embarrassments come upon him he will double and quadruple his efforts to surmount them. He has intelligence of a high order, coupled with an integrity of character, stainless and blameless before the world. In his intercourse with men of whatever station in life, he is always truthful and irreproachable. He is dignified, and yet a man of marked sociability and cordiality. Modest in his demeanor, he has always about him the demeanor of the true gentleman, and from this standard never lowers himself. Where he is best known he is most appreciated as a man, a neighbor and a public-spirited citizen. He is in the prime of life, and the very heyday of his vigor and usefulness. With health guaranteed to him, that indomitable perseverance that has characterized him through life, will be sure to carve out for him a name and a place among the noted men of the oil regions at once distinguished and enviable.

A. D. ATKINSON.

NEW BRUNSWICK, N. J.

ASHER D. ATKINSON is a Pennsylvanian by birth, born in the city of Philadelphia, on the 30th day of September, 1821. When he was seven years of age his father removed to the city of New York, where he engaged in the retail drug and medicine business. Here young ATKINSON grew up and was educated. While from early childhood he had more or less to do with the drug store his father conducted, it was not until he was sixteen years old that he entered the establishment as a clerk, and gave his undivided attention to it. From his earliest remembrances, he was called "Doctor," but he did not entitle himself to the distinction—even if he did then —until after he reached his majority. He commenced the study of medicine at the age of eighteen, and continued some months after he was twenty-one—but he did not complete his preliminary preparations. It is questionable, therefore, if he be, even now, entitled to the affix of "M. D." to his name, for he abandoned all idea of adopting medicine as a profession soon after attaining his majority, from an early and later cherished aversion to it. He acquired, however, a commendable proficiency in surgery, and a thorough knowledge of anatomy, ere he gave up his studies, and these have been serviceable to him upon many occasions, but he has never practiced medicine outside his own family circle.

He continued his connection with the drug and medicine traffic, in conjunction with his father, until 1861, when he surrendered it to enter into the just then developing petroleum enterprises of the Western Pennsylvania oil region.

In the spring of 1861, Dr. ATKINSON visited Titusville for the first time. His father-in-law, Mr. John Barnsdall, had preceded

Woodburytype. A. P. R. P. Co., Phila.

ASHER D. ATKINSON.

him some months to look after interests he had acquired by purchase in 1860, and with whom Dr. ATKINSON had associated himself as a partner.* Their first leases were upon the Parker Farm
flats, near Titusville, and these were in course of development
when Dr. A. reached there, in 1861. Here let us retrace our steps
to show just how. Dr. ATKINSON became interested in the development of Petroleum.

During the summer of 1860, Mr. William Barnsdall, who resided then, and now, at Titusville, visited Dr. ATKINSON, in New
York City, taking with him specimens of the oil obtained from
" the Drake " and " the Barnsdall Wells." During his visit, the
subject of Petroleum was fully discussed—its development, value,
extent, and in fact every conceivable phase of it was elaborately
examined. Dr. ATKINSON became deeply interested in the subject, and made several experiments with the small specimens before
him, and finally " determined to put some money in it." He subsequently joined Mr. John Barnsdall in his purchase of interests,
and, as before stated, in the spring of 1861, disposed of his drug
and medicine business, and practically removed to Titusville to
give his attention wholly to this new enterprise.

Great activity in drilling wells, leasing lands, and buying oil
tracts, were visible upon every hand at the date of Dr. ATKINSON'S
arrival at Titusville. The developments were as yet confined to
the Parker Farm and the Watson Flats, though operators had ven-
tured still further down " the Creek "—as far as Rouseville, and
below. The daily product of the wells at and near Titusville was
limited to a few hundred barrels, and the price was uniformly $10
per barrel. Later in the summer and fall of 1861, the great flowing
wells of the lower McElhenny Farm were struck, and in a few days
thereafter oil sold for twenty-five and even ten cents per barrel !

* Mr. John Barnsdall was among the early operators of 1860, upon the Watson and
Parker flats, below Titusville. He died in that city in 1863, universally regretted
and respected. He was a man of worth and integrity, and lent to the early development of petroleum a vigor and enterprise that won for him distinction and general
prominence.

The scores, and perhaps we should say hundreds of operators upon the Watson Flats, the Parker Farm, and other tracts in the immediate vicinity, found their small wells of five, ten and twenty barrels daily product, almost valueless, and the territory next to worthless for paying oil wells—and they were generally abandoned for the "flowing well region," further down the Creek.

Dr. ATKINSON was among the first to go. His first investment was upon the Foster Farm, below Shaffer Farm. Joel W. Sherman had leased a single acre upon this farm, and was busily engaged in "kicking down" the Sherman Well. Mr. John Barnsdall had met Mr. Sherman at Titusville soon after the "Fountain Well," lower McElhenny Farm, had been struck, and subsequently purchased a "one-quarter working interest," in his Foster Farm lease, paying him $250 for it—the purchasers to pay one-quarter of the cost of putting down the well. Long before this well was completed, Mr. Sherman's patience became exhausted, as did his working capital. He was ready, nay anxious, "to abandon the whole thing," upon very liberal conditions. He, over and over again, offered to sell his interest—one-half the working interest— for $500, and $400, and even $300! Mr. Barnsdall would reason with him, encourage him to "push ahead," and offered to loan him the amount he demanded for his interest, if he required it, to be repaid when the well was completed.

In giving these incidental facts connected with the history of "The Sherman Well," it is not intended to question the industry or the determination of Mr. Sherman. His was only the case of scores and scores who became disheartened, and were willing to give up their enterprise if they could realize money enough to get to their homes!

" The Sherman Well " was finally completed on the 10th day of March, 1862, and immediately started off at a product of 1,000 barrels per day! She continued to flow, gradually lessening in volume for about twenty-two months, and was thenceforward pumped until exhausted in 1865. As a pumping well, the Sherman netted her owners between $57,000 and $58,000. While her producing life held out, oil was sold from her great wooden tanks as low as fifteen cents per barrel, and as high as eleven dollars per barrel!

During the same summer, (1862,) and upon the same farm, the "Barnsdall Well" was struck, and flowed 100 barrels per day. Barnsdall & Atkinson owned one-third of the land interest of this well, and all the working interest. Having thus begun in 1862, the development of their own leases, they followed it up, and in 1863, they completed a well on the Fleming farm, adjoining the Miller farm, on the north and west. This well flowed from the commencement, 300 barrels per day. On the very day the Fleming farm well was struck, Dr. ATKINSON completed a fifty barrel pumping well, on the bluff, in the rear of the Sherman Well. Both these wells, we may add, were struck, and in operation the day previous to the striking of the famous "Noble Well," on the adjoining farm.

During the summer of 1862, Mr. Barnsdall, for the firm, contracted to deliver at their wells, over 50,000 barrels of oil, at prices ranging between thirty-five and sixty cents per barrel. While this contract was in force, and only about two-thirds of the oil had been delivered, Mr. Barnsdall died, and Dr. ATKINSON, though not legally bound to do so, fulfilled the agreement to the letter—delivering the oil at "the Sherman," "the Barnsdall" and Atkinson wells, and this, too, when $4 and $6 per barrel could have been realized for it! But the great loss sustained by this single transaction was subsequently fully atoned for. The money realized from this sale of oil, was re-invested in lands in the vicinity, and from their later development, Dr. ATKINSON laid the foundation of his present ample fortune.

In 1864, Dr. A. purchased the John Fleming Farm, located on the high lands, in the Shamburg district—two or three miles north and east of Miller farm station. For this property he paid $20,000. "A refusal," for the purchase of this farm had been obtained by

Dr. Potter, of Tidioute, at $5,000. Dr. ATKINSON expected to buy it for that sum, but upon "interviewing" Dr. Potter, he learned his error, and thereupon offered him $20,000 for his chance to purchase—a clear profit of $15,000! This generous offer, for such it was at that time, the territory being three miles from "the Creek," and as far from any developments, was accepted, the whole amount paid down, upon the execution of the deed from Fleming, and the title to it passed to Dr. ATKINSON. We may add here, that the Doctor still owns and operates this farm, and that its product, though limited, yields a goodly monthly income.

From and after the purchase of this property, in 1864, Dr. AT-KINSON, to a considerable extent, withdrew from the oil business, removing to Brooklyn, N. Y., with his family, and taking up his residence there—though occasionally returning to "the Creek" to look after his yet profitable oil interests.

In the early spring of 1867, Dr. ATKINSON began the development of the Fleming, or Atkinson farm—although a derrick and engine-house, boiler and engine, had been put upon the property in 1865. In July, 1867, however, the first well was completed and operated,—and we may add, demonstrated the value of the farm for oil purposes. This first well—a pumping one—produced seventy barrels per day. In December following, on Christmas day, the second well was completed, and this was a "flower" of *four hundred barrels* per day! About the 1st of January, following, the third well was struck, and this "New Year's gift" flowed from the start *three hundred barrels per day!*

During the summer of 1868, developments were continued without interruption, and with uniform success. "The Atkinson farm," became famous for its productive wells, and it was rapidly developed during 1868,–69, and '70. About thirty wells in all were drilled upon the farm, averaging in depth, 952 feet, and all or nearly all were profitable oil producers—many of them largely so. During the summer of 1868, a well was struck upon the farm which had this peculiarity. The first FOUR days she flowed eleven hundred barrels of oil, each day—making FORTY-FOUR hundred barrels during her flowing life. Ever after, the well was known, as "4. 11. 44!"

The purchase and development of this farm was a source of great profit to Dr. ATKINSON. During the years 1868, and '69, the average daily product of the "Atkinson farm" was fully 2,500 barrels per day. Oil sold readily for from $2.50 to $6.50 per barrel during these years—making the average price during the producing life of the property $3.50 to $4.00 per barrel. Dr. ATKINSON owned the farm, and paid no "royalty to the land"—and while he had much of it leased to, and developed by others, an average of one-half the oil produced from all sources, was turned over to him. We have no correct statement of the total product of oil from this prolific farm, but the amount realized IN CASH, to all interests, exceeds $1,600,000!

A. H. CHENEY—and who among the early operators "on the Creek," from Titusville to Oil City, does not remember rollicking, always happy "Lon. Cheney"—and JAMES MOORE were the first Superintendents of the Atkinson farm, to whom Dr. A. gave, "out and out," various interests from which each realized largely. Both are since deceased, and both left considerable means to surviving relatives. F. E. Hammond is the present Superintendent—a position he has held since 1869.

We have stated, that soon after the purchase of the Atkinson farm in 1864, Dr. ATKINSON suspended further operations until 1867. He removed with his family from Titusville, to Brooklyn, N. Y., in 1865, and soon after engaged in real estate operations there. Among his purchases, was one hundred acres of land located at various points surrounding the City of Brooklyn. This investment was made for his children, and this he will retain for their sole benefit.

In 1868, Dr. ATKINSON purchased a summer residence at New Brunswick, N. J., and here free from the turmoil of a great city, his summers have since been passed. This purchase included a tract of land lying within the city limits, and containing ninety-four acres of land. For this property, he paid $25,000. Subsequently, in 1869, or 1870, he disposed of all but TEN acres of this tract of land, realizing from the sale, sufficient to pay the original cost—$25,000—and, at the same time, provide for the erection and completion of one of the finest private residences in the City of New Brunswick, or in the County of Middlesex, in which the city is situated. We may add, that this elegant residence has cost him not far from $50,000—nearly all of which, with the ten acres reserved from the original purchase, may be set down as clear profit! Since the completion of his residence, he has become a permanent resident of New Brunswick.

Dr. ATKINSON, while he retains his interests in the Oil Region, may yet be regarded as practically withdrawn from active participation in the later developments. He, however, makes annual visits to his old tramping grounds, and cultivates a lively interest in all that pertains to the good of its people and the growth and prosperity of the industry. In later years, he has become prominently identified with real estate transactions in Brooklyn, N. Y., at New Brunswick, and upon the Raritan River, in New Jersey. In all these, he has been successful, and largely so ; and so real estate operations, involving large capital, absorb much of his time and large means.

In 1847–'8, Dr. ATKINSON resided nearly a year in the south, principally at New Orleans. Since the war, with the exception of last year, he has passed his winters in the southern portion of the Union, and during these annual journeys, has visited every legislative assembly in the South.

" The Fashion Plantation," near New Orleans, and once the property of Gen. Dick Taylor, son of President Z. Taylor, was confiscated by the Government just after the close of the war. An acquaintance of Dr. ATKINSON'S purchased the property, and, requiring money to pay for it, he applied to Dr. A. for a loan of $20,000. The loan was granted, the doctor securing himself by bond and mortgage. This he held a year or two, and was finally enabled to realize upon his investment, suffering only a moderate loss!

In 1853, Dr. ATKINSON married Miss EMMA BARNSDALL, daughter of Mr. JOHN BARNSDALL, then a resident of the City of New York, and a brother of WILLIAM and JOSEPH BARNSDALL, of Titusville, Pa. This marriage relation doubtless furnishes the motive and incentive to Dr. ATKINSON's subsequent prominent and successful connection with the development of Petroleum in Western Pennsylvania, Mr. BARNSDALL, the father-in-law, being a large owner in the second well struck in the Oil Region, upon the Watson flats below Titusville.

Personally, Dr. ATKINSON is a man of many excellencies, and marked private worth. Large-hearted, so to speak, he is generous to a fault. Attractive in manner, he is always companionable, and sure and unvarying in his friendships and attachments. A single instance will illustrate his generosity and liberality. Col. E. L. Drake had, in 1860, or 1861, purchased a tract of land lying in the then village of Titusville, consisting of twenty-six acres. He paid a nominal sum for it—$2,000 to $3,000—giving a mortgage for the amount unpaid.* In 1863, this mortgage became due, and the holder pressed Col. Drake for its payment. The property had meantime quadrupled in value. Col. Drake had applied to several personal friends to aid him in his embarrassment, but had obtained no relief. Some one suggested that he make his application to·Dr. ATKINSON. "But I don't know Dr. ATKINSON!" said Col. Drake. "That'll make no difference," said his friend; "the doctor will help you out, I know." Col. Drake presented the matter to Dr. ATKINSON, giving him a full detail of his threatened disaster. The property was offered to Dr. A. for $10,000. "I will give you $12,000 for it," said Dr. ATKINSON, "and I will hold it subject to

* This property has since become part and parcel of the City of Titusville, and is worth to-day, half a million dollars. It is that part of the city lying east of Martin Street.

your future efforts to dispose of it at a still further advance." The
doctor added: " If at any time during the next twelve months, you
can sell it for more than $12,000, you may have all the profits!"
The sale was made, and Col. Drake was saved from impending
bankruptcy, and by a total stranger. The property was subsequently sold, and within the time named by Dr. ATKINSON, for
$16,000; but neither Col. Drake, nor Dr. ATKINSON profited by the
advance. The individual who purchased the property from Dr.
ATKINSON, of whom a deed had been obtained for $12,000, with
the understanding *that all over that amount*, obtained for it, should
go to Col. Drake, neglected to make good his verbal agreement to
do so, and retained the $4,000 profit as his own! But the trans-
action, so far as Dr. ATKINSON was connected with it, did honor
to his generosity and liberality.

Dr. ATKINSON is in middle life, the picture of health, and seems
to enjoy the very personification of an unbroken constitutional vigor.
He is just what he seems to be at first acquaintance, a plain, substantial, good man, without guile, ostentation, or modern "airs."
A man of clear conceptions, and rapid comprehensive grasp, he
has business qualifications of a high order. Honor, honesty and
integrity of character, make up his private worth, and endear him to
all who know him. In his domestic relations he is indulgent and
devoted—his little family circle being his " Heaven upon Earth."
As a citizen he is enterprising, liberal, and cordial with all. As a
neighbor he is obliging, and full of those little attentions and accommodations so generally appreciated, and we may add, indispensable
in all communities. As a friend he is cordial, devoted and steadfast.
Such men make the world better for their living in it—and this is
the estimate we put upon Dr. A. D. ATKINSON, coupled with the
hope that he may live many, many years, to enjoy the bounties now
surrounding him, among the least of which, in our judgment, is
the ample fortune he has amassed by his own industry.

Woodburytype. A. P. R. P. Co., Phila.

JOHN L. McKINNEY.

JOHN L. McKINNEY.

TITUSVILLE, PA.

JOHN L. McKINNEY was born at Pittsfield, Warren Co., Pa., on the 21st day of June, 1842, of thrifty, industrious, well to do parentage, and is the second son of a family of seven sons and one daughter. He had the advantages of an excellent district school education until he attained the age of 15—from the age of 12, aiding his father in the conduct of his lumbering and farming inter- ests, as he was enabled to do, during the school vacations of the summer months. At sixteen he assumed almost entire charge of his father's books and accounts, involving considerable amounts of property which was constantly changing in value, location, &c. ; often superintending the sales of lumber, farm products, and whatever else made up the bulk of his father's business. All this responsibility young McKINNEY discharged with fidelity and an ability rarely exhibited in one of his years and experience. He remained with his father in this responsible position three years, when at the age of nineteen, he resolved to grapple the world in his own behalf.

The discovery of Petroleum oil, near Titusville, and at other points on Oil Creek, had already attracted the attention of the pub- lic, and thither young McKINNEY bent his steps. His father strongly opposed this new enterprise on the part of his son—not yet come to man's estate—but visions of wealth, and a desire to "strike a blow for himself," outweighed all else, and in the sum- mer and fall of 1861,we find him at Franklin, Pa., with less than $500 cash capital, looking for an opportunity to invest. He was not long in search of a "a good thing," as he deemed it. His first venture was upon a lease and well, located on the Allegheny River, south of Franklin. Here he risked a large portion of his cash capital, drawing upon his father for much of the lumber and timber needed in the erection of a derrick, engine house, &c.

This well was a failure—made so through the bad management of the parties engaged in its drilling and subsequent testing, and young McKINNEY's loss was total.

This first investment proved to be disastrous, but Mr. McKIN-NEY, with a wise foresight, had partially provided against such a contingency. He had invested in one or two "interests" on "Oil Creek," both which proved to be successful, and left him sound financially.

In the spring of 1862, he purchased an undivided one-third interest in a well and lease on the Jno. McClintock farm, near Rouseville. At the date of this purchase, the well was down about 125 feet, and had been drilled thus far with a "spring-pole." This process was continued to the depth of 300 feet, and then a "horse-power" was substituted, which exhausted itself at 400 feet, and a steam engine was found to be indispensable. With this steam power, the well was drilled to the depth of 512 feet, and at once began to flow, at the rate of 500 barrels per day! The well continued to flow for nearly two years, gradually lessening in product, but yielding to its fortunate owners, thousands and thousands of dollars in profits.

The history of this well, and Mr. McKINNEY's connection with it, may be set down as the history of very many others of the earlier developments "on the Creek," and may not be without interest to the general reader. After the purchase of his one-third interest in this lease, he gave to its development all his individual effort, and every dollar of money he was able to command. He worked his regular "tower," "kicking the pole," and followed the business of "a driller," with all the industry and devotion of an interested owner. Weeks, had run into months, in labors upon this well, and yet all was in doubt. Hope alone kept the operators to their

task. It might, nay, many prophesied it would be a "dry hole"—
a summer's work lost, to say nothing of the ready cash required
"to prosecute with due diligence." But the sturdy industry of the
owners did not flag. They labored night and day, surmounting
obstacle after obstacle, and embarrassment after embarrassment,
looking confidently at times, and doubtfully at others, upon the
results of their sacrifices. After the steam-power had been applied,
Mr. McKINNEY improved the opportunity to visit his home in
Warren County, little anticipating that during his absence—"not
to exceed two weeks,"—the well would be down, and a golden
stream gushing from its mouth, destined to enrich every man owning interests in it. But this consummation was realized. Mr.
McKINNEY heard of his good fortune a few days after the well was
struck, and immediately set out across the country to Rouseville,
then called "Cherry Run." When he reached the vicinity of the
well, he ascertained that his partner had sold the entire property
for a good round sum, taking no account of HIS interest!

When Mr. McKINNEY purchased his share in the property, he
drew his own contract for the sale, and upon its execution, laid it
away among his papers, neglecting to have it filed in the Protho-
notary's office at Franklin. The parties purchasing from his partner, had made a thorough search, and found no record invalidating
his claim to ownership, and, thereupon, closed the trade, paid the
consideration, and were actually in possession of the property, when
Mr. McKINNEY returned! Here was indeed a threatened calamity.
What to do, or whither to turn for relief, was the absorbing question
with young McKINNEY—who, while yet a minor, possessed the
will and determination of a full-grown man. He promptly noti-
fied the purchasers of his ownership of an undivided one-third interest, and his willingness to take care of his portion of the product
of the well. *They* in turn refused to acknowledge his rights, insist-
ing upon their *own*, by absolute purchase. While these interviews
were transpiring, Mr. McKINNEY had recourse to his contract, and
found the "six months recording clause" still in life! About two

days were left him, to put his claim on file. He was not long in reaching Franklin, when upon consulting Mr. C. HEIDRICK, one of the able attorneys of Franklin County, he ascertained that his interest was safe and that his ownership in the well, was beyond cavil or doubt.

The purchasers from Mr. McKINNEY's partner, upon presentation of his incontestable proofs, acknowledged his claim, and very shortly after paid him a large sum for his interest. The property, however, was worth, or proved to be worth, during the life of the well, fifty times the money Mr. McKINNEY received for it. The purchasers realized handsome fortunes from its product, some of whom are to-day rolling in the wealth it brought them.

After disposing of his interest in this property, he made good use of the little store he had acquired, buying interests in some ten or twelve other leases on Cherry Run and elsewhere, and pushing developments as rapidly as possible. The "Baker well," which produced one hundred barrels per day, was the first struck, in which he was interested. Others, many of them without name, but abundant in product, followed, and Mr. McKINNEY was really upon the flood-tide of success. The celebrated "Mountain (or Phipps) well," which produced three hundred barrels per day, was among his acquisitions. These continued successes ran through 1862–3–4, and '65, when he began the sale of his interests, here and there, having determined to change somewhat the character of his operations. He disposed of many of these at fabulous prices, receiving in payment considerable sums of money, and larger amounts in certificates of stock in various oil companies, then "as plenty as the leaves of the forest"—many of which never reached the stock boards of eastern monied centres. A few were "not worth the paper they were printed upon;" but a fair fraction proved to be valuable. In all, Mr. McKINNEY realized in cash and stocks, more than $500,000 for his oil interests. While his stocks had a nominal value, it is safe to say he left the oil region $200,000 richer than when he risked his all in 1861.

In 1864, still retaining several small interests in his early investments, Mr. McKINNEY, in connection with Mr. Wm. C. Duncan, of Pittsburgh, and Mr. George Work, of Philadelphia, and others, visited the then developing oil belts of Green Co., Pa., and leased large tracts of land along Duncard Creek. This leased land was afterwards sold to other parties, the original lessees realizing a considerable profit. In 1864–5, Mr. McKINNEY took up his residence in Philadelphia, still holding a business relationship with the oil regions—occasionally visiting the scenes of his early successes, purchasing interests here and there, and steadily adding to his worldly possessions. During the spring and summer of 1866, he purchased interests in several leases and wells on Benninghoff Run, and subsequently, consolidating various other productive and valuable oil properties, among which was an oil farm, owned by him *in fee*, "The Benninghoff Mutual Petroleum Company" was organized, and Mr. McKINNEY was elected its vice-president. This organization was upon a substantial basis, having valuable property and some of the very best producing wells " on the Creek " upon which to pay dividends. It did for a time pay very largely, but "the crash of 1866 and 1867," swept away values, obliterated oil companies, and "gobbled up" oil producers and their labor of years, leaving the oil region, if not a desert of industries, at least a section almost wholly given over to bankruptcy and ruin.

In addition to his oil traffic, Mr. McKINNEY dealt largely as a general Stock Broker, and in the disasters of '66 and '67 he found himself deeply involved. The accumulations of his earlier years, seemingly ample for a life-time, were rapidly exhausted, and he, left as empty-handed as when, six years before, he made his first venture in the oil fields of western Pennsylvania. But this "complete ruin" did not discourage or dishearten him. He resolved to " try again," and to begin as before, " at the top of the ground," and work out of his financial embarrassments. He returned to the region in 1868, and took some leases at Pleasantville, just then the point of attraction for good producing wells. Here he met with moderate success, and in a few months began to see clear sky and mended fortunes. His industry and devotion to his affairs soon placed him upon comparatively independent ground, and marked out his future.

He remained at Pleasantville until June, 1869, when an opportunity offering, he disposed of his interests at a fair profit, and resolved to enter the new oil fields at Parker's Landing. Admonition and friendly advice, and in some instances strong protests were indulged in by disinterested friends against this "suicidal enterprise," as nearly every one termed it. Mr. McKINNEY had, however, "put his hand to the plow," and would not turn back.

Late in the summer of 1869, he made some leases and purchases at Parker's Landing, and completed two wells, one of which was a kinsman of the "dry hole" race, and the other produced from four to six barrels per day. The cost of wells at Parker's Landing was nearly double that of other oil districts, and these first investments were not "strong arguments" in favor of Mr. McKINNEY's enterprise in fixing upon this district as the point of his future operations. He, however, did not falter or abate one jot of his confidence in the ultimate success of his now redoubled efforts. He made more and larger leases, and commenced the drilling of numerous wells, in that since wonderfully developing region.

Early in the spring of 1871, success began to crowd upon him, and through that year, and up to this time—August 1st, 1873— scarcely anything he has consented to identify himself with, but has served to swell his bank account, and add to his repute as a successful oil producer.

While conducting his large oil business, he managed an agency for the sale of Gibbs, Russell & Co.'s engines, boilers, well tools, &c., and during this connection, which was incidental rather than actual and positive, he sold not far from $500,000 worth of this great establishment's work, and suffered a loss from "bad debts," of less than $500!

In the spring of 1872, Mr. McKINNEY, in connection with R. H. Sterritt, purchased the one-third interest Jno. T. Russell owned in Gibbs, Russell & Co.'s Novelty Iron works, at Titusville, and at Nunda, N. Y., paying therefor a princely sum. While he is not prominently active in the conduct of this mammoth industry, his sound judgment is uniformly sought by the remaining members of the firm. An enterprise of this magnitude, involving so large a capital, certainly requires the best business talent to be commanded, and Mr. McKINNEY is in all respects, up to the requirements of the responsibility imposed upon him. It is needless to add that the present firm of Gibbs, Sterritt & Co., is among the most successful in the oil regions, or in Western Pennsylvania.

When in the fall of 1872, "The South Improvement Company" began to foreshadow its objects and aims, Mr. McKINNEY stood, with the large body of producers, opposing its schemes, as the effort of a monopoly, to gather into its embrace the producing interests of the oil region. He weighed carefully, every new phase of this all-important and all-absorbing movement, and gave to each and all its later developments, an unbiased examination. When the region "rose as one man," to oppose and bitterly denounce "The South Improvement Company," he held aloof, and was unwilling to join in the universal clamor—not that he abandoned, or in the least became lukewarm in the interests of producers, but that he deemed a calmer judgment than was manifested, indispensable to an adjustment of all questions at issue. With the subsequent action of "The Producers' Congress," and its efforts to control or monopolize the product, shipment and sales of oil, Mr. McKINNEY had little to do, and less confidence in as a means of relief. He regarded the proposed "plan of operations" as unsubstantial and wanting in the essential elements of probable financial and commercial success. The later history of these transactions, shows how clearly he comprehended "the situation," for the results were a full warrant for his judicious action from beginning to end. He was, however, among the first to suggest ways and means to arrest what was generally regarded as an impending calamity. He urged first, a suspension of the drilling of wells for six months; and subsequently, the shutting down of all pumping and drilling wells for thirty days. The first despatch that passed over the wires from Parker's Landing to other oil centers, electrifying the entire region, advising and demanding this last great sacrifice at the hands of producers, was dictated, if not actually written by Mr. McKINNEY, and signed by the firm of which he was the head—McKINNEY & NESBITT. Having dealt this blow, he earnestly followed up his convictions with unremitting labors to unite the Parker's Landing district in the movement, and in ten days thereafter, saw the fruits of his efforts in more than 4,500 wells shut down for thirty days!

In the early months of 1873, Mr. McKINNEY, with his brother, J. C. McKinney, who for a year past has been his sole partner, purchased the two Hemphill farms, and the Barnhart farm, at Millerstown, now the great oil field of the Parker's, or lower district, consisting of about 250 acres. For this property they paid $70,-000. Subsequently Mr. JOHN H. GAILEY, of Parker's Landing, became a purchaser from the McKINNEY brothers, of an interest in these oil lands, and they are now being developed under the firm name of McKINNEY BROS. & GAILEY.

Mr. McKINNEY, besides the business connections heretofore named, and his manufacturing interests, at Titusville and Corry, Pa., and at Nunda, N. Y., as a partner of Gibbs, Sterritt & Co., is a stockholder and director in two of the soundest banking institutions in the lower oil region—The Parker's Savings Bank, at Parker's Landing, and The Millerstown Savings Bank, at Millerstown, Butler Co., Pa.

We may be permitted to mention here, what should have been stated before, that Mr. McKINNEY on the 14th of February, 1866, married Miss IDA D. FORD, of Pittsfield, Penna., a lady of such womanly and motherly excellencies and attractions as are sure to adorn the home circle, and become a jewel in the crown of her husband's triumphs and worldly renown.

Mr. McKINNEY has the elements of a successful business man largely developed. He possesses a well balanced mind, is a con-

stant "worker," and a thorough, earnest "thinker" upon all matters pertaining to his business affairs. While there is nothing
penurious or small in his dealings with men, he yet goes to the
bottom of every proposition engaging his attention or challenging
his favor. Details and their bearing upon results, are carefully
examined, and rarely with faulty conclusions. In his intercourse
with business men he is reserved yet positive, and always influential. His opinions are grounded upon a substantial basis, and he
is never without ample language to render them intelligible. A
man of positive convictions, he possesses the ability to defend his
opinions and carry men with him. In his relationships to the
world he is frank, zealous, open-hearted, in the strongest acceptation
of these terms. Possessing the characteristics of the true gentleman, he bears about him the dignity of a true manhood. In private
life he is companionable and sociable, beyond the power of appreciation until wholly and really known. Retiring in manner, and
reticent in habits, he rarely obtrudes himself upon others. Cordial
and confiding in his friendships and attachments, his estimate of
men is at once prompt, and usually correct.

In business circles Mr. MCKINNEY takes rank among the first.
He is reliable, and prompt under all circumstances. His industry
and devotion to his private affairs, and his consequent success, is
known of all men, and hence it is that his obligations are regarded
everywhere as "first-class" and "gilt-edged." He is at this writing a few months past his thirtieth birthday, and it is no stretch of
probability to say, that a most prosperous future is opening to him,
and that he will advance to it with the same realizing sense of his personal power, responsibility and duty, that has characterized him
thus far through a life of remarkable activity and conceded usefulness and excellence.

FRANK W. ANDREWS.

THERE are few prominent men now living in the oil region of Western Pennsylvania, who, identifying themselves with the early development of that great staple, petroleum, have not carved out for themselves a history in many regards remarkable. When the discovery was made, great numbers came, saw, and strived manfully to conquer, but after a brief career, dropped out of the ranks and were heard of no more. Many came, it is true, and in a few weeks or months, by sheer "luck," gathered up their one, two, three and five hundred thousand dollar fortunes, and abandoned the field to others. But the men who have *made* the oil region of Western Pennsylvania, and have been made by this wonderful phenomena of nature, are those who began early, and have remained through ten or twelve years of adversity and prosperity, to enjoy the fruits of their industry in a final triumph over all obstacles and embarrassments. Among this class of men, the subject of the following sketch has a deserved distinction, creditable alike to his industry, his will and his determination.

FRANK W. ANDREWS is a native of the Green Mountain State, born in the town of Vernon, Windham County, Vermont, on the 30th day of May, 1838. In May, 1840, his father, a successful farmer, removed with his family to Ohio, and settled in Geauga County, engaging extensively in stock-growing and the cultivation of a large farm. Here young ANDREWS grew to manhood, working upon the farm summers, and attending a district school winters, acquiring a tolerable common school education. When nineteen years old, he assumed the role of a teacher in a district school in the vicinity of his father's residence, and acquitted himself with

Woodburytype. A. P. R. P. Co., Phila.

F. W. ANDREWS.

credit. Subsequently he attended Hiram College, at Hiram, Ohio. At the age of twenty he determined to commence the battle of life, singly and alone, and with about $75 in money, he made his way west and south, arriving at Cairo, Ill., undecided as to a southern or western destination. While at Cairo, he fell in with a couple of gentlemen, who were publishers or owners of a map of the "Great West, and the States of the Union," together with a book entitled "The History of the United States and Territories." He was tendered an agency for the sale of these publications, his field of operations embracing Southern Illinois and the State of Missouri. For each book or map sold, he was to receive one dollar. He entered at once upon his enterprise, and for a time his expectations were fully realized. It was, however, a business he was scarcely adapted to, and we next hear of him as a school teacher, in a small provincial town, in the State of Missouri. He applied for the position, and after passing a creditable examination as to qualification, was questioned by the commissioners upon various points of school policy, and among others was asked if he kept a "loud" or a "silent" school. Young ANDREWS was extremely anxious to put in a correct answer to this interrogatory, and after a hesitancy of a few seconds, believing he comprehended the situation, he promptly answered, "a silent school, of course." This did not meet the approval of the learned commissioners, whom traditionary practice had impressed with the efficiency and desirability of schools of the "loud" sort. They intimated to the young pedagogue that "he wouldn't do!" Young ANDREWS, however, resorted to diplomacy, and suggested that he would accept the position upon a week's trial, at the end of which time, if *his* "school policy" did not meet the approval of the commissioners, and scholars as well, he would resign his charge. Upon these terms he commenced his labors. Among the attendants were many young men and young ladies his senior in years, but not in acquirements. His thorough discipline, his superior ability as a teacher, his self-control and perfect reserve, together with his industry and devotion to his duties, won for him a respect and admiration on the part of parents and students, never before extended to a teacher in that locality. Of course he was retained; Mr. A. maintaining his reserve, and preserving a dignity, so to speak, that challenged the respect of his scholars and commanded the approbation of the commissioners and patrons of the school. He continued his charge of this school for five months, and when he left, he bore away with him the confidence and good will of scholars, commissioners and the community generally. Had he consented to remain, as he was urged to do, his salary would have been $50 instead of $40 per month.

Soon after the close of his school in the spring of 1859, Mr. ANDREWS, full of the spirit of adventure, and a commendable determination to hew his own way to fortune, resolved to join the throngs just then gathering upon our western borders, bound for the lately discovered gold and silver mining regions of Colorado and Pike's Peak. Securing the co-operation of three or four friends and acquaintances of like ambition, they promptly began the preparation of an outfit. They purchased four yoke of cattle, a heavy emigrant wagon, which they loaded down with provisions, clothing, etc., for a lengthened expedition, and in a few days were upon their wearisome journey across the plains to the new Eldorado, where it was said gold in abundant quantities could be had for the picking up, and where untold wealth in other precious metals awaited the persistent and industrious pioneer.

As the little band of adventurers travelled " westward ho !" their company rapidly increased, and long before one-third their journey had been accomplished, they numbered over one hundred persons, all destined for Pike's Peak and the golden mountains of that far off country. Day after day they met large and small parties, who had turned back, discouraged and heart-sick of gold hunting. They told fearful tales of suffering and disappointment, which sadly demoralized the ranks of Mr. ANDREWS' party, and as a consequence, two, three, five and ten of their number from time to time " gave out," and joined those who were " homeward bound." The result was disastrous. The "faint-hearted" increased with alarming rapidity, and there were only thirty left of the first one hundred recruited! At last a majority of these "struck, and would go no further!" There was *one*, however, who had resolved to "stick." Mr. ANDREWS had set out for Pike's Peak, and if life and health were spared him, he "would go through if'he went alone!" Turning back with the rest, for he could not release his teams or their load, he travelled a portion of one day toward home, when he encountered another party, bound for Pike's Peak, and he gladly joined in this new expedition, and turned his face once more toward the west.

We shall not follow this band of resolute men in their toilsome journey across the plains to the point of destination. Suffice it to say, they were a little more than three months in accomplishing their journey, encountering all the vicissitudes, trials, accidents and incidents of an emigrating party upon the great plains lying west of the Missouri River, and east of the Rocky Mountains. Their numbers, as in the first instance, were largely augmented, until their rolls showed 125 persons. Daily as they toiled and travelled on toward the setting sun, they passed returning parties who had "seen enough and knew enough of Pike's Peak to satisfy them!" and had turned their faces homeward. They had become disheartened, and had resolved to return to civilization. All told the same tale of disappointment, disaster, and misfortune. This did not, however, appal the brave men who made up this second expedition with which MR. ANDREWS had cast his lot. They determined to see the end of their journey.

In June following, this company entered the city of Denver, Colorado, one hundred and twenty-five to one hundred and thirty strong. A few days spent here in replenishing their reduced outfit, and in recuperating their wasted strength, and they were again on the road over the mountains to the gold diggings of Pike's Peak. Reaching finally their destination, prospecting began in earnest, and for the purposes of brevity in this mining venture, we may add, that startling success did not follow in the track of any one or more of Mr. ANDREWS' party. Many who had come to regard the expedition as a failure, joined in the numerous parties making up and returning to the States, while others abandoned the mines and betook themselves to employment in the mining towns of the region. The result of all this "thinning of the ranks" was just TWO of the original or subsequently recruited mining party, and MR. ANDREWS was ONE of these two, and Capt. BARNES, a hardy old California miner, was the other. Mr. ANDREWS determined to remain. He had suffered too much of privation and hardship to surrender upon so slight an experience in "gold hunting." He had come too many hundred miles, and had sacrificed too many of the comforts of life to entertain the idea of abandoning all upon so slender a pretext. With his sole remaining companion, he took up his line of march to the rich placers of Clear Creek— operating at the "Spanish diggins';" and here their prospecting and mining operations were really begun. The first day's return to MR. ANDREWS was "washings" of the pure metal amounting to $16! Claims were "staked out" at various points, and labor expended on each sufficient to make good their rights as discoverers. Early in August, however, Mr. ANDREWS was stricken down with fever, and for a long time his life was despaired of.

During his illness he sold a few of his claims, receiving in exchange some personal property and a small amount of money. He owned other and more valuable mining interests, which he retained, having determined to develop them the following season.

He recovered his health and strength slowly, and after a confinement of nearly two months, fearing the rigors of a winter in the mountains—snow to the depth of eighteen to twenty inches having already fallen—and this in October—he determined to close up his affairs and return to the East. He began his journey eastward soon after, and reaching the Platte River at Denver, he constructed a skiff or small boat, intending to follow its course to its junction with the Missouri, six hundred miles away. For some days he made satisfactory progress, but the farther he went, the more shallow and unnavigable the water, became, and long before he had made one-half his journey, he was compelled to take to the dry land! Briefly, he reached the mouth of the Platte River, late in the fall of 1859, and soon after bent his steps toward St. Joe, Missouri, where he arrived in the month of November. Remaining here a few days, he sought and obtained employment as a teacher, and entered upon his trust, determined with the means thus realized, to return to his mining interests at South Pass as soon as spring opened.

The winter passed, and the $150—the amount received for his three months' services in teaching—in hand, he began to gather up his outfit for a second expedition. Before its consummation, however, ill health compelled him to change his determinations, and he returned to Ohio. This was in the early spring of 1860. He reached home in due time, and turned his attention to the supervision and conduct of his father's business. In January, 1862, he married Batpih L., eldest daughter of Anson Reed, Esq., of Newbury, Ohio, a very estimable lady.

Late in the fall of 1863, he resolved to visit the Oil Region of Western Pennsylvania, and during the winter of 1863–4, we find him engaged in the lucrative business of "hauling oil," from Tarr Farm, to Oil City and Franklin, and oftentimes to Titusville, employing several teams for this purpose, and when "the Creek" admitted, "boating" the same commodity to the two former points. This "means of livelihood" he followed into the fall months of 1864, when he determined, having accumulated, as he believed, means sufficient for the undertaking, to try his fortunes in operating. He secured leases on Cherry Tree Run, then undeveloped territory, and put down *four dry holes!* As a commencement, this result was far from flattering to his zeal and industry. But he did not weaken in his faith, or slacken in his resolve to succeed.

In the spring of 1865, he put down a single well on the McClintock farm, and later in the same year, drilled three on Pit Hole Creek, all which were like his first *four,* dry !

The Pit Hole excitement was about looming up, and thither Mr. ANDREWS went, full of resolution, quickened and intensified by his previous failures. On the day of his arrival there he secured five half-acre leases, agreeing to pay a bonus of $4,000 for each, "and one-half the oil." He paid $250 down upon each of the five leases —$1,250—and bound himself to pay the balance of the sum due —$18,750—in 60 days thereafter! After completing these contracts, his "cash account" was about balanced! He, however, contracted for drilling the first well, Mr. W. W. Thompson, of Titusville, since so famous as a successful oil operator, being the contractor. Mr. Thompson purchased a small interest, to apply on the drilling account, and immediately commenced the work of sinking the well. Of the remaining interests in the first venture, Mr. AN-DREWS disposed of sufficient to pay the cost of putting it down, retaining the balance. He continued to sell interests in the remaining *four* leases at excellent prices, and in six weeks from the date of his contract, sales had been made sufficient to pay off all his indebtedness and leave him a handsome margin! To this good fortune, was added the new well Mr. Thompson had meanwhile completed, and it was flowing 400 barrels per day!

During the same summer, with other parties he purchased a reserve of $3\frac{1}{2}$ acres of the Rooker farm, in the Pit Hole basin, paying $75,000 for it. This property was prolific of good wells, and among them was the famous "Fisher well," which flowed from the start, one thousand barrels per day.

Soon after securing the four half acre leases, above-mentioned, Mr. Andrews purchased one acre of the Ball farm, one of the large producing tracts of Pit Hole, and for this "bit of land" he paid $5,000,—$1,000 down and the balance in monthly payments of $1,000 each! Before the second payment became due, or within thirty days after the purchase, he sold the fee of this single acre of land to Hopewell, Parker & McLaughlin, for $6,000, reserving one-eighth of the oil!

Pit Hole continued its wonderful developments, and as a sequence, the value of leases, lands and farms, went to fabulous prices. The "Hyner farm" may be taken as a fair record of all the surrounding territory. Mr. ANDREWS met the owner, Mr. Hyner, one evening, while the excitement was at simple "fever heat." "What will you take for your farm?" asked Mr. Andrews. Mr. Hyner hesitated a moment, and replied: "Fifty thousand dollars—cash!" "I will take it," said Mr. ANDREWS, "and here are $500 to make the contract binding!" "Never mind that," said Mr. Hyner, "you come down in the morning, and make the papers!" The morning came, and Mr. ANDREWS was promptly on hand—but Mr. Hyner didn't seem to be as eager as the night before —he talked less, and appeared to be in deep study! The papers were being prepared, and as they approached completion, Mr. Hyner wandered a little way off, and beckoned to Mr. ANDREWS. "I told you fifty thousand dollars cash would buy my farm, didn't I?" said Mr. H. "Yes," replied Mr. ANDREWS, "and the papers are nearly ready." "Well," said Mr. Hyner, "I meant fifty thousand dollars in gold!" This was a stunner to the enterprising purchaser. Gold was then selling at $2.15, and the new proviso made a difference Mr. ANDREWS had not calculated upon! Another negotiation was entered upon, based upon $100,000 in green-backs, and a ten days' refusal of the farm at these figures—$500 forfeit! Later, Mr. Hyner, claiming he could not make a perfect title, "bought" out of this contract, by refunding the $500 paid for the refusal for ten days, and gave Mr. ANDREWS a lease upon any part of the farm he might select, within sixty days thereafter. This farm was afterwards divided into halves or thirds, and sold, Mr. Hyner realizing $175,000 to $200,000 from the sales.

Mr. ANDREWS' Pit Hole ventures continued to grow more and more remunerative, and he steadily enlarged his operations. At one time he could have disposed of all his interests there, at a clear profit of $250,000! He added new engines and boilers, erected new derricks and engine-houses, and rapidly increased his leases and liabilities. Some months later, when "the bottom went out of the Pit Hole basin," Mr. A. was left like scores of others in that once attractive locality, with considerable Pit Hole City property, numberless oil well rigs, boilers and engines, tools, tubing, casing, &c., upon his hands, with a scant stock of oil, and less money! His embarrassments might have disheartened and crushed out a less determined man. Indeed large numbers theretofore successful operators at Pit Hole and elsewhere, did actually "give up the ship," and returned to their homes "in the States," with cash and bank accounts either badly shattered, or showing balances upon the wrong side. Mr. ANDREWS, however, saw in his disaster only greater inducements for effort. He returned to "the creek" and secured territory on Pioneer Run—this and the Benninghoff farm territory, just then being developed successfully. He again secured the services of his friend, Mr. W. W. Thompson, to drill his first well, Mr. T., as before, assuming a small interest to apply on his contract.

The well was first put down upon a lease adjoining the " Union," and proved to be a 300 barrel flowing and pumping well. The opening at Pioneer Run was so flattering and remunerative that he prosecuted his operations there, with redoubled vigor, through 1866–7 and '68, and with large success, and here, we may add, he laid the foundations for that ample fortune he has since amassed and now enjoys.

In the fall of 1867, he purchased five-eighths of the land interest of the Tallman farm at Shamburg, paying for the property a little over $40,000 cash. This farm at the date of purchase, was rated " Wild Cat Territory "—and the first and second, and even the third well drilled upon it came near establishing its character as " dry diggins." The fourth, however, atoned for the first three, for from the start it flowed and pumped over 300 barrels a day. This farm was thoroughly developed through the summer of 1868, and into the early months of 1869, and proved to be among the best oil tracts of the region, netting to the fortunate owners more than $ 500,000 ! and it is yet a source of revenue. At one time during the summer of 1868, there were four wells upon this farm, the product of which was 1,200 barrels per day. Oil at this time, and throughout the year, brought readily $3.50 to $4.50 per barrel, making the average daily income to the owners of the Tallman farm, from these four wells alone, quite $4,000! Add to these some thirty others upon the same farm, all producing bountifully, and some idea may be formed of the great volume of greenbacks daily dropping into the cash boxes of the owners.

During the summer of 1868, Mr. ANDREWS purchased the interest held by Wm. A. Byers, in the Tallman farm and property adjoining, paying him $55,000 cash, which proved a good investment.

Early in the spring of 1869, he returned to Cherry Tree Run, "in search of satisfaction," as we infer, for his first failures in that locality. He contracted for drilling FIVE wells upon his own account, and in connection with other parties, put down THIRTY others. All these were remunerative investments, though not largely so. It is, however, safe to say, his "dry holes" of 1865, were paid for, and much more! During the same summer he drilled THREE wells on Shaffer Run, in the neighborhood of Charley Run, which were productive and paying wells. He also interested himself in the sinking of FIVE wells on the Allegany River, above Oil City, in the same year. These were small but paying wells.

In August, 1868, in connection with other parties, he leased forty acres of the McClintock Oil Company's land, near McClintockville, and twenty-five acres of the Robert Shaw farm, half a mile north of McClintockville, on the summit, overlooking Oil Creek. He leased, about the same time, seventy-five acres of the Corn Planter Oil Company's lands, adjoining the Shaw farm, with a portion of the Davis & Hukill tract, and soon after purchased twenty-six acres of the Robert Shaw farm, paying therefor only $300 per acre! Obtaining control of all this adjoining territory rapidly, he determined to develop it with his accustomed vigor. And here let us state one fact in regard to Mr. ANDREWS' mode of operating his territory. Having acquired possession, by lease or purchase, of any tract of oil territory, he promptly sets about developing and testing it. If a well is to be put down, he begins operations without unnecessary delay, and is the first to know the result. If the estimated cost of an engine and boiler, tools, tubing, casing, and the expense of drilling, amount to $5,000, he is only solicitous to see the "bottom dollar" and the bottom of the well at one and the same time. This we believe to be one important element of his success as an oil producer.

Immediately upon acquiring title to these *four* tracts—three by lease and *one* by purchase—he promptly contracted for drilling EIGHT wells, so located as to practically test the entire property. THREE of the EIGHT proved to be *dry holes;* but the remaining FIVE were abundantly productive. All this territory, embracing in the aggregate one hundred and forty acres, was rapidly developed, much of it producing largely and adding bountifully to the wealth of the enterprising owner. In the fall of 1871, he sold this property, real and personal, at $60,000, and this, after realizing from its product, in profits, more than $100,000 !

We have thus briefly sketched Mr. ANDREWS' career as a producer of Petroleum—and even this interesting record is incomplete, and shows but a small portion of his operations during the last five years, for he has purchased numberless interests in leases, and wells, and tracts of land, owned and developed by others. In *some* instances, these purchases were made for the sole purpose of assisting his less fortunate friends, but mainly as business ventures. The large majority of these latter investments brought him additional wealth, and have placed him in the front rank of producers of "the Creek," and region. During the five years covered by this portion of his history, there were months and months, that his DAILY income reached the sum of $4,000 from his oil properties alone! But under this "green-back shower," swelling into a flood that would have submerged and utterly destroyed some men, he remained the same careful, economical, judicious man of business, devoting to his affairs, his energies and constant attention.

He is still the possessor of valuable oil lands, and leases, and interests, in quite 200 wells, located at Brady's Bend, St. Petersburg, Petrolia, East Sandy, Bredensburg, the Milton, Huff and Henley Farms near Oil City, Fee Farms, Shaffer Run, Shamburg, Red Hot, Church Run and Colorado. He is the owner, *in fee*, of more than 5,000 acres of undeveloped oil territory, lying in various sections of the oil region, from Brady's Bend to Tidioute, on the Allegany River, and "on the Creek," from Cherry Tree Run to Titusville. These will add still further to his worldly wealth when developed; but we venture the prediction, that no amount of prosperity will change the general characteristics of the man as we have indicated them in this imperfect sketch of his history.

One fact in regard to the SYSTEM which Mr. ANDREWS has adopted in the conduct of his large oil business, deserves mention here. His oil interests, developed and developing, are scattered widely over the oil fields of the region. It would be an impossibility to visit them all, even once a month. He has therefore adopted a system of " checks and balances," that meet exactly the emergency. Every well is provided with blanks, which the person in charge is required to fill up once a day, giving the product of the twelve hours of his control. This record shows the product of the well at any date required. If delays have occurred, the cause is set forth. If expenses have been incurred, the amount is given, and the reasons are assigned fully. If the product has fallen off, it is either from neglect of duty or from natural causes, which is readily shown from these daily check reports. In this simple mode and manner, Mr. ANDREWS has from week to week, and oftener, if desired, the product of each and all his wells, and maintains as perfect control, and has as complete knowledge of his interests as he could have by daily or weekly visits.

Before closing this brief resume of Mr. ANDREWS' business career, we should add that he has large and increasing manufacturing interests at Titusville, at Corry, Pa., and at Nunda, New York, requiring a heavy capital to conduct. He is Vice-President of " The Gibbs and Sterrit Manufacturing Co.," one of the largest iron manufactories in North Western Pennsylvania. He is also a stockholder in four of the principal banks of the oil region, in three of which he is a director.

Mr. ANDREWS, is eminently a man of business, of acknowledged honor and unsullied integrity. He possesses a rare faculty of comprehending the intricate problems surrounding or entering into any matter challenging his attention. To this universally conceded acquirement must be added a more than commendable reticence as to his business affairs and ventures. SELF-MADE, he has become thoroughly wedded to a SELF-RELIANCE worthy of imitation. Quiet and unassuming in manner and habit, he is yet regarded by all who know him as a man of solid worth and real purity of character. Upright and honorable in all his business affairs, no " bonds " or " seals " can add to the binding force of his word, once given. Amiability, of the manly sort, is a prominent element of his character. Indeed, he seems always to be of an even temper, rarely ruffled or ill-humored.

Financial operations, involving thousands of dollars, are grappled with the same apparent ease and almost indifference, that accompany the simplest transactions. He, however, never loses sight of details, and his active mind is sure to work out successful results.

In his intercourse with the world, he is reserved and retiring, and yet he is cordial, and cheerfully so, with those who know him best. In his friendships he is warm and steadfastly attached. Possessing the finer sensibilities of a mature manhood, he cultivates all the social virtues and excellencies that render HOME attractive and inviting. A liberal giver to every worthy charity, and to the needy poor about him, he gives for the sake of giving, and without ostentation or a desire for publicity. Rigidly " temperate in all things," he is, besides, as blameless in his private life as he is unspotted and unblemished in his public career.

For a year past he has been in impaired health, partly induced, we doubt not, by the death of his wife, a most amiable and refined lady, a devoted companion, and a kindly, tender Christian mother. The poignant grief over such a bereavement, no man may appreciate or fully comprehend until the bitter cup has passed his own lips. Latterly, however, release from the cares of business, together with travel in a southern clime, have combined to reinvigorate and build up anew, a constitution heretofore preserved by an entire freedom from excesses of every character.

Mr. ANDREWS is now in middle life, full of vigor, full of enterprise, and in possession of ample ability, financial, as well as mental and physical, to accomplish desired results in the future, and we hesitate not to prophesy for him a career of great usefulness, and an individual prosperity sure to keep even pace with the careful, the determined, the economical, and the industrious.

HENRY HARLEY.

NEW YORK CITY.

THE subject of this sketch, one of the marked, and we may add, remarkable men of the oil region, was born of most excellent family, in Canton, Stark County, Ohio, on the 28th day of April, 1839. He is the eldest of a family of four—three sons and one daughter.

Young HARLEY had the advantages of an excellent common school education, and at the age of 16, entered the Rensselaer Polytechnic Institute, at Troy, N. Y., whence he graduated in 1858, with all the honors of the institution, as a civil engineer. He was soon after appointed to the position of Assistant Engineer, upon the Troy and Boston Railroad, and Hoosac Tunnel, of which Gen. Herman Haupt, formerly Chief Engineer of the Pennsylvania Central Railroad, was Chief Engineer. The responsibilities of this charge were assumed with so much confidence, and discharged with such rare ability, that a few months later, Mr. HARLEY received the appointment of Principal Assistant Engineer of the entire work, and was placed in the immediate control of the great work of the Hoosac Tunnel. This important and responsible position he continued to hold, discharging its duties with fidelity and ability, until the breaking out of the war in 1861, when the State of Massachusetts withdrew her aid from the road, and the work was suspended.

In 1862, Mr. HARLEY married, and soon after removed to Pittsburgh, where he became interested in the petroleum trade, in which he was very successful. He was the active partner of the firm of RICHARDSON, HARLEY & Co., at that day, one of the largest petroleum commission houses in western Pennsylvania. In 1863,

Woodburytype. A. P. R. P. Co., Phila.

HENRY HARLEY.

branch houses of this firm were established in New York and Philadelphia, and Mr. H. removed to the latter city, assuming the management of their large and increasing business at that commercial centre.

The excitements, and necessarily the speculative spirit engendered by and through them, which marked the history of petroleum developments in 1863–4, found Mr. HARLEY a willing participant in their attractions and prospective profits. He associated himself with Hon. Augustus Schell, James McLean, the late Benjamin Nathans, so mysteriously murdered in his own house some years later, John Bloodgood, Esq., and other prominent New York gentlemen and capitalists, purchasing large tracts of land in West Virginia, for oil development. This new enterprise required so much of his time and attention, that he was forced to relinquish his connection with the firm of RICHARDSON, HARLEY & Co., which he reluctantly did in the fall of 1864.

The results of his West Virginia enterprise did not meet his expectations, and in the latter part of 1865, we find him engaged in a comparatively new enterprise on Oil Creek, in Venango County, Penn., laying pipe-lines for the more rapid and less expensive transit of petroleum from the wells to railway facilities.

The first pipe-line, in the interest of the public, was laid from the old "Noble Well" on the Farrel farm, below Shaffer farm, to the then termini of the Oil Creek Railroad, at Shaffer. In this enterprise the Oil Creek and Allegany River Railroad was largely interested. It was known as "The Western Transportation Company," and was organized under a charter obtained from the Legislature of the State of Pennsylvania. Its projectors, however, "reckoned without their host." The pipe itself was $5\frac{1}{2}$ inches in diameter, and when completed, was about two miles and a quarter in length. It was laid upon a regular grade, and in the ordinary mode of laying water-pipe. Connected by lead points, it could not be made to hold its contents. It leaked at every joint, almost. Suffice it to say, this first attempt at piping oil was a disastrous failure, and it was abandoned.

The Pipe Line from Pit Hole to Miller Farm, was constructed in the fall of 1865. Mr. Van Syckle, of Titusville, was its projector and builder. Its construction required a large amount of money, involving Mr. Van Syckle in debt and embarrassments. The First National Bank of Titusville, then in the height of its power and prosperity, came to the relief of the enterprising gentleman, identified with the construction of this new mode of transporting oil to the railways, advancing to him as his needs required, more than $30,000. The line was finally completed, and after a few months of trial and disappointment, Mr. Van Syckle was forced to give it up as a losing venture. The enterprise, with its property, machinery &c., subsequently fell into the hands of the First National Bank, its largest creditor. This was in the fall of 1866. Shortly after, Jonathan Watson, Esq., a creditor of the Bank, came into possession of it, and subsequently disposed of portions of his interest to W. H. Abbott, Esq., and Mr. HARLEY. At this time—and since its completion we believe—J. T. Briggs and Geo. S. Stewart, of Titusville, had been running the line in the interest of the Bank.

Messrs. ABBOTT & HARLEY, who at this date owned one-half the Pit Hole and Miller Farm Line, and Mr. HARLEY, who owned all of the Benninghoff Run Line, believed the Pit Hole Pipe Line could be made to pay, in connection with the Benninghoff Run Line, already in successful operation, (of which we shall speak hereafter,) resolved to purchase Mr. Watson's interest, and consolidate the two lines, which was done, and thenceforward, for sixty or ninety days, the business was conducted under the firm name of ABBOTT & HARLEY.

And here we digress, in order to bring up the history of Mr. HARLEY's Pipe Line, constructing and completed about this date, from Benninghoff Run, to Shaffer Farm.

In the fall of 1865, Mr. HARLEY, began the construction of a Pipe Line from Benninghoff Run, to Shaffer farm, at that date, the termini of the Oil Creek and Allegany River Railroad. In this undertaking, his skill and experience as a civil engineer, were of great value to him. The line was constructed amid all sorts of threatening demonstrations, from a combination of oil teamsters, and lawless men, bent upon destroying whatever seemed inimical to their interests, or that ceased to minister to their wants. The threatenings of these men found open utterance at all points, and these were either privately or publicly endorsed and approved by those who assumed to have, or really cultivated an interest in the well-being of the men whose cause they espoused. Mr. HARLEY, however, with the same vigor and determination, which has characterized him through life, "pushed things," and in the spring of 1866, his enterprise was an acknowledged success, and his efforts universally commended.

While this Pipe Line was in process of construction, the teamsters of Shaffer farm and vicinity, then forming a large fraction of the population of "the Creek" between the Lower McElhenny farm, and Titusville, came to regard Mr. HARLEY's Pipe Line, and indeed all Pipe lines, as legalized robbery, and as infringements upon their rights as common carriers—as a fatal blow at their means of livelihood—or as many of these exasperated, ignorant men expressed it—"an effort to take the bread from the mouths of their children." Of course these lawless combinations had plenty of sympathizers, aiders and abettors, who saw, in the destruction of the business which had drawn so many thither, to engage in "hauling oil," a loss of their own sources of revenue, and hence it is, or was, that Mr. HARLEY had few friends who dared to avow themselves such, and so he may be said to have been compelled to "go it alone." Threats of vengeance, torrents of abuse, and a wild clamor for what these men termed their "rights," greeted him upon every hand. His movements were watched with daily increasing frenzy, on the part of the large body of teamsters, all about him. There were street brawls and bar-room rows, growing out of these Pipe Line affairs, which finally culminated in an attempt to impede their construction by acts of violence, and a resort to mob-law. These infuriated, but misguided men, set fire to the wooden tanks belonging to Mr. HARLEY, many of which were filled with oil. They sought to destroy the Pipe Line, by breaking the joints, and by every conceivable device, determined to thwart the designs of the enterprising and plucky projector. They sent him letters threatening his certain assassination, if he did not abandon his scheme for impoverishing them by ruining their business! Several assaults were made upon the men employed in constructing the line, in which pistols, clubs and stones were freely used, but we believe no lives were sacrificed.

While these lawless demonstrations were transpiring, Mr. HARLEY was not idle. He secured several detectives from New York and elsewhere, made teamsters of them, and they in turn made confidants of their fellows. In less than a week after the adoption of this scheme for the capture of the leaders, Mr. HARLEY had more than twenty of their number under arrest, and conveyed to the county jail at Franklin! This strategetic movement demoralized the remainder of the gang, and in ten days after, more than three hundred teams and teamsters had shaken "the dust from their sandals," and gone hence to return no more. The twenty or more leaders arrested and committed to the county prison at Franklin, were kept there for two or three months, when, no one appearing against them, they were discharged, wiser, and we doubt not, better men!

Mr. HARLEY completed his line from Benninghoff run to Shaffer farm, and it was a positive success. It had cost him months of toil and a large amount of money, but all this was of little account when compared to the fearful trials and almost insurmountable embarrassments and determined opposition he had encountered from the people we have mentioned. Going to and from his works with his life in his hands, a price put upon his head, assassination threatened, overt acts committed, and the general voice, openly or secretly against his enterprise, one can, in a measure, at least, comprehend his unenviable surroundings. He, however, found himself ample for the emergency. Turning neither to the right nor to the left to appease the wrath of those who sought his life or threatened the destruction of his property, he completed his enterprise, and met his reward in its unqualified success.

" *The Western Transportation Company*," held the only charter granted by the Legislature of the State of Pennsylvania, at that period, for piping or transporting oil from the wells to railway stations. This was the charter used for constructing the 5½ inch pipe-line, from the Noble Well to Shaffer farm, before referred to. As opportunity presented itself, Messrs. Abbott & Harley purchased the stock of this company, and in the course of the summer and fall of 1867, found themselves in possession of sufficient of its script to control it all, and their lines were subsequently organized under the old " *Western Transportation Company's* " charter, and took the name of " *The Allegany Transportation Company.*" The following gentlemen were elected the first Board of Directors, held January 25, 1869 :

Henry Harley, W. H. Abbott, Joshua Douglass, J. P. Harley, and Jay Gould. At a subsequent meeting of the Board, Mr. HARLEY was elected President; Mr. Abbott, Secretary; T. W. Larsen, Treasurer, and William Warmcastle, General Superintendent. Mr. Warmcastle has been associated with Messrs. Harley & Abbott, in one capacity and another, ever since these gentlemen commenced their pipe-line enterprises, and laid, or superintended the construction of the first two miles of the Benninghoff Run line. Upon the consolidation of Mr. Abbott's and Mr. Harley's interest, and the organization of " *The Allegany Transportation Company,*" in 1867, Mr. Warmcastle was appointed General Superintendent, and has continued to hold the position with that uninterrupted relationship of confidence and rare ability always characterizing a faithful discharge of responsible duties.

In 1868, " *The Allegany Transportation Company,*" having grown into almost colossal importance, and hence under-

stood to be " a power " in the oil regions, the attention of the various railway lines to the sea board were attracted to it, and the President of the Erie Railway Company, Jay Gould, Esq., succeeded,
in advance of all negotiators, in obtaining control, by purchase, of
the valuable interest.   Simultaneously with this purchase, which
comprised little more than one half the stock, Mr. HARLEY was
appointed to the Superintendency of the oil traffic of the Atlantic
and Great Western, and Erie railways, with the title of " General
Oil Agent." This responsible position he continued to fill, system-
atizing its cumbersome and unwieldy proportions, and reducing
them to practical business comprehension, as susceptible of control
as the simplest problem in mathematics—through the administration of Jay Gould, and he was re-appointed to the same position,
upon the accession of the new management, of which Gen. John
A. Dix, present Governor of New York, was made President.   In
May, 1872, he tendered his resignation, his private affairs requiring his attention and presence in Europe. Soon after the acceptance
of his resignation, Mr. HARLEY visited Europe, spending some
months abroad, business affairs in the main engrossing his time and
attention.

Early in the spring of 1871, the Oil Creek and Allegany River
Railway, impressed with the belief, doubtless, that " The Allegany Transportation Company" had organized itself so fully and
completely in the interests of the Erie Railway Company, as to
compel it to stand in a position of armed neutrality, if not open
hostility to the interests of the former road, threatening seriously
their freighting facilities, resolved upon a new pipe line, to meet
the emergency.   A company was organized, under the title of
" THE COMMONWEALTH OIL AND PIPE COMPANY," in the interest of the Oil Creek Railroad, and its construction promptly
entered upon.

In August of the same year, Mr. HARLEY opened negotiations
with the Oil Creek road, for the purpose of effecting a combination
of interests, and shortly after terms were agreed upon, resulting in

the organization of "THE PENNSYLVANIA TRANSPORTATION COMPANY," with a capital of $1,700,000, owning and operating nearly 500 miles of Pipe Line, running hither and thither upon the surface of the earth " over mountain and gorge, over rock and plain," here, there, and everywhere, in the almost double triangle, made up of Tidioute, Triumph, Irvineton, Oil City, Shamburg, Pleasantville and Titusville, with the apex at Miller Farm.

The new organization at its first meeting elected the following Board of Directors: Henry Harley, W. H. Abbott, Jay Gould, A. R. Williams, J. Douglass, C. B. Wright, U. S. Lane, Geo. K. Anderson and W. H. Kemble. Subsequently Mr. HARLEY was elected President, Mr. Abbott, Treasurer, T. W. Larsen, Secretary, and Mr. Warmcastle, General Superintendent.

Mr. Thos. W. Larsen, the first Treasurer of The Allegany Transportation Company, and latterly, Secretary of The Pennsylvania Transportation Company, deserves mention in this connection. The position now held by Mr. Larsen, like that of Treasurer of the Allegany Transportation Company, is one of great responsibility, requiring ability of a high order, and integrity of an unquestioned character. Mr. Larsen's connection with this important enterprise from its commencement to the present day, is proof positive of his entire fitness for the position he occupies.

It may be stated here, without fear of contradiction that " The Pennsylvania Transportation Company," under its present organization and control, is among the wealthiest and most substantial institutions of this commonwealth. Its stock owners are comparatively few, but they are of a character to warrant unqualified confidence in its soundness and stability. Among the most prominent of these may be named Col. Thos. A. Scott, the great railway king of the continent, Jay Gould, C. B. Wright, U. S. Lane, HENRY HARLEY, W. H. Abbott, Geo. K. Anderson, and Mrs. James Fisk, Jr.

Since Mr. HARLEY's return from Europe he has employed himself mainly in closing up his business connections in New York

City, to make his home in Titusville, where he now resides, honored and respected by a wide circle of friends, who know and appreciate him for his worth, integrity and capacity as a man of business.

Personally Mr. HARLEY is a gentleman of rare excellencies. In his friendship he is devoted and constant. He belongs to that class of men who win their way to private and public esteem through a sort of magnetic channel, founded really upon "good will toward all." Few are blessed with this gift, but Mr. HARLEY seems to be its possessor in an abundant and enviable degree. In all his business connections he is recognized as the possessor of a clear head, with an ample comprehensiveness of detail, giving assurances of success to whatever enterprise he undertakes. In private life he is blameless, without high-sounding professions—charitable without ostentation, always ready with word and purse to relieve the needy and aid the unfortunate. In all his public and private relationships he is recognized as a man of worth, integrity and rare business capacity. His word is his bond, and his bond is gold, or its equivalent. He is now in " the flush of his manhood and the years of his usefulness," blessed with a vigorous constitution, guaranteeing bodily health equal to the responsibilities which his large and constantly augmenting interests rigidly impose. Let us hope these may be vouchsafed to him for many years, and that his career so auspiciously begun may be a "steady series of brightening out-looks," and of brighter achievements, through " a thousand moons yet unfulled."

Woodburytype. A. P. R. P. Co., Phila.

COL. R. B ALLEN.

COL. R. B. ALLEN.

PARKER'S LANDING, PA.

COL. ALLEN is a native of Delaware County, Pa., born in Upper Darby Township, on the 22d day of September, 1838. He comes of good old Pennsylvania stock, and is the second son of a family of nine children—four sons and five daughters. His father was a carpenter and joiner and master builder by trade, and this calling he followed successfully and industriously, rearing a large family, and giving each and all his children a good common school education. The subject of this sketch was kept steadily at school until he was sixteen years old, acquiring a thorough knowledge of mathematics, grammar, and the rudiments of a higher grade of scholarship. At sixteen he went to his trade with his father, and was counted both an industrious apprentice and a skillful workman. During his three years of apprenticeship he attended school during the winter months, storing his mind with a knowledge of the principles of mechanics, mathematics, &c. He followed his trade industriously, until the war broke out in 1861,—then in his 22d year. In April, soon after Fort Sumter had fallen into rebel hands, he volunteered for three years, in a Cavalry company, organizing in the vicinity of his home, and called "Mad Anthony's Boys." There were vexatious, if not needless delays, in getting into the service, for weeks and weeks after their organization, but the "boys" made good use of their time and leisure. When they came to be mustered, as they were, in July following, they were comparatively well drilled and disciplined for the tented field. Upon their muster-in, they were united to and formed a part of the 5th Regiment Pennsylvania Volunteer Cavalry. The Regiment was promptly marched to the front, and in August, 1861, became part and parcel of the

Army of the Potomac.   In 1862, the Regiment participated in the Peninsula Campaign, and the advance upon Richmond.   After the fearful disasters of that year, it was detailed for special duty in the Dismal Swamp region, then infested with marauders, Ku-klux, and spies.   They remained here until the early months of 1864, when they were united to Gen. Kautz's Division, and ordered to march over land, to City Point, charged with the duty of severing railway communications, south of Richmond, and making a junction with Grant's lines at Bermuda Hundreds.   The Division accomplished its mission, destroying the railways between Petersburg and Richmond, and all other railway communication, south.   Subsequently, the Regiment took an active part in the great Col. Wilson raid, upon the Danville and South Side Railroad.   In all these advances, skirmishes, raids, and engagements, young ALLEN bore a prominent part, in upholding, defending, and maintaining the flag of his country.

His term of enlistment ended in the fall of 1864, when he was honorably discharged the service, and returned to his home in Delaware County, Pa.   Here he again took up his trade of a carpenter and builder.   The swamp fevers of Virginia, and the severity of the campaigns of 1864, had left their traces indelibly upon his otherwise rugged constitution, and only a few months after his return home his health failed him, and he was obliged to abandon the severe labor of his profession.   He spent a year or more as a book-keeper in Philadelphia, and subsequently, engaged in the manufacture of patented articles, labor-saving machines, &c. This enterprise did not meet his expectations, and after a few months of profitless labor, he disposed of his interest, and was again " upon his oars."

About this date a personal friend, suggested the acceptance of a Superintendency of an Oil Company in the oil regions of Western Pennsylvania.   The suggestion met the hearty co-operation of Col. ALLEN, and very soon after, he was appointed Superintendent of " The Clarion and Allegany River Oil Company," with head-quarters at Parker's Landing, and in October, 1868, he assumed the responsibilities of the trust confided to him. At the date of his arrival at Parker's, not more than four or five producing wells were in operation there, and two or three of these were the property of The Clarion and Allegany River Oil Company, whose superintendent he was. He was not long, however, in making himself familiar with his duties and responsibilities, nor in apprising the company of the exact condition of their valuable property. From the outset, he won the confidence of those who had entrusted him with their interests, and this relationship has been strengthened and intensified by five or six years of contact, for Col. ALLEN, is still the superintendent of " The Clarion and Allegany River Oil Company,"—a fact that speaks volumes for his efficiency, devotion and trustworthiness.

In the spring of 1869, Col. A., commenced operations upon his own account. His capital was not large, but his determinations were equal to his requirements. His first lease consisted of half an acre of land on the Fullerton Parker farm, at Parker's, at a graded royalty. At the completion of this, his first well, its product reached scarcely FIVE barrels per day! His second venture was upon the Robinson farm, same locality, and this proved to be a dry hole! Upon the heels of these seriously deplored failures, Col. ALLEN became thoroughly in earnest, and as rapidly as possible contracted for the drilling of ten or twelve new wells in the Parker oil field. Very many of these were largely productive, and the profits of the plucky proprietor in the aggregate were largely remunerative.

The fall, winter and spring of 1869, and '70, witnessed a very marked advance of the lines of development, to the northeast and southwest—from the Clarion river to Bear Creek, and beyond upon the same belts. Col. ALLEN, moved with the picket line, and continued his operations vigorously and with profitable results.

Without going into a detailed statement of his enterprises during the past three years, we may justly claim for him, prominence and an honorable distinction as a producer in the Parker's District. He has been interested as principal or partner in many of the new oil fields, from Parker's to Bear Creek, Fairview, Petrolia, Karns City. &c., and north-westerly to the Clarion River, and into the St. Petersburg District. He is steadily increasing his operations, extending his developments and adding to his investments, and these are sure to add to his worldly possessions, and work out for him a name and fame, which he may justly, and without boast, claim as his own.

Col. ALLEN, is President of "The Grant Pipe Line Company," having its termini at the Parker's Railway station, of the Allegany River Railroad, directly opposite Parker's Landing. The responsible duties of this position he discharges with consummate ability, and to the entire satisfaction of stockholders and patrons. He is also, a director in "The Exchange Bank of Parker's Landing."

Whatever has promised good to the interests of the producer of Petroleum oil, in western Pennsylvania, has had the hearty concurrence, and unqualified approval and support of Col. ALLEN. When, in the early months of 1872, "The South Improvement Company" sought a foothold in the oil region, and gradually unfolded its fearful demands for a monopoly of the Petroleum trade, not alone in its production, but in its transportation and absolute control, at home and abroad, Col. ALLEN was among the first, foremost and most uncompromising of its opponents and denunciators. By voice and purse, and unremitting effort, he accomplished the work of a score of less determined men, and did it promptly, cheerfully and well.

Col. ALLEN, wherever he is known, is recognized as a rigidly upright and honorable man—upright in all his business relations, and honorable in all his personal and private transactions. A man of quick perceptions, he unites in himself administrative ability of a high order. Socially, he is by instinct and education, a gentleman. Of commanding presence, and yet of modest pretense, he has about him a weight of character, rarely possessed and sure to be observed and felt by all who make his acquaintance. Of warm friendships and unmistakable convictions, his opinions and judgment of men, are based upon thorough acquaintance, and are never given hastily or without mature deliberation. A man of generous impulses, he gives freely of his means for all worthy objects. He is universally respected, and admired for his worth, integrity, and for the tenacity with which he clings to his convictions and purposes.

Col. ALLEN is a bachelor, and has surrounded himself with all the luxuries of a bachelor's home. Books, paintings, and whatever contributes to his enjoyment and comfort, make up the attractions of his sumptuous apartments, and these he shares with his troops of friends, dispensing his hospitalities with bountiful generosity. In closing this incomplete sketch of the man and his past, we may say, without being open to the charge of meddlesome interference, that while he seems to enjoy his bachelor-hood and single blessedness, in the fullest degree, no man of his social excellencies and rare personal good qualities, has the right—the Scriptural right we mean—to thus "hide his light under a bushel," and so we admonish and beseech him to marry, and that too, without delay, and repent, if repentance becomes necessary, at his leisure.

COL. E. A. L. ROBERTS.

TITUSVILLE, PENNA.

COL. E. A. L. ROBERTS, the inventor of the Torpedo, for Oil Wells, was born in the town of Moreau, Saratoga County, New York, on the 13th day of April, 1829. The early years of his boyhood were without incident, save that at the age of thirteen, he boldly struck out for himself. In 1846, then in his seventeenth year, he enlisted as a private in Col. Pitcher's company recruiting at Sandy Hill, N. Y., for the Mexican War. Though a mere lad, he served his country faithfully, and was counted one of the best soldiers in the company, often receiving the commendations of his commanding officer. After a service of twenty-two months, at the close of the Mexican War, he was honorably discharged, and returned to his home in Saratoga County, and soon after entered the Academy at Amenia, Dutchess County, N. Y., where he remained one year. After leaving the Academy, in 1851, he entered the dental office of C. H. & W. B. Roberts, at Poughkeepsie, N. Y. Here his natural bent for mechanics, found an ample field, and a few years later we find him a full partner with his brother, Dr. W. B. Roberts, in the city of New York. This partnership continued one year, and then Col. R. resolved to " go it alone." He disposed of his interest to his brother, W. B. Roberts, and soon after opened a dental depot in Bond street, New York, where he entered largely into the manufacture of Dental material, and in a very short time brought to perfection a Mineral Compound, now used by the Dental profession for making what is known as " Continuous Gum Teeth." For his many improvements made in den-

Woodburytype. A. P. R. P. Co., Phila.

Col. E. A. L. ROBERTS.

tal science and dental operations, he was awarded three gold and silver medals by the American Institute.

In 1857, he obtained letters patent for a Dental and Coupeling Furnace, now in general use by Dentists, and by United States Assay officers.

In 1858, he perfected and obtained letters patent for a Vulcanizing Machine, which proved to be of great value to the Dental profession, and which is now in use in all parts of the world. This Vulcanizing Machine was infringed, and he was compelled to bring suits in its defence. These proved to be very expensive, and he was forced to sell his patent for the paltry sum of $2,000. The purchaser subsequently realized from the invention over $100,000.

In 1859-60, Col. Roberts constructed and brought to successful completion, his large Oxhydrogen Blow Pipe, and made experiments in melting 60 ounces of platina. This scientific achievement is referred to with considerable elaboration in Appleton's Cyclopedia, under the index word, "Blow Pipe and Platina." A fuller description of this valuable discovery is given in Hodge's work on American Inventions.

The Rebellion of 1860-61, found Col. ROBERTS in the full enjoyment of health and a robust, vigorous constitution. Promptly he lent his individual aid to the government, and actively engaged in raising Regiments and forwarding them to the scene of conflict. In 1862, he was appointed Lieut. Col., of the 29th New Jersey Volunteers, and remained with it—oftentimes its Commanding officer—until the battle of Fredericksburg, under Burnside, had been fought. Soon after, failing health compelled him to resign his command and return to his home in New York.

In 1863, he assisted Col. F. A. Conkling of New York city in recruiting the 84th Regiment, New York National Guards, and was complimented with a Captain's Commission, and assigned to Company C., and soon after was placed in charge of the Center St. Arsenal, and was in command of that post when the riots of that year in New York, threatened such serious consequences.

In July, 1864, Gov. Seymour of New York called for volunteer regiments of the National Guard, for 100 days, and Col. Conkling tendered the 84th Regiment, and was accepted, and ordered to march without delay.   Col. ROBERTS still held the command of Company C., and was ready, and went promptly to the Capitol with his command.   From Washington, the Regiment was ordered to Martinsburg, thence to follow up Sheridan's Division, through that part of Virginia since made famous by being the scene of " Sheridan's Ride," with all its glorious consequences and invaluable results.

On the expiration of the one hundred days which the Regiment had volunteered to serve, Col. ROBERTS returned to New York, and completed drawings of his Torpedo, for Artesian and Oil Wells, together with specifications, which he had commenced in 1862.   This done he applied for a patent.   This was in Nov. 1864.   He then constructed six Torpedoes, and came to Titusville, with a view to test their adaptability to the uses he intended.

It was the labor of days and weeks to induce owners of wells to permit him to use his new invention for increasing the production of oil wells.   Some said it would do no good, others feared it might do injury, and the majority believed the process would absolutely destroy the wells.   In January, 1865, however, he obtained permission to explode two Torpedoes in the ' Ladies' Well," on the Watson Flats, which largely increased the product of the well, creating a great deal of excitement, and establishing beyond cavil, the feasibility and practicability of the process.   From that day, the virtue and necessity for Torpedoes in oil wells may be dated.

These successful experiments with Torpedoes in oil wells, prompted others to lay claim to originality of discovery, and within a few months not less than, six applications were made at the Patent Office for the same invention.   Interferences were declared by the different claimants, and a contest of more than two years was had before the patent was issued.   Appeals had been taken from the

Examiner's decision, to the Examiner-in-chief, and from the Examiner-in-Chief, to the Commissioner of Patents; and from the Commissioner of Patents, to the United States Supreme Court of the District of Columbia. In every instance, decisions were rendered in favor of ROBERTS. These several suits embarrassed Col. ROBERTS seriously, and delayed the issuance of his patent for Torpedoes until the 20th of Nov., 1866—more than two years after filing his specifications.

Infringements upon this patent became frequent and vexatious; and to protect his rights, Col. ROBERTS was compelled to commence suits against several parties. The cases of Roberts vs. Nickerson, and Roberts vs. Hammar, were among the first tried—the Court, after a patient hearing, sustaining the ROBERTS patent without qualification. Other suits were brought, and a like result was obtained in each case. The suits of Roberts vs. Dickey, and Roberts vs. The Reed Torpedo Company, followed, and were contested to the last. Decisions were rendered in ROBERTS' favor, fully sustaining the validity of his patents, by Judges McCandless, Grier, Strong and McKennan.

"The Producers' Association," an association embracing among its membership a large majority of the Oil Producers of the Pennsylvania oil fields, having a President, Secretary, Treasurer, and Board of Directors, Executive Committee, &c., in order to define their own rights under the Roberts' Torpedo Patent, determined to contest the validity of Col. Roberts' claim to originality. From $40,000 to $60,000 was raised to this end, and thereupon the Torpedo war became general, and at all points, determined and uncompromising—Col. Roberts maintaining his position with a consciousness of the justice of his claim, and "The Producers' Association" only desirous, as they alleged, to arrive at such conclusion as would establish the rights of all parties.

In January, 1871, a final hearing was held before Judges Strong and McKennan, in the U. S. District Court at Washington, and the Roberts Patent was sustained—the Court granting a perpetual injunction against all infringers in the following language: "*The complainant is therefore entitled to a perpetual injunction, and to a decree for account.*"

The expenses attending the litigation of suits brought by Col. ROBERTS to defend his rights and franchises, have been very great. He has disbursed more than $100,000 in these cases, numbering not less than two hundred in all. Many convictions and imprisonments have followed, and more than $60,000 in judgments have been obtained against infringers here and there, to the present time.

On the 7th of May, 1873, he secured Letters Patent for a new and useful "Improvement in treating Explosive Compounds, to render them safe for blasting and other purposes." Under this Patent, Col. ROBERTS has demonstrated that he can treat Explosive Compounds, especially those mixed or combined with water, so as to render them absolutely harmless, and to be safely handled, or stored for any length of time, "without any decomposition occurring that would give rise to dangerous results."

On the 3d day of June, 1873, Col. ROBERTS obtained a re-issue of his patent for torpedoes, granted Nov. 20, 1866, and ante-dated May 20, 1866. By this re-issue, he claims "a new and useful improvement in process of increasing the capacity of oil wells, and also of restoring oil wells to productiveness." He also secured by this re-issue a patent for the apparatus used in lowering the torpedo to the oil-bearing rock.

Just what Col. ROBERTS claims by his "specifications," as his invention, upon which letters patent have been granted, is thus concisely stated in his application:

1. The method or process of increasing or restoring the productiveness of oil wells by causing an explosion therein, at or near the oil-bearing point.

2. The method or process of increasing or restoring the productiveness of oil wells by causing an explosion at or near the oil-bearing point, in connection with superincumbent fluid tamping, substantially as set forth.

3. In combination with a torpedo adapted to deep wells, the employment of a weight and guiding cord for the purpose of exploding the charge, substantially as described.

4. The means herein described for suspending, adjusting, and exploding torpedoes in oil wells at any desired or predetermined point, consisting of a wire or cord, attached to such torpedo and through which, as a medium of communicating ignition, the torpedo is exploded.

Col. ROBERTS, following the natural bent of his genius and rare mechanical skill and ingenuity, is now engaged in perfecting a new and novel mode of propelling water craft of every conceivable character and burthen, both inland and at sea. If this new motive power proves to be a success, and it looks very like it, a revolution in navigating our inland lakes, rivers and canals, and indeed the ocean, is sure to follow. We regret the experiments now making are not sufficiently developed to warrant a full description of this new mode of propelling vessels of all kinds, in whatever waters they may be found.

Col. ROBERTS is at this writing, a little over forty-three years of age. He is blessed with a strong constitution, a clear head and a mind filled with original ideas. His personal appearance denotes activity of both mind and body. A man of iron will and indomitable perseverance, he exhibits a capacity equal to any emergency which may surround him. In all public enterprises he is liberal, open-handed and proverbially generous. To private charities he gives with a bounty and cheerfulness worthy of emulation. His private friendships are not many, but they are lasting and close. In the city of his adoption may be seen many monuments of his enterprise, erected under his own eye, and with a view to adding beauty and thrift to all about him. We add the hope that he may be spared to a green old age to enjoy the fruits of his genius, his labors and his enterprises.

MARCUS BROWNSON

TITUSVILLE, PENNA.

MR. BROWNSON is a New Yorker by birth and rearing, born in the County of Delaware, on the 23d day of May, 1822. He is the next youngest of a family of seven children—three sons and four daughters. His father was a farmer, and young MARCUS remained under the parental roof, assisting, as he grew up, in the labors of his father, until he was nineteen years old. At this age, his father died, leaving him in sole charge of the farm. His educational advantages were limited to the district school of the neighborhood, and thus he was enabled to improve a portion of each year, three or four months at farthest, until the death of his father, which occurred, as before stated, when he was nineteen years old. He, however, obtained a very good knowledge of the common branches, and was considered a proficient scholar.

Upon the death of his father, young BROWNSON assumed the entire responsibility of the conduct of the farm, providing for the support of the family, and the payment of a heavy incumbrance upon his father's estate, divided by will among the heirs. He continued to reside upon the farm, and to conduct it industriously and successfully for nearly fifteen years, steadily working out of the embarrassments that surrounded him at the commencement; and, we may add here, that long before these years had passed, he had provided for all the liabilities upon the property, and had, by economy and careful attention to his affairs, been able to count up a goodly sum as the profits of his industry and thrift.

On the 1st of January, 1844, Mr. BROWNSON, then twenty-two years of age, married Miss MARY A. WALLING, a lady of New

Woodburytype. A. P. R. P. Co., Phila.

MARCUS BROWNSON.

England birth and parentage. It is both just and proper that we make special mention of Mrs. Brownson in this connection; for her husband attributes to her, as he ought, the possession of womanly qualities, sure to gladden the household, lighten the burdens of her help-mate, be a guide and example to her children, and a jewel in the family and social circle.

In 1855, eleven years after his marriage, Mr. BROWNSON disposed of his farm, and found himself the possessor of $3,000 in gains, which, with the additional sum of $500, devised by his father's will, gave him a small fortune of $3,500, to embark in any new enterprise that might offer itself.

Some months prior to the sale of his farm, Mr. BROWNSON received a visit from a younger brother, who had long been a resident of Oregon. Many times during these months of re-union, the brother had expressed a wish, amounting almost to a prayer, "that MARCUS would return with him to Oregon," and make that his future home. Mr. B. finally determined to accede to his brother's wishes, if he could make advantageous sales of his property. His purpose settled upon, it was only the work of a few weeks to get ready. The farm was sold, everything settled up, preparation for the journey completed, and at the end of three months from the younger brother's arrival, the twain, with their families—the younger brother had left his family when he first migrated to the then territory of Oregon—set out for that far distant country.

This was in the early spring months of 1855. The party sailed from New York, via the Isthmus route, and in due time landed at San Francisco. About the date of Mr. BROWNSON's arrival on the Pacific Coast, the Indian tribes of Oregon and Washington Territories were in arms against the whites, and massacres, murders, and depredations of every conceivable horror were being committed by these dreaded savages. Gen. John E. Wool, then in command of the Department of the Pacific Coast, was fitting out transports with army supplies and munitions of war to subdue these " barbarian wards of the government!" Mr. BROWNSON sailed in the same steamer which conveyed Gen. Wool and retinue, to the scene of slaughter. The steamer, with its convoys, was bound for Portland, Oregon, and without accident or delay, reached that port.

The headquarters of the United States forces was at Corvallis, a point of some importance on the Williamette river, seventy-five miles distant from Portland. This "settlement" Mr. BROWNSON resolved to make his future home, and thither he removed soon after his arrival at Portland. He had scarcely made a home for himself and family at Corvallis, ere he was tendered the appointment of Commissary of Subsistence in the army, and upon the discharge of these responsible duties, Mr. BROWNSON entered at once. Subsequently he was commissioned "Purchasing Agent," in the Quarter Master General's Department, thus adding to his burdens, and, at the same time, enlarging his usefulness and power. This latter appointment covered the purchase of horses, mules, cattle, and indeed almost every needed article required by the Army of the Oregon Indian Campaign. This position he held, discharging its varied and oftentimes hazardous responsibilities, with faithfulness and fidelity to his trust, until the close of the war, which continued about eighteen months.

After his retirement from government service, he engaged in the business of purchasing and driving cattle to Portland, whence they were shipped to Van Couver's Island, Puget Sound, Victoria, &c. At this time the Frazier River gold mining excitement was at its height, and some questions of boundary had arisen between the British government and our own. British troops were stationed near Van Couver's Island, and our own defenders were "on the look-out" at several points on the coast, at Bellevue Island, Fort Townsend, etc. Added to these bodies of men were "forty thousand" gold miners scattered along Frazier River, and to these camps as well as the several military stations, cattle were shipped for a market. We have thus particularized, that the reader may form some idea of the enterprise with which Mr. BROWNSON was identified.

This business connection and important traffic—important in more ways than we have time or space to mention, was continued, increasing in magnitude and augmenting in profits, until the spring of 1860, when Mr. BROWNSON determined to re-visit his home in the State of New York, and soon after sailed thence, where he arrived in the early days of June of that year.

The Presidential canvass of 1860, was just opening, and the country was filled with mutterings of an approaching national crisis, never before experienced. The election of Abraham Lincoln followed in the fall of 1860, and thenceforward the threatened rebellion and secession of the Southern States of the Union, began to assume alarming proportions. The "overt act," was committed, firing upon Fort Sumter, in April, following, and the North and the South sprang to arms. Mr. BROWNSON promptly espoused the Union cause, and did his whole duty in aiding the government in the terrible struggle for its own maintenance, through four years of bloody civil war. He had resolved to remain at the North, and "survive or perish" with those who would preserve the government of our fathers from rebellious hands. At one time during the war he held minor relationships to the armies of the republic, furnishing troops to the thinning ranks, and whatever else he might do, to sustain "the flag of the free." As a member of the Sanitary Commission, he traversed the lines of the Army of the Potomac, visiting the sick and ministering to those requiring assistance of any character.

During these years of civil strife, he engaged in no settled business enterprise, until some time in the fall of 1863, when he removed to Binghamton, N. Y., where he remained only a few months, for he had already determined to visit the oil regions of Western Pennsylvania, and try his fortunes in that marvelous field of industry and prodigal wealth.

In June, 1865, Mr. BROWNSON removed to Titusville, then the great distributing centre of the oil fields of "the Upper Creek region." The Pit Hole "dazzle" was at its zenith, and thither, a few days subsequent to his arrival at Titusville, he bent his steps, and not long after, was induced to purchase a small interest in a well going down, upon the Holmden farm. Without stopping to give details of this investment, we may add, that the interest cost him $500, and that the well proved an utter failure! Some months later he purchased a brick building and lot, upon the site now occupied by "The City Bakery," on Spring street, Titusville, and this he rented, for the purpose it is now used—a bakery. At the end of six months the building was destroyed by fire, and so this investment "turned to ashes"—for it was only partially insured. Mr. BROWNSON rebuilt it of brick, in 1865, and soon after re-sold it to the original owners.

These, and other unfortunate ventures, had to a degree crippled Mr. BROWNSON financially. His entire capital, when he arrived in Titusville, in 1864, did not exceed $7,000. And this sum, however carefully handled, and invested, without some gains, would hardly "stick to a man through the Pit Hole campaign, and stand one fire with a loss of $3,000 to $4,000!" Added to his embarrassments, was the fact that he had loaned to a personal friend, whom he sought to aid in his enterprise, $2,500, taking some "wild cat" security for its payment. This debtor betrayed his confidence, took "French leave," and departed hence, leaving Mr. BROWNSON and other creditors to get their honest dues as best they could. It was full three years before the absconder was heard from, and four years before Mr. B. met him face to face. Meantime, a year or two after his departure, the security left to Mr. BROWNSON became valuable, and the $2,500 were realized. About the close of the year, 1865, however, the balances of Mr. BROWNSON'S $7,000 capital, balanced each other—he had next to nothing in cash, and he owed no man a dollar! And here we may remark, *en passent*, that during a business career of more than thirty years, he has paid one hundred cents on the dollar of his indebtedness, and has never asked rebate or discount from any sum he was honestly or legally liable for.

In 1866, Mr. Brownson resolved to try his fortunes as an oil producer. He had little or no cash capital, but he had an unblemished name and fame, and this was capital sufficient for an industrious, determined man like him. He removed to Pioneer, in the spring of that year, and soon after, in connection with others as partners, secured three half acre leases upon the Benninghoff Farm, (Western Run,) just then developing quite largely. For these leases about $1,000 each was paid, "and half the oil." As before stated, his own resources were very limited. His friends at Titusville had tendered him accommodations, by way of endorsement, &c., to prosecute his enterprises, but he thankfully declined them, preferring, as he avowed, "to fail, if fail he must, without embarrassing his friends, and dragging them down with him."

After procuring the leases above named, interests were disposed of in each, sufficient to repay the bonuses, and complete the wells. In this safe way he began his operations as a producer of petroleum, retaining in each lease a "free interest," which in the end, proved to be quite productive, and in each instance were sources of revenue and profit to him.

We should state here, that before his removal to Pioneer, he had acquired a small interest in a well on the McClintock Farm, and another at Benninghoff Hollow, the former of which was a failure, and the latter a paying investment. The wells, three in number, upon the leases first mentioned, were each good producers, and went far towards mending the broken fortunes of Mr. Brownson. Later, in 1867, he was interested in two other wells upon the Benninghoff Farm, one of which paid good dividends, and the other proved valueless.

During 1867, and '68, the line of developments extended across Pioneer Run, to the Foster Farm, and here Mr. Brownson continued his operations during these years with flattering success.*

---

* During the years of 1868, and '69, Mr. B. was also extensively engaged in oil operations at Shamburg, and Red Hot, and held interests in eight or ten good producing wells. This property he sold, in two different quantities, after realizing very acceptably from their product for nearly two years.

In 1869, continuing his developments upon the Foster Farm, he extended his enterprise and operations to the Pierson Farm, west of Pioneer Run and the Benninghoff Farm. Success crowned his efforts, and later in 1869, and '70, in conjunction with F. W. Andrews and others, developments were inaugurated upon the Shaw Farm above Rouseville. This farm was, during its producing life, prolific of good wells. The enterprising lessees, while they disposed of their product, in 1868, and '69, at very low rates, were yet enabled, before the farm was exhausted, to realize large gains, and it is safe to say, that here Mr. BROWNSON laid the foundations of his future prosperity and substantial wealth.

It was during these three or four years of prosperity, that in extending his enterprises and enlarging his investments, Mr. BROWNSON became interested in the marvellous "Venture Well," at Fagundas, that produced from the start between three and four hundred barrels per day! In 1870, he disposed of his interest in the "Shaw Farm," together with his Fagundas and Hickory property, to F. W. Andrews and Charles Bly, for $55,000.

He had the year previous made sale of his Pierson Farm property, together with some other small interests in that vicinity, for $10,000.

In 1869, or '70, Mr. BROWNSON, in connection with F. W. Andrews and T. C. Joy, of the one part, and D. B. Benson & Co., of the other part, purchased four hundred acres of land lying in what is known as the Colorado oil field, eight or ten miles north-east of Titusville. This property is very productive, and is being developed steadily, though not as vigorously as it would be if the price of oil was nearer its remunerative value.

About the date of the above purchase, Messrs. Brownson, Andrews and Joy—Mr. Joy being the original owner of a portion of the tract—became joint proprietors of two hundred and twenty-five acres more, *in fee*, adjoining their other landed interests at Colorado. This property is also being developed judiciously, and is proving itself excellent oil territory. We give elsewhere a more complete history of the Colorado oil field, and omit further reference to it here, beyond the fact that it is a source of steady gain to its proprietors, Messrs. Andrews, Brownson and Joy, who still own and operate it.

In the spring of 1869, Mr. BROWNSON visited Parker's Landing "upon a tour of inspection." This was his first visit, but he became thoroughly convinced, that this was oil territory, and that its limits were yet to be ascertained. Before leaving, he had secured—with other parties—a ten acre lease, southeast of Lawrenceburg, and between that point and Bear Creek. Upon this lease, four wells were drilled during the summer of 1869. These wells were paying, but not largely so. This property Mr. BROWNSON held, and operated for a year or two, and finally disposed of it, at a nominal sum.

In 1870, Mr. BROWNSON, in conjunction with Mr. A. N. PERRIN, a producer of clear record, deserved prominence, and we may add, deserved success, purchased a five acre lease, upon the Robinson farm, above Parker's Landing, upon which were two wells, one producing 140 barrels, and the other 20 barrels per day, with a third to be completed, paying therefor, $45,000. This large transaction was followed by other investments by Mr. B., in fee, and in developments,—in connection with other parties—upon the Thomas and Parker farms, at Parker's Landing, and additional enterprises of like character on the Clarion River. We omit a detail of these varied operations. They were all flattering successes, however, and added to the wealth of the owners. In the spring of 1871, the several investments named in this connection, were sold to Fisher Bros., for $60,000. How much of this princely sum should be regarded as clear profit, may be approximated, when we state the fact, that during Mr. BROWNSON's part ownership of them, they paid bountifully, and over and above first cost !

On the very day this sale was consummated with Fisher Bros., Mr. BROWNSON purchased the western half of the "Walker farm," situated four miles, southwest of Parker's Landing, and two miles north-east of Petrolia city—containing ninety-two acres. For this property he paid $45,000. This farm adjoins the Shakeley farm, upon which developments had already been made, establishing its character as good oil territory. Upon the eastern half of the Walker farm, had been drilled the " Gold Dust Well," which started off at 150 barrels, and maintained this great volume for many months. Mr. BROWNSON, owned an interest in this well and lease—about seven acres—and was identified with its development from the first.

Immediately upon acquiring title to this property—the Walker farm—the work of developing it began, and during the summer, fall, and winter of 1870, and '71, four wells were put down.

When Mr. BROWNSON, purchased this farm, one lease had been given, and soon after he came into possession, another was made, to other parties—the first named, being the first drilled upon the farm. Just before these two wells were completed, however, he purchased both, paying $18,000 for them. With the exception of one other lease and well, which he put down mainly himself, but upon an agreement to divide the " working interest " with a valued friend, he has developed this property entirely as his own enterprise. He, has now, twelve producing wells upon " The Walker," or "BROWNSON Farm," the monthly product of which is between five and six thousand barrels. The average cost of these wells is $8,500. So that in prosecuting these later developments, he has expended more than $100,000. To this may be added, the expense of an iron tankage capacity of 10,000 barrels, at a cost of $8,000.

Mr. BROWNSON has recently purchased additional oil territory, on the new oil field, near Millerstown, so lately demonstrated, and denominated the " head centre " of the lower oil region. He is the owner in fee, of one half of the " Boyle farm," containing one hundred acres of land, and is the possessor in fee, of a one-third interest in two other contiguous farms, containing, about one hundred and seventy-five acres of land. This property is in the line of developments, at and near Millerstown, and is regarded upon all hands, as eligibly located, and sure to be valuable for oil purposes. Should these anticipations be realized, and there seems to be little doubt in this direction, Mr. BROWNSON, has in the near future a mine of wealth, which his indomitable industry and perseverance will surely open and secure, adding to his fame as a producer of Petroleum, and greatly augmenting his wealth.

Mr. BROWNSON, as this hurriedly prepared sketch of him indicates, is a gentleman of superior business qualifications and unsullied public and private repute. He is a man of generous impulses, and marked liberality both in his business relations, and in his intercourse with the world about him. A man of social excellencies, high moral culture, and Christian practices, he commands the respect and warm friendship of all who know him. Prompt in the discharge of every duty imposed, he bears with him an unblemished record, and an unspotted commercial name and fame. Whatever he agrees to do—no matter when or where, or whether under "Seal" or not—he will do, faithfully and surely to the end of the contract. By this emphatic language, we mean to convey to the reader the impression, unqualifiedly, that Mr. BROWNSON, is an honest man.

He is in middle life, hale, vigorous and full of determination. His course is yet onward, and his ambition is yet unflagging. A man of quiet manner, he wins the confidence and bears off the good opinion of all who may fortunately be thrown into his society. Without guile, or even the ability to deceive, his frankness, like his manhood, never forsakes him. Upright and honorable in all his transactions, he is a man to cultivate, whether in the walks of private life or in the more rugged and tortuous paths to wealth and its acquirement. MARCUS BROWNSON is among the men of the oil region whose industry, sagacity and tried honesty have won him distinction, and an ample competency. This he has richly deserved, and none will grudge him the enjoyment of his good fortune now, or in the future.

JOHN C. BRYAN.

TITUSVILLE, PENNA.

JOHN C. BRYAN, is a native of the county of Cork, Ireland, where he was born on the 25th day of December, 1831. He was one of a family of twelve children, five boys and seven daughters, all whom came to man's and woman's estate, in the enjoyment of health and vigor. Young BRYAN, had the advantages of an excellent school, until he reached the age of eighteen, when he resolved to try his fortunes in the new world, and soon after set sail for America. He landed in New York, the possessor of very little of this world's wealth, and after a few days spent in the Metropolis of the Western hemisphere, he went to Erie, Pa., where soon after, he entered the employment of Sennett Barr & Co., iron workers and machinists, with whom he remained three years, mastering his business, and becoming thoroughly wedded to his profession. Leaving Erie, he entered the employment of Sidney Shepherd & Co., at Buffalo, New York, with whom he remained one year, and then engaged with "The Buffalo Steam Engine Works," for another year. In the winter of 1855, he was tendered, and accepted the position of foreman in the establishment of Teesdell & Cole, at Conneautville, Crawford County, Pa., remaining here one year. The following year he spent in the employ of Sennett, Barr & Co., at Erie, in a supervising capacity. In the summer of 1857, he entered the employment of Brown Bros. & Co., iron manufacturers and machinists, at Warren, Pa., and remained with them until May, 1858, when he returned to Ireland upon a visit to his relatives, and in pursuit of health and relaxation from the exacting demands of his profession.

A year spent in the land of his birth, found him recuperated in

Woodburytype. A. P. R. P. Co., Phila.

JOHN C BRYAN.

health and strength, and he resolved to return to the United States, and in 1869, we find him again in the service of Brown Bros., at Warren, Pa. He remained here until the fall of 1860, meantime, superintending the construction of the first steam engine manufactured in Warren county.

The discovery of Petroleum Oil, in apparently inexhaustible quantities, was made at Titusville, Pa., in the fall of 1859. Mr. BRYAN was not long in determining his course of action. With his accustomed business foresight, he saw clearly the magnitude of this new development, not only to the world at large, but to that branch of mechanics which had been his pride and study for years. Within a few weeks the oil region, from Titusville "down the Creek," was the scene of a most wonderful excitement. Men with, and without means, flocked thither, and speculations and developments were the rule of the hour. Wells were put down, with "spring poles," "horse power," and every conceivable mode was adopted to secure the "golden stream," from the rock below. Col. Drake, had used a small steam engine to drill his first well sixty-nine feet, but this was deemed quite too expensive a process, while oil brought only one, two or three dollars per barrel, or $25 to $40, for a flat-boat load of forty to sixty barrels. At this time "second sand rock oil," was all operators sought for, and hence "spring pole," water, and horse powers were deemed sufficient for the needs of drillers. This product was soon exhausted and deeper wells demanded, and to accomplish this, steam was deemed absolutely necessary. To employ this power, required capital as well as mechanical skill and genius for their manufacture. Mr. BRYAN comprehended the needs of the section, and proposed to Brown Bros., to enter at once into the business of supplying engines and boilers, of sufficient capacity to meet the requirements of the just then developing wealth of the oil region. They, however, did not at first accept the proposal of Mr. BRYAN, regarding their ability to compete successfully with eastern manufacturers as extremely problematical. Everything, in the *iron* trade, and required in the manufacture of engines and boilers, had to be transported by wagon, over twenty to thirty miles of unimproved roads, the nearest railway facilities, being at Corry, and Union.   In spite of all these embarrassing surroundings, Mr. BRYAN proposed to Brown Bros., the building of one steam engine, the cost of which was to be charged against his services, if not promptly disposed of. The result proved how completely Mr. BRYAN comprehended the matter, for not alone was this first engine sold, but orders were taken, and executed with promptness, for six additional.   These engines were all rated at six horse power, and sold for $600, to $800 each.

While constructing these engines Mr. BRYAN was engaged in the manufacture of *pole tools*, the rope or cable not as yet being used. The first set of drilling tools, after the discovery of oil by Col. Drake, in September, 1859, was designed and manufactured by Mr. BRYAN.

During the spring and summer of 1860, Messrs. Kingsbury, Ames, and Richards, from Brown Bro.'s establishment at Warren, Pa., erected a small iron foundry in Titusville, upon the site now occupied by the Titusville Manufacturing Co.'s extensive works, and tendered to Mr. BRYAN a leading position.   This was accepted, and the engagement lasted about six months.   Mr. BRYAN had from the first, made himself master of the situation, and by closely observing the operations of the new enterprise, arrived at the conclusion that the business was being conducted at a loss to the proprietors.   An examination of the liabilities and assets of the firm disclosed this fact very palpably.   Mr. BRYAN proposed to purchase the establishment, and soon after formed a co-partnership with Mr. William McMullen, of Warren, Pa., and the firm of McMULLEN & BRYAN became proprietors of the first iron foundry and machine shop in the Oil Region.   The amount paid for the property was $6,000, and, in Mr. B.'s judgment, all it was worth.

In December 1862, Mr. McMullen died, leaving the business to the care and conduct of Mr. BRYAN, alone.   At this juncture the outlook was not inviting or encouraging. Business was almost at a stand still. Oil was selling at one and two dollars per barrel, and even less, and there was little demand for it at these prices. Money was scarce, and the wants of oil producers in the iron trade, next to nothing. All these discouraging surroundings did not frighten or appal Mr. BRYAN. He had commenced his business in Titusville with the determination to work out his financial salvation, and come what might he resolved to "stick." The establishment employed but five hands, and there was often nothing for these to do. Business, there was none—stagnation was visible everywhere, and often Mr. BRYAN was advised by friends who seemed impressed with the idea of his utter destruction in a business way, to " pull up and go elsewhere." He, however, listened to none of these " croakers," but redoubled his efforts to adapt his manufactures to the wants of all classes of people. When the demand for drilling-tools flagged, he turned his attention to the manufacture of plows, scrapers, or whatever else would be required by farmers, contractors, &c., &c. When the plow and scraper market was overstocked, saw-mill irons, gearing, &c., were turned out in fair quantities and excellent mechanical attraction. When nothing else offered, he applied himself to the construction of patterns for a new steam engine, which during his leisure hours he built and set up in the establishment, anticipating orders for more. In this he was not disappointed. From this time onward his business rapidly increased, to such a degree that in the summer of 1864, thirteen machines were required (employing over thirty men,) to meet the requirements of the oil trade. Prosperity, thrift and the busy hum of a score and a-half of operatives, greeted him upon every side. The summer of '64 was indeed a season of profit to him, due alike to his industry and determination to succeed over all obstacles.

But disappointment and disaster lay in his path. The earnings of years were destined to become the sport and sacrifice of the devouring element. On the 22d of November, 1864, following his previous months of prosperity, a destructive fire swept away his all. The foundry and machine shops were in ruins! The loss was not less than $42,000, and fell entirely upon Mr. BRYAN. It included his all!

On the morning succeeding the fire, friends gathered about him tendering sympathy and assistance, if needed. "What are you going to do, John?" said one to him. His answer was characteristic of the man and his indomitable will. "What am I going to do? *I shall carry on the Foundry business right here: I am not dead yet!*" The reply was an order for clearing away the debris, and the commencement of the erection of new buildings of ample proportions and greater adaptability to the growing demands of his extended and extending business. The promptness with which his determinations were set in motion and executed may be comprehended when it is added, that early in March following—a little over three months after his disaster—found him again in full tide of successful operation, with greatly increased facilities for the execution of any and all orders made upon him. The completion of his new establishment in so short a time was such a "gilt-edged" certificate to his energy and perseverance that old and new customers flocked to him from all quarters of the oil belts, crowding his manufactory with their work, and his counting-room with their orders. Mrs. McMULLEN still retained the interest owned by her husband at his death.

Mr. BRYAN was again upon the flood-tide of prosperity. The works gave employment to forty men, testing the full capacity of the machinery in use. Prosperity was wafted to the new establishment upon every breeze. The mechanical excellence of all its work was known of all men. But there was still another disaster in store for the proprietors. On the 17th of April, 1865, the rains descended, and the floods came, carrying destruction and devastation in their track. The oil men of that day will remember the fearful character of that terrible scourge. It was the most disastrous blow before or since received by this industrious region. Everything—trade, money, product, credit, confidence and effort seemed to be paralyzed, or entirely swept away.

In this emergency, the ever-fertile brain of Mr. BRYAN was again taxed to its utmost limit. While failures, bankruptcy, stagnation, and almost a total destruction of values were upon every hand, he bated no jot or tittle of his energy and manly perseverance. He maintained the credit of his establishment through all this crisis of the oil region, and came out brighter and prouder, so to speak, than ever. As the storm-clouds cleared away, he took careful note of his surroundings and found a clear sky and an open sea, in which to sail his sturdy craft, and he kept her from the " breakers."

In May, 1866, following the reverses of the previous year, he sold a one-third interest in the establishment to Capt. John Dillingham, whose accession to the enterprise proved to be of great value. In the latter part of the same year—1866—Mr. BRYAN added to the increasing business of the establishment a commodious boiler-shop, and this, contrary to the judgment of Mr. Dillingham, whose opinion was, that it would require more capital than they could command.

The manufacture of boilers in the Oil Region, at this time, was so much of an experiment, and its success so doubtful, that the firm not only furnished all the machinery and stock, but they contracted with Mr. Ackerman, a practical boiler-maker, to give him one-half of all that could be made out of the undertaking for one year. The business was so limited for the first twelve months, that the contractor did not realize day-wages. Operators preferred to go abroad for their boilers, regarding those of home manufacture as unreliable. The firm, however, continued the manufacture of boilers, disposing now and then of one or more, so that at the end of two years, their character was substantially established, and from that day to this, BRYAN, DILLINGHAM & Co.'s boilers have grown in public confidence, and are now regarded as second to none manufactured in any part of the country.

In 1868, a brass-foundry department was added to their extensive works, to enable them to supply the increasing demands for brass boxes, gibs, and every kind of brass fixtures used in the manufacture of engines, gas pumps, &c., &c., heretofore dependent upon remote and foreign industries.

We have thus traced the personal history of Mr. BRYAN from his boyhood, to "point a moral and adorn a tale" of industry under every conceivable discouragement, and a deserved success under many perplexing embarrassments. Few men, encountering the seemingly overwhelming disasters which have been the lot of the subject of this sketch, would have proven in so eminent a degree, superior to them all. The instances we have given of the business calamities falling to the lot of Mr. BRYAN, are not in detail, nor are they elaborated in the least. The reader may, with reflection, form a faint idea of the character of a business man, who can, from the ruins of three sweeping losses, in six years, rise to success and to final affluence. The great iron establishment of BRYAN, DILLINGHAM & Co., now operating under the auspices of " *The Titusville Manufacturing Company*," is the product of the indomitable will and perseverance of JOHN C. BRYAN, built up to success under disasters that would have disheartened most men. To-day this great industry ranks second to none in Western Pennsylvania, involving a capital of $350,000, and giving employment to more than two hundred and fifty men. The establishment has built over 300,000 barrels of tankage in the oil regions. They turn out half a dozen boilers and engines a week, and derrick rigs, and drilling tools without number, every thirty days. To these departments must be added their brass foundry, manufacturing all the brass work for their own needs, and immense quantities for the wholesale and retail trade of the region. Their great iron tank transportation works, recently put into successful operation, adds largely to their importance as manufacturers, and is a guarantee of the bountiful success of this great mart of industry and mechanical excellence.

Within the past year and a half the establishment, of which Mr. BRYAN was so long the successful head, prepared and put into successful use, some of the best tapping machines in the United States. These are necessary to meet the wants of refiners, latterly a growing and all-important industry for Titusville and other refining centres. Indeed, it is not too much to say, that this establishment is capable of, and really does a larger diversity of work for oil purposes, than any other in the State of Pennsylvania. The success of this great industry, as we have before said, its wealth and world-wide reputation are mainly due to Mr. BRYAN. Titusville has long been the radiating centre for our heaviest oil operations, and it is safe to say, the wants and necessities of producers have not been in advance of Mr. BRYAN's ability to devise and execute—and to his genius and mechanical skill must be awarded much of the perfection now apparent in the manufacture of tools for drilling, pumping, and the clearing out of oil wells, and indeed for improvements of every character for oil well uses.

Titusville is a city of enterprise and enterprising business men, and Mr. BRYAN, from the date of his residence there, may be, nay must be, regarded as among the foremost in all improvements of both a public and private nature. While others, in the struggle for success, have fallen by the way, his pertinacity and determination to surmount all obstacles stands out in bold relief, the true index to his character as a man, a neighbor, and a public and private citizen. As a mechanic he has few equals. This fact is conceded by all who know him. He has been awarded contracts for building some large refineries and tanks in New York city, and these have been executed with such fidelity and satisfactory results, that others of like nature have been tendered in such numbers, that it was impossible to fill them—and this, when other and rival establishments proposed to do the same work at lower rates.

These details of the growth of the iron interest of the oil regions, as exemplified in the sketch of Mr. BRYAN's life, have been given as forming an important part of the history of the vast mineral wealth of this section, and the mechanical means adopted for its development. Without this our history would be incomplete,—in the language of another, who recently sketched the personal history of Mr. BRYAN for the pages of the *"Petroleum Monthly,"* "We of to-day are alone interested in the business and successes of to-day. Looking back ten years, we can scan the entire history of Petroleum in Western Pennsylvania. It seems but yesterday, that Colonel Drake opened this mine of wealth, now so successfully worked, and whose bounty reaches to every civilized nation on the globe. But in the future, when this generation has passed away, the men who are now the supporting pillars of the grand edifice of commercial industry and prosperity here erected, will live in the respectful memories of their successors."

Mr. BRYAN'S personal characteristics are those of an active, untiring business man, wholly given to his profession, and thoroughly identified with all its details. In person he is six feet in height, and when in robust health, the picture of activity, and perseverance. He is clear-headed, logical, and almost invariably correct in his mode of reasoning, and his conclusions upon any subject he has in hand. He has been a member of the City Council of Titusville, and occupied from first to last a prominence in that body, at once honorable and commanding. His perceptions are quick and generally accurate, and he estimates men, not from their wealth, but upon the broader basis of honor and honesty. His personal friendships are many and real. He is known and appreciated for his business probity, and for fidelity to all his obligations, public or private. For a few months past he has been in delicate health, and has necessarily withdrawn from the more active pursuits of a thus far busy life.

Woodburytype. A. P. R. P. Co., Phila.

GEO. H. DIMICK.

GEORGE H. DIMICK.

PETROLIA, PA.

MR. DIMICK is a western man, born near Kenosha, Racine County, Wisconsin, on the first day of April, 1839. He is the third, of a family of five children, four sons and one daughter. His father was a farmer, and the sons were reared to this independent calling, until they reached years of maturity. GEORGE was kept at the district school until he was fourteen years old. At this age he was sent to the High School at Kenosha, Wis., for a few terms, and subsequently, attended the High School at Milwaukee for two years and a-half. At the age of seventeen he taught a district school in his native town for a single term, and at nineteen was engaged as principal of one of the Ward Schools of the city of Milwaukee. This position he held two years, or until the spring of 1860.

In the fall of 1860, he came into the oil regions, and located at Rouseville, then known as the "Buchanan Farm," and subsequently as Cherry Run. Not long after his arrival at Rouseville he received the appointment of Superintendent of the two Buchanan farms—one of which was being operated under a lease to Rouse & Mitchel, and the other by Rouse, Mitchel & Brown. This position he held about one year.

During the latter part of 1861, he began operations upon his own account, and upon the two Buchanan farms. At this date the "spring pole," water and horse power, were alone employed in drilling oil wells, and the "first sand" was the lowest depth reached, and this was obtained at from 180 to 200 feet. The first well drilled upon the Archie Buchanan farm, and the "first sand" was struck at 180 feet. The drilling, which required months of time, was accomplished wholly by "spring pole" power.

This well produced from twelve to fifteen barrels, and was owned by Rouse, Mitchel & Brown.

Mr. DIMICK's enterprises were measurably successful during the summer of 1861, but an unlooked-for affliction came upon him, in the tragic death, by burning, an account of which is given elsewhere in these pages, of his friend, Henry R. Rouse, which changed for the time at least the current of his ambition. Allied to Mr. Rouse by ties of consanguinity, and from his business connections, fully acquainted with the private affairs of the unfortunate gentleman, he was selected to settle his estate. This responsible duty he entered upon during the fall and winter of 1861, in connection with the administrators, named by the deceased just prior to his death. This labor required some months of his time, and while in its performance he marked out his future. The war for the Union, was at this time assuming colossal proportions. Communities, Towns, Counties and States were rivaling each other in arming and equipping Companies, Regiments and Brigades for the bloody contest. Mr. DIMICK soon after the closing up of his responsible duties to his dead friend, proceeded to New York City, and there enrolled his name as a private in "Scott's Nine Hundred," afterwards known, as the "First Regiment U. S. Volunteer Cavalry." Col. Thomas A. Scott, since so widely known as the Railway King of this continent, was mainly instrumental in procuring the passage of a law authorizing the Regiment, and when it was ready for service, was named in honor of him, "Scott's Nine Hundred," and was known during its three years active duty by this title.

The regiment reached Washington in the early spring of 1862, and was detailed for special duty, acting when required, as escort to President Lincoln, and members of his cabinet, and keeping watch and ward over the federal capitol.

Mr. DIMICK entered the regiment as a private, and was at its organization elected first sergeant. Subsequently he was promoted to First Lieutenant and Quarter Master, and then to a Captaincy.

He was detached from his regiment, and served as Post and Division Quarter Master, and later was honored with the position of acting Assistant Adjutant-General.

The regiment, after eighteen months' service about Washington, was transferred to the Department of the Gulf, under Gen. Banks. For a considerable time the command remained in New Orleans, and was afterwards stationed at Baton Rouge, having charge of points on the river intermediate to New Orleans. During this time Capt. Dimick was acting Assistant Quarter Master. After New Orleans and the lower Mississippi had been brought back to their allegiance to the Union, the regiment was transferred to Memphis. This was in the fall and winter of 1864. The regiment had already served three years and three months, and in March, 1865, Capt. D. was honorably discharged. After he returned to his home in Wisconsin, he received his commission as captain, a promotion he had earned by gallant and meritorious conduct, and which ought to have been awarded him months before his term of enlistment expired.

In May, following his discharge from the army, he returned to the Oil Region. This was about the date of the Pit Hole development, and thither Mr. D. bent his steps. He at once identified himself with the rapid growth and fabulous wealth of this newly discovered oil-producing locality, entering largely into real estate sales and putting down wells upon his own account. "The United States Petroleum Company," the first to lease and successfully test the Pit Hole basin, appointed Mr. D. an agent for the sale of leases upon their property. Large sums of money were paid for half acre leases, near the old "United States Well," and Mr. D. had the agency of most of this property. His transactions were heavy, rapid, and very profitable. During the summer of 1865, and while Pit Hole was the centre of attraction to operators and speculators, small fortunes were eagerly expended in the purchase of half acre leases at a quarter, and one-half royalty! One day, while Mr. Dimick was busily engaged in his office, a sedate, well-dressed, elderly gentleman entered, and inquired for the "United

States Oil Company's agent?" Mr. D. was pointed out to him, and he promptly approached. "What figures do you put upon lease No. 49, Holmden Farm?" asked the stranger. "We have concluded not to part with that half acre," said Mr. DIMICK, "but to operate it ourselves." The clerical gentleman was evidently disappointed, but not wholly satisfied. He pressed Mr. D. to fix a price upon the coveted half acre plot, and before he gave up the effort to secure it, offered $20,000 for it! This sum was refused, and the stranger departed. It was afterward ascertained that he was the representative of a body or company of "oil smellers," who for some days had been examining the surroundings of the "United States," the "Grant," and other large flowing wells in the vicinity, and had fixed upon this lease as one sure to rival all others. They declared that this lease covered the great jugular vein of the deposits. The plot happened to be one of several the Company had reserved for their own development, and so could not be sold at any price. When it was subsequently operated, it proved to be only nominally productive territory!

Throughout the summer and fall, and into the winter of 1865, Mr. DIMICK operated largely, and carried controlling interests in *eleven dry wells* on the McKinney farm alone! These were, however, more than balanced by interests in many large producers,— among which were interests purchased, after they were down, as ventures, in the great "Pool Well," that tanked out daily, 1,200 barrels—the "Eureka Well," which gave out a golden stream of 800 barrels per day—the famous "Grant Well" which flowed, for many months, (as did the others,) 700 barrels per day—and the "Burchill Well," which produced steadily 350 barrels per day. He held paying interests in very many other large producing wells of less note than those we have named, all which added largely to his resources and wealth. His business steadily augmented, and was mainly successful through the fall and winter of 1865-6. But in spite of all these evidences of accumulating riches, which at times, amounted to thousands of dollars a day, when in January, 1867, Mr. D., shook the dust of Pit Hole from his feet, his Bank balances were far from flattering. Like hundreds of others, he had remained quite too long, and when he came to gather up his resources, their value was easily computed. Indeed the balances were largely upon the wrong side.

In the spring of 1867, Mr. D., found himself physically reduced, and he resolved upon a sea voyage. He proceeded to New Bedford, Mass., and took passage in the good ship "Maywood," bound for Cumberland Inlet, away up in the Arctic regions. Without giving a detailed account of this somewhat remarkable trip, to these far-off ice bound-shores—and it was full of incident —we may add that the voyage was accomplished in about fifteen months, when Mr. D., returned to his home in Wisconsin, greatly improved in health. Soon after, or in the fall of 1868, he again returned to the oil region, locating at Pleasantville. He remained here, a year or more, operating to a limited extent at Hickory, Rouseville and Pleasantville, buying and selling lands, leases, interests, &c. In the spring of 1871, he removed to Parker's Landing, and became a partner in the extensive oil well supply establishment of Messrs. McKinney & Nesbitt, then representing on the one hand the great iron manufactory of Gibbs, Sterritt & Co., of Titusville, for boilers, and engines, drilling tools, &c., and the Messrs. Morris, Tasker & Co., of Philadelphia, with the National Tube Works of Boston, and the Crescent City Tube Works of Pittsburgh, upon the other, for casing, tubing, and well fixtures generally. Eight months after his connection with this firm, Mr. DIMICK purchased the interests of his partners, and thence forward, with a single change, that of DIMICK & McCORMICK—he conducted the constantly growing business, with success. During his year and a half control of this branch establishment at Parker's Landing, it is safe to say that his sales and receipts amounted to $500,000, every dollar of which was scrupulously accounted for and paid over.

In June, 1871, Mr. DIMICK made a thorough survey of the

country lying five to seven miles south and west of Parker's Landing, extending his observations and examinations to what is now
known as Petrolia City—between Fairview and Argyle—in Fairview Township, Butler County, Pa. Reaching what seemed to him
to be a natural basin, at a point where South Bear Creek and
Dougherty Run unite, he settled into the conviction that "*this
must be oil territory.*" Near by, and upon the high ground overlooking this basin, a large number of men, farmers and neighbors,
were engaged in raising a barn. There may have been fifty or
sixty of them. Mr. DIMICK, had previously met one of them
and had "interviewed him," as to the possibility of the vicinity
being oil territory, and he was not long in interviewing others at
the "raising." Subsequently he offered to drill a well somewhere
in this basin, provided the owners of the land would give him
leases of portions of their farms, sufficient to warrant the undertaking. Terms were finally agreed upon, and before Mr. DIMICK
left the "raising party," he had obtained leases upon five farms
in the immediate vicinity, for himself and partners. He took a
lease of twenty-nine acres of the Blaney farm—sixteen acres of the
Jameson farm, and ten acres of the W. A. Wilson farm—all which
covered or were located in the basin before referred to. He also
obtained a ten acre lease on the Graham farm, west of Argyle, and
another of ten acres upon the James Wilson farm, south of the
Blaney farm, about seventy-five acres in all, and all at one-eighth
royalty. We have been thus particular in our detail of this
transaction, because it illustrates better than we can express or ex-
plain it, the mode and manner of "Wild Catting" as it is termed,
for new developments. The first well drilled upon this property,
was located upon the line dividing the Jameson and Blaney farms.
It was arranged that the one-eighth royalty, should be equally
divided between the owners of the land, each receiving one-sixteenth. This well was completed on the 17th of April, 1872, and
started off at 100 barrels per day ! When these leases were secured, there were no developments within four miles of Petrolia

or Argyle. Other parties, at a later day obtained leases, a mile or more below Mr. DIMICK, and were a day or two in advance of him in getting their wells down, but the "pioneering" about Petrolia, we think belongs to Mr. DIMICK, and the firm of which he was a member, and which was formed after the first well was commenced, (DIMICK, NESBITT & Co.) The developments just mentioned, were the signals for emigration to the new Oil-dorado. "Petrolia City," became a famous oil centre. Its growth was rapid, and in many respects substantial. The first wells in the vicinity were struck in April, and before the close of the following month, May, the moving mass began to arrive, and by July, a city with metropolitan pretensions and cosmopolitan make up, sprang into existence. Early in 1872, a borough charter was obtained, and in February following, an election of borough officers took place. Mr. DIMICK was named by the respectable portion of the community for Burgess. The "roughs" and "slums" of the oil region, named one of their own number as an opposing candidate. When the election came on, many of the best citizens of Petrolia, feared the "roughs" might succeed with their infamous representative, who all day long possessed the polls, desiring, if they did not do so, to "vote early and often." When the vote was canvassed, it was found that Mr. DIMICK had been elected by a large majority—the slums and "riff-raff," being able to muster less than one fifth of the whole number cast. At the regular election, which followed late in March, Mr. DIMICK was unanimously nominated and re-elected without a dissenting vote. This is a compliment rarely paid to officials of this character, but we doubt not, Mr. D., merited this endorsement and compliment, at the hands of his fellow-citizens, for upon all hands he is denominated a faithful, vigilant public officer.

Mr. DIMICK's career, as a successful producer, may date from his Petrolia City investments and developments. He has, of his own, and in connection with Mr. Nesbitt, many other oil interests, and tracts of oil territory, in process of development, or soon to be put to the drill.   These properties are located in different portions of the Parker oil field, at St. Petersburg, Slippery Rock, and in the immediate vicinity of the wonderful developments in and about Greece City.

Personally, Mr. DIMICK is a gentleman of active temperament, excellent executive ability, and sturdy industry.   He has dignity of character, and yet ample social characteristics to relieve him of a seeming stiffness of manner.   He is a man of quick perceptions, and superior business qualifications.   Reliable in all his relations, a man of integrity and private worth, he adorns alike the private circles of life, and is amply qualified for the sterner conflicts of a public career.   As Burgess of Petrolia City, he is deservedly popular with all classes, and he discharges his varied duties with a clear conception of his responsibilities, without fear or favor.   He is at present upon the high road to deserved success, and this will assuredly be his future; for to a commendable perseverance, he adds an industry and a will, always sure to overcome obstacles and whatever of embarrassment that may stand in the way.

Woodburytype. A. P. R. P. Co., Phila.

GEO. H. NESBITT.

GEORGE H. NESBITT.

PETROLIA, PA.

MR. NESBITT is a native of Erie Co., Penn., born in Washington township, on the 21st day of July, 1847. His father was, and is at this date, a well-to-do farmer, residing upon the same farm upon which he settled many years ago. GEORGE is the youngest of a family of nine children, six sons and three daughters. In his earlier years, he had the advantages of a good common-school education, acquiring a thorough knowledge of the common branches, and was known and acknowledged, before he attained the age of twelve years, as a bright, if not a brilliant scholar. He early became impressed with the idea of self-reliance in the battle of life; and at the age of fifteen, left his father's house, resolved to work out his own success. He made his way to the Oil Regions, reaching Titusville late in the fall of 1862. Here he soon after entered the employ of BURTIS BROS., oil-refiners, intending to learn the business. He continued in this position a few months, when, in consequence of a general stagnation of the trade, induced by the beggarly prices of oil, the refinery closed, and GEORGE was thrown out of employment. The following spring, 1863, he engaged with DENSMORE BROS., extensive and successful operators on the Tarr and Funk farms. Here he remained into the summer of 1863. During 1863, in connection with his brother-in-law, W. H. DOUGHERTY, he built a refinery at Bull Run, Shaffer farm, and this enterprise was continued with success until the spring of 1864, when Mr. NESBITT disposed of his interest, and soon after entered into the business of shipping oil to Pittsburgh, and points intermediate, by barges, flat-boats, &c. When the waters had subsided, his occupation was gone; and during the remainder of that year, he turned his attention to real estate operations. He dealt successfully in property of this character, buying and selling oil leases, oil well interests, oil lands, &c.

In the spring of 1865, then scarcely eighteen years old, he began to operate upon his own account, and drilled wells on lower Benninghoff Run, and upon the Dempsey and adjoining farms on Cherry Tree Run. These enterprises were, taken as a whole, only partially successful; but Mr. N. was enabled to pay all his indebtedness, dollar for dollar, and keep his balances upon the right-side. While these several wells were being put down, "the Pit Hole glory" came upon the region, and startled everybody. Mr. NESBITT was among the first to reach the new developments, and secure leases. He leased, and during 1865, put down, or was interested in, wells upon the Holmden, Rooker, Hyner, Ball, McKinney and Morey farms. In conjunction with his oil operations, he entered into real estate speculations, dealing in leasing, interests, &c., and occasionally erecting buildings, and leasing or disposing of them at fabulous prices to the daily augmenting population of Pit Hole City. It is safe to say, his two years' residence at Pit Hole were marked by as many changes of fortune as commonly fall to the lot of men of his years and experience. There were days and months that his profits and prosperity and rapid accumulations of wealth were simply wonderful. He counted his gains by the tens of thousands. But when in 1867, Pit Hole "gave out," and its glory departed, he found himself, like hundreds of others, able to get away with only a fraction of his earlier acquisitions.

During the latter part of 1866, and while endeavoring to dispose of his Pit Hole interests, he received the appointment of Assistant Assessor of Internal Revenue—his district comprising Pleasantville and its vicinity—and this was his residence and headquarters during 1866. In the spring of 1867, he returned to "the Creek," locating at Pioneer. Here he operated extensively, on the Benninghoff, Upper McElhenny, and Foster Farms, and with considerable success. The low price of oil during the summer of

1867, was a serious check to rapid money-making, but Mr. NES-
BITT was fortunate in obtaining good wells, and while he sold his
product at $1.50 per barrel, and upwards, he was enabled to count
his profits at a good round figure. In the spring of 1868, he
moved with the mass to the then developing oil deposits of Pleasantville. Here he again entered largely into real estate investments, dealing as before in oil lands, oil wells and oil interests
generally. Besides this, his chief pursuit, he put down several
wells upon his own account, and nearly all his ventures were
sources of profit. He remained at Pleasantville until the summer
and fall of 1869, when his health failed him, forcing him to withdraw from active pursuits. He, however, formed a business connection with Mr. G. H. Dimick, and together they purchased some
abandoned wells at Rouseville, with a view to resuscitating them.
This undertaking did not prove a profitable one, and during the
summer of 1869, they turned their attention to prospecting for oil
belts and new territory. They located the famous Goodrich lease,
on the Scott Farm, north-east of developments on Hickory Creek,
and established the "belt" upon which the celebrated "Venture
Well" was subsequently obtained, and which gave Fagundas City
and its marvellous production so much notoriety. During the same
season they located a large lease on the Cope Farm, near the
McGrew developments, south of Oil City. All these located leases
afterwards proved to be very productive, demonstrating the correctness of the surveys and examinations of Messrs. DIMICK &
NESBITT.

During the winter of 1869–70, Mr. NESBITT removed to Parker's
Landing, and shortly after formed a business connection with Mr.
J. L. McKinney, in the sale of oil well supplies, machinery, &c.
This business had already grown into large proportions, involving a
heavy capital, but the firm extended their operations as opportunity offered, dealing extensively in oil lands and oil interests at
various points in the Parker region. Upon the opening of operations in 1870, and throughout the spring and summer of that year,

they were known as leading operators and producers, extending their developments to Bear Creek, and upon the Black and Dutchess Farms, and wheresoever a fair return was promised for their enterprise. By judicious investments, good management and careful, pains-taking industry, the firm prospered largely. Indeed, their oil investments and varied operations absorbed so much of the time and required so fully the attention of both the partners, that in the spring of 1871, Mr. G. H. Dimick was admitted a partner in the oil well supply branch of their trade, under the firm name of McKinney, Nesbitt & Co., and to him was confided the care and conduct of this important portion of their interests.

In the early part of 1871, Mr. NESBITT was again afflicted with ill health, the results of exposure to storm and cold, and overtaxed energies, and the greater part of the subsequent months of that year were spent in eastern cities, the better to secure medical advice and treatment. But though ill, he was not idle or relieved wholly from the cares of his business. In conjunction with his partners, he gave much attention to the sale of oil lands and interests to eastern and home parties. The oil excitement at Parker's Landing ran high during all the summer of 1871, and oil lands were in great demand. The firm of McKinney & Nesbitt was among the boldest and most enterprising and active, and as a sequence, had an extensive trade and realized large profits.

In January, 1872, he returned to Parker's Landing, greatly improved in health. During the spring and summer following, the firm of McKinney & Nesbitt entered largely into oil operations, extending their developments from St. Petersburg, to the *then* extreme south-western limits below Parker's Landing. They secured leases, purchased lands, and put down many wells, largely increasing both their expenditures and their profits. Among their earlier developments during this year, was the first oil producing well at Petrolia, since so prolific of oil wells of fabulous production.

In the early part of 1872, Messrs. McKinney & Nesbitt, in connection with, and we may add, at the suggestion of Mr. G. H.

Dimick, who prospected the new territory, obtained a large lease, including parts of the Blaney, Jameson, W. A. Wilson, James Wilson, and Graham Farms, four or five miles in advance of other developments, and upon the site of what is now known as Petrolia City. Their first well at this point was completed about the middle of April, 1872, and flowed from the start one hundred to one hundred and fifty barrels per day!

Soon after this first well was down, Messrs. Dimick & Nesbitt, and Mr. William Lardin, the contractor for drilling their first well, purchased the interest of Mr. McKinney in the property, and Mr. L. has since been a member of the firm of Dimick, Nesbitt & Co. The new proprietors of this very valuable property continued their active operations, and before the close of the summer of 1872, it had been thoroughly gone over and tested. The history, in brief, of its development is a most remarkable one, and is beyond question without parallel in the Parker District.

Let us take "the first SIX" of these wells put down upon this territory, in proof of our assertion:

No. 1, flowed and pumped from the start over one hundred barrels per day! No. 2, about eighty barrels per day. No. 3, over one hundred and eighty barrels per day. No. 4, two hundred and fifty barrels per day. No. 5, one hundred and fifty barrels per day, and No. 6, tanked two hundred and twenty barrels per day. It is questionable if the entire oil region, in later years, will afford an instance of such wonderful product, and in such rapid succession, at this date, April, 1873. There are fourteen wells upon this lease, one of which has been pouring forth its golden stream for quite twelve months, and the majority of the remaining thirteen have been producing largely for months past, and bid fair to hold out for months to come.

Mr. NESBITT has no business, or financial connections other than we have mentioned, if we except one, he holds in "The Parker's Savings Bank." He is a Director in this institution, and has been so honored for two years past.

GEORGE H. NESBITT is yet a young man, but he has left an impress upon the history of petroleum developments in Western Pennsylvania, which may well be his pride and boast. Without capital or influential friends, he made his way into the oil region, when a mere lad, and by his industry, frugality and fidelity to every trust confided to him, has hewn his way to prominence and prosperity. He is to-day the possessor of a handsome competency, accumulated in the earnest and honest discharge of all his obligations. No man can say truthfully that GEORGE H. NESBITT has knowingly wronged or designedly over-reached him.

Personally, Mr. NESBITT is a gentleman of modest and retiring manners. He never obtrudes himself or his opinions upon any. He is, nevertheless, a man of convictions, and does not hesitate in the avowal of them whenever and wherever occasion requires it. In business circles, he stands the test of ability, capacity, and unquestioned honor and honesty. He is a man of irreproachable private worth, and unblemished integrity, and such a man must achieve, as he deserves, success in all his business transactions.

Woodburytype. A. P. R. P. Co., Phila.

WILLIAM D. ROBINSON.

WILLIAM D. ROBINSON.

KITTANNING, PA.

MR. ROBINSON'S ancestors were among the early settlers of Perry Township, in the County of Armstrong, Penna., his father having removed thither, and to what is now known as Parker's Landing, in 1810, where he soon after entered largely into the manufacture of leather, erecting a tannery, and necessary buildings for this purpose.

The subject of the following sketch, WILLIAM D. ROBINSON, was born in this new home in the wilderness, near the present borough of Parker's Landing, on the 2d day of October, 1820. He was the eldest of nine children—seven sons and two daughters. Schools were not as plentiful as now, in this new Petroleum field, but such as they were, young ROBINSON enjoyed and improved, as soon as his age would permit. He was schooled until he was twenty. Meantime, he acquired a thorough knowledge of the business his father carried on. He was known as a first-class tanner and currier, and followed this calling until he had nearly reached his majority—attending school winters, and assisting his father from seven to eight months of each year.

When he was in his twentieth year he sought and obtained a clerkship in Shippen's Furnace, Venango county, Pa. This position he held about a year, when he was tendered a like responsible place in an iron manufactory at Brady's Bend. Here he remained three years, faithfully discharging his duties, challenging and receiving the confidence and esteem of his employers.

In 1843, he embarked in business for himself, opening a country store at Parker's Landing, for the sale of dry goods, groceries, crockery, hardware, etc., etc. This enterprise he continued for nearly twenty-six years, establishing a character for uprightness and honorable dealing never questioned. He retired from mercantile pursuits in 1869, having acquired a competency, accelerated doubtless, by his successes in developing and producing petroleum oil.

In the winter of 1864,–5, the oil excitements of the upper and lower Oil Creek region were at their height, and Mr. ROBINSON very earnestly conceived the idea, that oil deposits existed in the region of his third of a century's residence. He had examined and carefully noted the then generally received opinion of "surface indications," and soon reached the conviction that oil could be found there. He purchased thirty-six acres of the old Homestead farm, lying on the Allegany River, and now forming a portion of Parker's Landing. This thirty-six acres of land he made the basis of a stock company, called " *The Clarion and Allegany River Oil Company*," with a capital of $100,000.

The stock was all subscribed—mainly by Philadelphia capitalists, who knew Mr. ROBINSON as a truthful, honest man—and upon the organization of the company Mr. R. was made the Superintendent, and assumed charge of the work.

In the spring of 1865, he commenced his first well, under the auspices of this company, *and this was the first oil well drilled at Parker's Landing.* The embarrassments attending this first effort to find oil at Parker's Landing, may be estimated by those familiar with new territory. All the machinery for the well had to be boated from Pittsburgh and Oil City; and there was neither derrick nor development between these two points. Fifty and sixty miles from a machine shop, if a break occurred, Pittsburgh, or Oil City, or Titusville, were the nearest points for repairs. It required the entire summer of 1865—nearly six months—to complete this well. In October, 1865, the sand-pump brought up the unmistakable evidences of a "third sand," or oil rock. The well was tubed, and started off at about ten barrels per day. It averaged the first year nineteen barrels per day, and oil was sold from it, during its first two or three months' product, at $8 per barrel. The well continued to produce for a long time, and was a source of much profit to the company.

During the same year (1869,) he began and completed, upon his own account, a third well, on the Fox farm, just above his first developments. This proved to be another success, the well producing from the start, and for more than a year thereafter, seventy-five barrels per day.

Mr. ROBINSON has been a resident of Kittanning, Pa., thirty miles below Parker's Landing, since 1859. Here he has provided himself with a commodious residence, furnished with all the modern luxuries of a sumptuous home. During the past thirteen years he has given his personal attention to his large business at Parker's Landing, where he is still the fortunate possessor of some valuable oil property, spending much of his time in that locality.

In all his business relations he has maintained a high character for promptness, probity, and unflagging industry. No man stands higher in the esteem of those who know him best. He is clear-headed, honest to a fault, and possesses the financial ability to execute to the letter, every contract he may enter into. He is a man of generous impulses, and does good without ostentation, or desire for notoriety. Of simple tastes and habits, he enjoys his ample fortune without worry or apparent anxiety. Possessing to the fullest extent the confidence and esteem of all who know him, his declining years are sure to be his truest and happiest. He has a well preserved constitution and vigorous health, the results of correct habits of life, and he bids fair to live to a good old age, to enjoy the fruits of his industry, and the friendship and devotion of a wide circle of friends.

JAMES S. McCRAY.

FRANKLIN, PA.

OF all the remarkable phenomena of Petroleum, among the most singular have been those which attended the development of the McCray Farm, at Petroleum Centre. Nearly half surrounded by the Creek, along the margin of which, directly at the base of that hill on which the farm is situated, were several of the largest flowing wells of the Pennsylvania Oil Region, which, together with the infinite number of good pumping wells that perforated the ground in the immediate vicinity of the territory, at Wild Cat, Petroleum Centre, and the Egbert Farm, for eight years, sapping the sources of supply to an unheard-of extent, yet in October, 1870, six months after oil was struck on the hill, this farm produced over twenty-seven hundred barrels a day, yielding to its owner the enormous income of more than six thousand dollars a day, and yet but little more than half the production was his!

To gratify an honest interest which the public are sure to feel in the history of men whose lives have been connected with remarkable events, we pause in the description of the farm to glance at the career of the man. A career it has been which however much wanting in the wild interest of adventure, is not without a healthful lesson in industry and perseverance, to restless, surging, feverish, fortune-seeking young America.

He was born November 16, 1824, about a mile below Titusville, on what is now known as Watson's Flats. Here, till his father removed to a farm near the Allegany, several years of his childhood were passed. The meagre education in that day deemed requisite for a farmer's son, was acquired at the log school-houses, here and there maintained for the enlightenment of country youth during

Woodburytype. A. P. R. P. Co., Phila.

JAS. S. McCRAY.

the winter months. Up to that time, in the history of American evangelization, there were still a few who clung to the antique supposition, now quite obsolete, that children owed the parents who had nourished them in their tender helplessness, some debt of gratitude that demanded a practical recognition. Consequently, after giving his father two or three years of service, we find J. S. McCRAY, at the age of twenty-two, starting out for himself, with only two dollars in his pocket.

For two years he worked at one of the numerous saw-mills in the lumbering region, along the Allegany, above Oil City, and by the practice of those economies which the hardy back-woodsmen, who are now the bone and sinew of the land, commonly observed, at the age of twenty-five, he was able to buy a team and take up the land which now constitutes the McCray Farm, and for which he paid about two thousand dollars.

Between farming and lumbering he continued to lay up the means of making a comfortable home, and at the age of twenty-eight, married Martha G. Crooks, of Venango County, a respectably connected young woman, of sterling good sense, who was a willing helpmeet in their humble beginning, and an invaluable guide when the flood of fortune, that might have unsettled a steadier head, rolled in upon him.

When the news of Drake's success spread like wildfire up and down the country, drawing crowds of curious and excited visitors, Mr. McCRAY, together with several others, formed a Company, and took a lease of two acres on the Buchanan Farm, at Rouseville. The lease proved a remunerative bargain, but before the first well was quite down, McCRAY embraced an opportunity which was offered, to sell out his interest, and he soon after secured a lease on the Blood Farm, adjoining his own.

It was here that fortune first smiled on his oil operations. He had obtained a most favorable lease from his old neighbor, who still held the farm, and in company with several others, he put down the famous "Maple Shade Well," which for a long time flowed eight hundred barrels per day, a quarter of which production, clear from all expense, belonged to Mr. McCRAY, who, besides owning the lease, had an interest in the well.

After realizing more than twenty-five thousand dollars by the sale of oil, he sold his interest in this property for fifty thousand dollars!

In 1863, Dr. Egbert, then as still one of the most enterprising of operators, having acquired possession of the strip of bottom land along the Creek at the base of the McCRAY hill, and having struck the renowned "Coquet Well" almost on the McCRAY line, took a lease of all that terrace—called the McCRAY flats—which adjoined his own property, giving three-eighths of the oil as royalty, and immediately began to develop.

The first three wells sunk, for a long time averaged about three hundred barrels of oil per day, netting of course a handsome revenue, and piling up wealth as fast as any man could reasonably wish. In all, there were seven or eight wells sunk on his land at the foot of the hill; and in 1865, when the production was at its height, Mr. McCRAY refused a clear half million in cash for the farm. Though he refused this offer, not because he thought it less than the value of the property, but from reluctance to give up a homestead acquired by hard labor and constant frugality, and endeared to his family by cherished associations; subsequent events showed it to be a very fortunate decision.

In the possession now of an ample fortune he surrounded his family with such solid comforts as their education and former hardy lives enabled them to enjoy, and prudently determined to give his children such an education as would fit them for any social position.

Early in the spring of 1870, Keffer & Watson, operating on the Dalzell tract, on top of the hill, struck a well which began to flow four hundred barrels per day.

This well was near the McCray line, and when it was struck, immediately leases were in anxious demand on the McCray farm *on the top of the hill,* and operators were crowding to obtain them on any terms at all.

He, however, fixed a uniform price of one thousand dollars per acre bonus, and one-half the oil. Even at these unusual figures there were plenty of experienced operators, ready to take leases. He also began to operate himself on a large scale, and so remunerative was the territory that in the whole Oil Region of Pennsylvania there is probably not to-day another piece of land equal in size, on which are to be seen standing so great a number of derricks. So successful were the operations, and so rapid the increase in production, that six months after Keffer & Watson's strike—in October, 1870, the McCray farm piped twenty-seven hundred and thirty barrels of oil daily, about seventeen hundred of which were his own, yielding for a time, perhaps the largest revenue ever received by a single firm from the production of petroleum.

But when the magic smile of fortune had turned the rugged hill, over which they had toiled in the hopeful days of their younger life, following their quiet herds, dreaming only of a competency that would enable them to spend the evening of their lives in repose and plenty ; happy because industrious, honest and thrifty ; ambitious only to give their children a better chance in life than their's had been ; when fortune surrounded them with gold for the realization of their wildest dreams, the hand of death stole in upon the glittering scene, and with one rude crash shattered their fondest hopes. The eldest of their three children, a beautiful daughter just merging into womanhood, died, bowing their hearts in a grief that riches can never heal.

MR. McCRAY is in middle life, blessed with a vigorous, robust constitution, and as a sequence, enjoys most excellent health. He is a man of active temperament, thoroughly devoted to his business concerns,—and these he manages with an ability seldom witnessed in men of his educational advantages and experience. Latterly he has come to regard all enterprises, bidding for his approval and capital, with great caution, and is therefore not likely to become the dupe of sharpers, or the prey of speculators upon his bounty or generosity.

The product of the McCRAY farm from the commencement of operations in 1871, to July, 1873, must reach, if it does not exceed, 500,000 barrels of oil—one-half of which has been turned over to Mr. McCRAY. Nearly 150,000 barrels of this is now held in iron tanks by him for better prices.

A year or more since, Mr. McCRAY purchased the elegant private residence of Mr. BROUGH, at Franklin, Venango Co., Pa., and soon after removed thither, and is now a resident of that beautiful city. Here he is honored and respected for his integrity as a man, his enterprise as a citizen, and his worth, honor and honesty as a neighbor and friend.

CITIES AND TOWNS.

THE OIL REGION OF PENNSYLVANIA.

WITH the progress of Oil developments, cities and towns, with teeming inhabitants, arise. A great number of these remain to mark the industry and the fortunes of the business. We have only to point to Titusville, Oil City, Franklin, Parker City, Tidioute, and others of less importance, to prove that notwithstanding the hazardous character of the oil business at these points, it has built up for itself very many enduring monuments.

In the following pages we shall endeavor to describe the cities and towns of the Oil Region as they now stand. The reader should not entertain too exalted an idea of many of these "cities." They spring up as if from the touch of the magician's wand—are swept away by fire, or disappear, only to re-appear miles in advance of their last location. They are portable and ported. The owners of the buildings migrate with the greatest unconcern. The structures that shelter them to-night, may to-morrow, do a similar service miles in advance, at the scene of some new strike.

From the year succeeding Col. DRAKE's initial venture, the striking of a good well has time and again throughout the Oil Region resulted in the establishment of a flourishing town, with a celerity truly marvellous. Who is not familiar with the unpoetic name and eventful history of Pit Hole—the civic wonder of the century? Well might foreigners, accustomed only to the slow advance of national progress, exhibit astonishment at the story of this remarkable creation of the feverish times, when a few brief weeks served to accomplish the work of years, and hundreds of anxious men, in three short months, crowded into their lives the multiform experiences of a decade. Never before had mankind beheld a nearer realization of some imaginative tale of the Arabian Nights — nor is the world likely soon again to witness an ascent and a tumble so bewildering as the rise and fall of the once famous Pit Hole city, located in the north-east corner of Cornplanter township. But Pit Hole, and others of minor note in the upper district, have disappeared,—and he who would see thoroughly, representative "oil towns," must perforce visit the oily realms of Butler County, and interview the new oil "cities," springing up there.

TITUSVILLE, CRAWFORD COUNTY, PA.

LET us begin our review, or description of the chief Cities and Towns in the upper oil fields, with TITUSVILLE, located as it is, upon the extreme north-eastern boundary of the oil producing region, near the head-waters of Oil Creek, in the north-eastern corner of Crawford County. TITUSVILLE became a point of importance, upon the successful completion of Col. DRAKE'S well, in 1859,—though the Borough was incorporated in 1847. At that date—1847,—the settlement, contained less than 200 inhabitants. JOSEPH L. CHASE, was the first Burgess of TITUSVILLE, under a special law of the State, creating it an incorporation. WILLIAM BARNSDALL, S. S. BATES, J. K. KERR, and G. C. PETTIS, composed the first Council, under this enactment. In 1861, the Borough was re-organized under the general act of 1834, retaining as its corporate government, a Burgess, six Councilmen, six School Directors, and other officers required by its Charter. Year by year, since 1861, the city has grown in population and wealth,— and in 1867, a City Charter was secured, with a Mayor, Common Council, uniformed Police, School Board, &c., &c.

The population of TITUSVILLE, by the census of 1870, was a little short of 10,000. Its steady growth since, warrants an estimate of the population now, at 12,000. The city presents an attractive and solid appearance. The business streets are adorned with handsome stores, mostly brick structures, and the thoroughfares are well lighted, and are kept in good repair. During the past year—since July, 1872—more than a mile of wooden pavement has been laid in the principal business avenues—SPRING, DIAMOND, FRANKLIN, PINE, and WASHINGTON Streets*—an improvement greatly needed, and rendering these thoroughfares at once attractive and substantial. Many splendid private mansions have been erected during the past two or three years, both elegant in style and elaborate in finish. The city has besides scores and scores of beautiful residences. The principle streets are wide, and handsomely graded. The city is lighted with gas. It has some ten or twelve churches, representing almost every profession of Christianity. It has four public schools—one in each of the four Wards of the city,—two of these school edifices being large and substantial brick structures, that have cost not less than $50,000 each. It has a public park, which is beautified yearly by an expenditure of $2,000 to $5,000. Its city sewerage is nearly complete, and the new "Holly Water Works" will be finished this year.

The city has six banking establishments, all upon solid basis—a Board of Trade—"Oil Exchange"—two daily newspapers—*The Herald* and *The Courier*, the former the Republican, the latter the Democratic organ—and one weekly, *The Sunday Morning Press*— "independent, but not neutral." It has among its many industries, two large Iron establishments, employing hundreds of men in the manufacture of engines and boilers, oil well tools and machinery, car tanks, oil tanks of immense size and capacity, &c., &c. These are "THE GIBBS & STERRITT MANUFACTURING COM-

---

* The principal portion of this wood pavement is the product of local enterprise, Col. E. A. L. ROBERTS being the patentee, and ROBERTS BROS. the contractors.

PANY," and "THE TITUSVILLE MANUFACTURING COMPANY."
Besides, there are many others of less capacity—ADAM GOOD'S
Brass Foundry and Machine Shop, being among those de-
serving special mention. In addition to these noted industries,
TITUSVILLE has eight or ten Oil Refineries, involving the employ-
ment of a large amount of capital, and hundreds of operatives. Its
Hotel accommodations are second to no town or city in Western
Pennsylvania. It has an Opera House, which, for size and
beauty, is not excelled in this portion of the Commonwealth.

TITUSVILLE has two Railways entering her limits—"The Oil
Creek & Allegany River Railway"—connecting with the Atlantic
& Great Western Railway at Union and Corry—and at the latter
point, with the Philadelphia & Erie Road, East and West, and
with the Cross Cut, or Buffalo, Corry & Pittsburgh Railroad to
Brockton, on the Lake Shore & Michigan Southern Railroad.

"The Warren & Venango Road," terminates at TITUSVILLE at
present—connecting that city with Warren, Pa., and with Dunkirk, N. Y., on the Erie, and Lake Shore Road. This Road has its
prospective terminus at Oil City, and the road-bed is nearly completed to that point.

TITUSVILLE has all the advantages and attractions of a growing
Western city. It has wealth, enterprise, industry, thrift, and a
"go-ahead," about her citizens, that will bring prosperity and continued growth as surely as day succeeds night.

PLEASANTVILLE, VENANGO COUNTY, PA.

THIS is one of the most attractive and prosperous towns of the
upper Oil Region. Situated on the high land, overlooking the
country for miles in every direction, PLEASANTVILLE is a beautiful interior village of two to three thousand inhabitants, six miles
east of Titusville. The streets are wide, and are laid out regularly, beautified by substantial private residences, not costly or
luxurious, if we except a dozen or more of the better class of struc-

tures of this character, but are home-like, substantial, and inviting. Every street in the village is a model of neatness and order.

PLEASANTVILLE has half a dozen brick blocks, devoted to the purposes of trade and business. It has two Banking Institutions of financial solidity, Churches, an Opera House, and first-class hotels, ample for all the public requirements. In the midst of an oil producing region not yet exhausted, and surrounded by an agricultural territory second to no village in the upper Oil Country, PLEASANTVILLE will have sure prosperity and steady growth for years to come.

———

MILLER FARM, PIONEER, PETROLEUM CENTRE, COLUMBIA, TARR FARM, RYND FARM, and ROUSEVILLE, were once busy oil centres, with large populations. They each have, even now, their attractions, and are all located in the valley of Oil Creek, between Titusville and Oil City, on the line of the Oil Creek and Allegany River Railroad. PETROLEUM CENTRE and ROUSEVILLE may be said to retain much of their early thrift and later enterprise, being surrounded by good producing territory. PETROLEUM CENTRE is perhaps the most important point, located in the midst of productive oil farms, and yet a shipping station of note. It has two churches, a banking office, stores, machine shops, an opera house, hotels, &c., and a population of 2,000 to 2,500. It has also an excellent daily paper, small, but ably conducted,—"*The Record.*" ROUSEVILLE has all the advantages and conveniences of the "Centre," including churches, banking offices, hotels, and an opera house,—but it has no daily or weekly newspaper.

———

COLUMBIA, Story Farm, is a point of considerable note and importance—being the local habitation of one of the best oil companies in the upper oil field—"The Columbia Oil Co." It has yet hundreds of acres of undeveloped oil lands.

OIL CITY, Venango County, Pa.

Oil City is located upon both sides of the Allegany River and Oil Creek—Venango City—across the Allegany River—having been added to its corporate boundaries a year or more since, by special enactment of the Legislature. The population of the city, with its additions, must be 7,000 to 8,000.

"About twenty years prior to the discovery of oil," or in 1840—we quote a writer who seems to know whereof he speaks—"a small settlement was made on this spot, consisting of a grist mill, a furnace, hotel, and boat landing, but this soon fell into decay. In 1852, Mr. John Hopewell established a store, the first settler after him being Francis Halliday, who purchased the land where the city now stands from the Government. The amount of this purchase included several hundred acres. After a while, the greater portion of this tract was sold to Dr. John Nevins, who again sold it to Plummer & Drum, of Franklin. In 1860, the Michigan Rock Oil Company became the purchasers, and by them the foundation of the present city was first laid. The land on the east side of the Creek was once, if not now, owned by the United Petroleum Farms Association, of New York, having been purchased by them in 1865, of Graff, Hasson & Co.,—originally belonged to Cornplanter, the renowned Chief of the Seneca Indians, a branch of the famous Six Nations. Three hundred acres of this property was presented to the Chief, by Congress, as a reward for distinguished services rendered by him during the wars of the Revolution. Cornplanter, it is alleged, was very fond of ' fire-water,' and in one of his drunken frolics, it is claimed, he sold this property for a mere song."

Soon after the striking of Drake's well in 1860, the town began to increase rapidly, so much so indeed, that in 1865, its population was about 8,000. There are eight or nine churches, a public library, two large school-houses, several machine shops, three Refineries—one of which has the largest Still capacity of any in the

Oil Region—besides a large number of excellent stores, well stocked with every description of goods. The town has an ably conducted morning paper, *The Oil City Derrick*, and a weekly, *The Oil City Times*, issued from the office of *The Derrick*. A substantial iron bridge over Oil Creek unites the eastern and western portions of the city. The railroad bridge of the Atlantic & Great Western Railway crosses the Creek at its mouth, and a fine wooden bridge stretches over the Allegany, connecting OIL CITY with its recently added territory—Venango City.

From the commencement of the Oil business, OIL CITY has been the principal shipping point of the Petroleum Region. The Meadville branch of the Atlantic & Great Western Railway has its terminus at OIL CITY. The Oil Creek Railroad runs through it, having its connections at Corry with the Philadelphia & Erie Railroad. The Allegany Valley Railroad has its terminus here, and the Jamestown Railroad, (Lake Shore & Michigan Southern Railroad,) run trains into the city. In addition to these advantages, the Allegany River and Oil Creek, at certain seasons of the year, furnish cheap means of transportation.

The amount of iron tankage at this point is computed at more than a million barrels.

OIL CITY has within its limits a number of small producing wells, and it is surrounded by territory where wells, although not doing any thing very extraordinary—if we except the Sage Run developments—are steadily pumping from one to five barrels of oil a day. The agents of many of the largest refining and exporting houses are stationed here, and an oil exchange has been established upon the principal street, which is characterized by all the bustle and anxiety noticeable in similar places in larger cities. With its numerous Railroad connections it may easily be imagined that the depots present a very animated appearance, and the tracks seem always crowded with long trains of tank cars, bearing away the precious oil to all quarters of the country.

During the past two years OIL CITY has added largely to her

growing business needs, by the addition of very many splendid brick structures. Few cities in Western Pennsylvania can boast more substantial or more attractive blocks for the purposes of trade, than OIL CITY—and these are being constantly and rapidly added to. Two or three years of prosperity and enterprise, like that of the two years past, will go far to make OIL CITY the chief town of the Oil Region.

TIONESTA, FOREST COUNTY, PA.

TIONESTA, just above, on the opposite side of the Allegany, is a thriving village of about eight hundred inhabitants, many fine private residences, several stores, hotels, and churches. It is the county-seat of Forest county. A new court-house and jail were completed a few years since. Two newspapers—*The Journal* and *The Press*, representing each of the political parties—are published here. There is also a banking office. Tionesta Creek here debouches into the Allegany.

TIDIOUTE, WARREN COUNTY, PA.

TIDIOUTE is in Warren county, still above Tionesta, on the Allegany river, and is the centre of an excellent oil producing region. It is one among the oldest towns in this part of Pennsylvania, is well built, has several churches, school-houses, an opera house, hotels, stores, three banking offices, " Grandin Bros. Bank," " People's Savings Bank," and a private banking institution—all in excellent repute—machine shops, and a daily and weekly newspaper, *The Journal*. It contains about 3,000 inhabitants. At present, a very heavy iron bridge is being constructed over the Allegany river at this point, to connect the village with an addition to its boundaries.

TRIUMPH, FAGUNDAS and NEW LONDON, in the neighborhood of Tidioute, are emphatically Oil Towns, located in the midst of Oil Fields that have been producing oil in abundance, during the past four or five years. TRIUMPH and NEW LONDON seem to have an exhaustless oil product, and are, therefore, the more thrifty and prosperous. Mr. E. E. CLAPP, of President, is the principal owner of the New London oil territory—Henderson farm and others—and is one of the largest and most enterprising operators in the upper oil region.

___

FRANKLIN, VENANGO COUNTY, PA.

FRANKLIN is the county seat of Venango county, and is situated at the junction of French creek with the Allegany river, and on the west side of the latter. The present town was laid out in 1795, by General William. Irvine and Andrew Ellicott, who were commissioners appointed by the State, under Act of Assembly to select a site for the county town. Having been chosen, Franklin, was laid out in what was then known as in-lots, out-lots and out-tracts. The town was first organized into a borough, in 1829, Judge William Connelly being selected as Burgess. Franklin received its chief impetus from the Petroleum excitement, and is now a substantial city, with a steady growth, and has a population of 6,000 to 7,000.

The streets are broad and well paved, and the principal business houses are located in elegant blocks of brick buildings. The residences of the well-to-do inhabitants are very handsome, and remarkable for the care and tastefulness with which the grounds are laid out in gardens and ornamental shrubberies. There are in the town several machine shops, two or three foundries and one large flouring mill. A new and very elegant and commodious court-house has lately been completed, which is of red brick with brown stone facings. This edifice would contrast favorably with many of the public buildings in more pretentious cities. The school-house, not far distant, is also a handsome structure, capable of accommodating 1,000 children. There are churches of almost every denomination, several private schools and two weekly newspapers, *The Spectator* and *The Citizen.*

On the high ground opposite the city, upon the banks of French Creek, several wells have been put down, some of which started out with a yield of as much as one hundred barrels per day. This fact, of course, soon became known, and the hill-side is now literally bristling with derricks. The oil found in this neighborhood is of more than ordinary quality, almost all of it being known as lubricating oil, and the wells along French Creek are almost all yielding fairly. On the Allegany side, on the contrary, the yield is small, and a great number of dry wells may be seen scattered about. (See chapter on Lubricating Oil.)

Two Lubricating Oil Manufactories are located here, employing a capital of $300,000 to $400,000. These distil nine-tenths of the lubricating oils produced in the vicinity of Franklin.

Franklin has three or four banks and banking institutions, ample hotel accommodations, an Opera House, and very many elegant and commodious stores and warehouses. The Meadville Branch of the Atlantic and Great Western Railway passes through the city limits, as does also the Jamestown and Franklin road. The Allegany Valley Railway is upon the opposite side of the river, a mile from the business centre of the city,—and all three of these railways have depots for the conveniences of passengers and freight.

FOSTER STATION, and SCRUBGRASS, are points of importance, and are located still further down the Allegany River. FOSTER, is an oil producing and shipping point, and beyond this has few advantages. SCRUBGRASS has an oil field, and two or three years ago, boasted a prolific yield. It has a railway eating-house, and is a point of some importance for shipping oil. The village may have a population of 300 to 500.

EMLENTON, CLARION COUNTY, PA.

EMLENTON is one of the oldest towns in this part of the State. It was a lumbering center years before the discovery of oil. Its rapid growth dates from the oil developments of the lower oil field, since 1865–6. It has a population of 2,000 to 3,000, with churches, elegant public school buildings, hotels, stores, and a weekly newspaper,. published by that veteran journalist, J. R. JOHNS. EMLENTON is seven or eight miles above Foxburg, on the Allegany river, and on the line of the A. V. Railroad. It is an oil shipping point of importance.

———

FULLERTON, three miles above Foxburg, is an oil-shipping station for the St. Petersburg district and points contiguous to FULLERTON, on both sides of the river. The tankage capacity here must be fifty to seventy-five thousand barrels.

THE LOWER OIL FIELDS.

PARKER'S LANDING, ARMSTRONG CO., PA.

"PARKER'S" is the largest and perhaps the most important town in the lower oil field. From this point diverge the great developments in oil which have astonished the world. Parker City is the base of operations, and may justly be termed the capital of the lower oil field.

The first settlement was made here about seventy-five years ago, by JOHN and WILLIAM PARKER,* and for upwards of forty years it continued to be a small village deriving support from an iron furnace situated about half a mile from town. This furnace was established 1825.

With the progress of oil developments, the territory around PARKER'S became the scene of active operations, the village assumed the importance of a busy, thriving town, possessing all the advantages and disadvantages, vices and virtues, of larger cities, known to fame. On the 1st of March, 1873, the charter incorporating PARKER'S LANDING, Lawrenceburg and Farrentown, (the two latter suburban towns,) into one city, was signed by the Governor of the State, thus consolidating, under one city government, these three towns, which had only been divided by name.

---

* It is worthy of note that WILLIAM PARKER, the first settler, came from the North of Ireland, bringing with him the pluck and energy of an Ulster yeoman, and from him are descended Parkers, of the firm of Parker, Thompson & Co., of Oil City, and Parker's Landing.

PARKER'S has twice been visited by fire, and on both occasions the great bulk of the town was swept away, but on each occasion it has re-appeared, built up in better shape, with increased dimensions, and more substantial buildings. Many of the lately erected buildings are of brick, and present tasteful architectural features. The city boasts a well equipped fire department, three banks, five hotels, and two churches.

The Allegany River at PARKER'S is spanned by an iron bridge, which has cost $100,000. On the east side of the river, opposite the city, is located the Allegany Valley R. R. depot. The Parker and Karns City R. R., now in course of construction, will materially aid the progress and prosperity of the place. This road runs along Bear Creek to Martinsburgh, thence to Petrolia. It will be ten miles in length, three foot gauge, and will cost $150,000. It is not unlikely this road will be extended to Millerstown, and thence to Butler.

PARKER CITY is noted for its men of energy and liberality— here are to be found the men of "the front." The city is peaceable, orderly, and well kept, being under the government of an active and efficient city council.

The population of PARKER CITY is about 4,000. The Oil Man's *Journal*, a weekly publication, of large circulation, is printed here, and it is very ably conducted.

FOXBURGH, CLARION Co., PA.

FOXBURGH is three miles above, and upon the opposite side of the river from Parker's Landing. Less than two and a half years ago there was nothing on the sight of this town but a freight shed, belonging to the Allegany Valley R. R. Now there are at least two hundred houses, and a population of over 1,000. It is the head-quarters of the Mutual Pipe Line Co., and an immense freighting business is done here. There is the usual complement of banks, stores, etc., and it is noticeable that the buildings are neater in appearance, and far more substantial than those seen in the more recently built oil towns.   It is situated on the Fox estate, and derives its name from the former owner, now deceased.   An iron bridge is being constructed here, to span the Allegany River.   Its cost will be not less than $100,000, and like that at Parker's Landing, will be the perfection of American skill and mechanism.

ST. PETERSBURG, CLARION CO., PA.

ST. PETERSBURG is of considerable size and importance, and lays claim to some antiquity.   A small village has existed here for nearly fifty years, but previous to the development of oil in its neighborhood, it was almost unknown.   There are now at least four hundred houses, and a population of two thousand inhabitants. The town enjoys many of the advantages of larger places, having a church, bank, and weekly newspaper, named *The Progress*.   In September, 1871, the Hulings' Well was struck on the J. J. Ashbaugh farm, which proved a 100 barrel well.   From this may be dated the revival of St. Petersburg.   The town is two miles from the Allegany river, west of Foxburgh.

PETROLIA, BUTLER COUNTY, PA.

THE attention of operators having been attracted to this quarter in the autumn of 1871, by the striking of good wells on the Sheakley farm, and subsequently (April 1872,) a strike made by Dimick, Nesbitt & Co.   This latter venture was attended with the usual results, and forthwith began a regular rush to the new field of operation.   With the first *rig* built, PETROLIA was founded, houses multiplied, and ere long the infant city presented all the bustle and rush characteristic of sudden oil towns.   Wells continued to be put down, PETROLIA grew apace, and in February of 1872, it was incorporated into a borough. The town possesses a most energetic population of about 1,500. PETROLIA justly ranks among the most attractive and prosperous towns in North-western Pennsylvania.

The town is built principally upon the Blaney, Graham, and Jamison farms, and stands in a valley running due north and south. The hill on the east side commands a view of Karns City, one and a half miles to the south, and Fairview, about a mile and a half to the west. At night the "*belt*" can be distinctly traced for miles by the hundreds of gas-lights which mark its outline. New buildings are constantly being built. At present there is in progress a church. PETROLIA is the central point for a number of the Pipe Lines. It has a bank, and other institutions necessary for a rising and prosperous town. PETROLIA utilizes the gas from the Fairview gas wells. It is supplied to many of the houses, and illuminates the streets.

FAIRVIEW, BUTLER COUNTY, PA.

FAIRVIEW is situated one mile south of Petrolia, on a commanding position, affording a delightful prospect of the country for miles around, from which circumstance the place derived its name. FAIRVIEW is of old date, a settlement having established itself here some seventy years ago, when the country around was nothing but a howling wilderness. Most of the buildings are located on one street, and generally present the appearance of considerable age.

FAIRVIEW has one good hotel, a tavern or two, one bank, a church and school-house, and numbers in population about 800.

ARGYLE, BUTLER COUNTY, PA.

ARGYLE is a suburb of Petrolia. It took its rise from a well put down in the fall of '71. At its best it had 600 inhabitants.

two hotels, two hardware stores and one or two machine shops. ARGYLE is but a wreck of its former self, and is a sad comment on the changeable character of our new petroleum towns.

––––––

KARNS CITY, BUTLER CO., PA.

KARNS CITY is about fifteen months old, pleasantly situated on a large plateau of ground about a mile long, and a quarter of a mile broad. Prior to the oil excitement, the site of KARNS CITY presented nothing to distinguish it from the average of farming lands. The McClymonds' homestead and an old saw-mill were its chief erections. The first well here was struck in June, 1872, and the second in September, and from the later period the place continued to improve rapidly. At the present time it has a population of about 1,000 inhabitants. A number of important business houses have established branches at this point. There are three large hotels—the *Empire House, Apollo House,* and *Exchange Hotel*—one bank, and the customary number of business houses of all varieties. The town is named in honor of S. D. Karns, Esq., a fortunate and extensive operator.

––––––

BUENA VISTA, BUTLER CO., PA.

Four miles south of Petrolia, on the road to Greece City, is the rising hamlet of BUENA VISTA. It is pleasantly situated at the intersection of roads leading in all directions. A settlement has existed here for at least sixty years. It has a Post Office, two taverns, and the usual supply of tradesmen. The population is about 400.

––––––

GREECE CITY, BUTLER CO., PA.

A little less than fourteen months since, GREECE CITY had neither "a *locale* nor a name." On the 24th of August, 1872, the first well was struck on the Jamison farm, and early in September the first building made its appearance in what is now GREECE CITY. GREECE rose rapidly—and fell with the same speed. It had, at the height of its glory, about 1,200 inhabitants, and has now four or five hotels, three banks, and the full variety of stores. The "city" is fast disappearing from the face of the earth.

The Morrison well at this place is the second largest yet struck—so far as the productive value is concerned—having at the present date produced $90,000 worth of oil, and still continues to produce fifteen barrels of oil per day

BUTLER, BUTLER COUNTY, PA.

The first settlement of whites in what is now Butler county, was made in 1796, when the four families of Harbison, Holt, Fulton, and Kennedy, located in Middlesex township. The town is pleasantly situated, and the surrounding country is rich in soil and mineral products. BUTLER is laid out with tolerable regularity—the streets being intersected at right-angles, and possessing ample width. A large proportion of the buildings are of brick, and the side-walks paved with the same material. Iron furnaces once existed here, to which is due much of the growth of the town. It was made a borough in 1817. The first Court House was built in 1807—the present building in 1853. The church (Presbyterian) was built in 1822.

With the advance of oil developments around its vicinity, BUTLER has taken a step in advance. New buildings are being erected, and all branches of business have become stimulated. A number of wells are now going down in the BUTLER district, upon which the further advance of the town for the present depends. BUTLER is a well-ordered town, possessing all the necessary adjuncts to civilization, such as a court-house, jail, churches, school-houses, two newspapers, &c. It has now a population of 5,000.

MILLERSTOWN, Butler County, Pa.

A village has existed here for upwards of half a century—and previous to the advent of the derrick and drill, presented that dead-and-alive appearance common to towns situated at considerable distances from railway connections, manufacturing or mining industries. Its silent streets wore a Sunday aspect. It lived in quiet peace, and could boast of a couple of taverns, a number of stores, tradesmen's shops, a church, school-house, and less than two hundred men, women, and children. Many of the dwellings gave evidence that the owners were well-to-do, as they presented a comfortable appearance.

In March of the present year, the first well—"The Shreve"— was struck, which gave ample proof of the value of the MILLERSTOWN territory, and so attracted many of our operators. MILLERSTOWN soon became crowded with people from all parts of the Region. New wells were struck—and with each strike the town swelled in population and importance. At the present writing it has two banks, two Pipe Line offices, four hotels, four livery stables, a number of hardware and drug stores, and some six or seven grocery and dry goods houses, and has a population of 2,000 to 2,500. The town, as it now stands, is well built, and has the look of permanence. It is situated on a side-elevation overlooking the valley, and is surrounded by a beautiful and fertile farming country—its countless hills and dales forming a picturesque landscape of rare beauty. The *Sand Pump*—a daily newspaper—is published here.

BRADY'S BEND and EAST BRADY.

The Rolling Mills at BRADY'S BEND have been in existence for about fifty years. Their full capacity gives employment to 800 men. At the present time there are about 500 men at work. The town is situated on Sugar Creek, about half a mile from the Allegany river, and has six churches built by the miners and mill men.

EAST BRADY has a population of nearly 1,500. There are five hotels and one church, one machine shop and foundry, a number of groceries and dry goods houses, and three schools. There is a large amount of freight business done here on the Allegany Valley Railroad, consequent upon the opening up of the Oil districts of Modoc and Millerstown.

EAST BRADY has a weekly paper—*The Independent*—which enjoys a large circulation—Col. YOUNG, one of the *live* men of the profession of Journalism, being its editor and proprietor.

MODOC, BUTLER COUNTY, PA.

Scarcely a year has elapsed since the success of operations a short distance north of Greece City, also in Concord township, attracted the attention of enterprising operators to an uninviting tract of land about midway between the many roads leading to Fairview and Buena Vista respectively. The property was secured, and the spot chosen on which to put down a well. The place was low and forbidding in its appearance, surrounded with a dense growth of underbrush, having a small run and extended ravine contiguous, and in no respect possessing any natural features likely to please an æsthetic taste. Neither did it seem destined to reward the adventurers with abundance of the greasy fluid of which they were in search, as the "belt" was generally believed to run in a direction that would leave the entire locality completely "out in the cold." A rig was put up during the fall and winter, drilling quietly went on, and on the twenty-second of last March the biggest strike since "sixty-five," was announced to the outside world. Such was the origin of the renowned Troutman Well, the most prolific yet discovered south of Oil Creek.

At the outset, the new "spouter" actually flowed at the enormous rate of over a thousand barrels per diem. It was only three feet in the third sand, and the tools were left suspended in the hole for greater convenience in agitating the well occasionally. It required two pipes to convey the product from the casing to the tanks, of which a number were provided in the quickest manner known to the business. The owners of the big strike were William Vandegrift, Joseph Bushnell, Warden and Bostwick, who had likewise secured a large tract of land in the immediate vicinity. For weeks the "geyser" continued to flow without serious diminution, frequently giving the pipe line lively exercise to convey the oil to Greece City, exactly three miles distant. Up to this date, it has yielded nearly *ninety thousand barrels* of petroleum, thus ranking it with the most productive wells ever struck in any portion of Oildom.

The great success of this venture was attended with the result that has invariably followed in such cases. Operators and speculators flocked to the latest Oil-Dorado, intent on acquiring some interest in the neighborhood. Rigs multiplied rapidly, and the nucleus of a town appeared as if by magic. The Starr and the Sutton farms, below the Troutman property, were generally considered with special favor, and upon these a village was at once located. Building lots were staked off that found ready purchasers or lessees at high figures, balloon houses were quickly erected along both sides of the principal street, even the fearful condition of the roads did not deter crowds from seeking this second Pit Hole, and within a month a town bristling with business of every description occupied the vacant fields of four weeks before. The embryo city received an Indian epithet, and henceforth Modoc became a familiar name in petroleum circles. Many oleaginous sages wisely shook their heads at the spectacle of activity everywhere presented.

MODOC, now contains one leading hotel, two or three hardware stores, a dozen groceries, a livery stable, two banks, and a score of unlicensed local-option dispensaries. The place is built without special regard to architectural arrangement, its appearance and surroundings are decidedly picturesque, and the newest tyro would not be long in discovering he had "fetched up" in a genuine oil city! With the exception of the three old homesteads of the Troutman, Starr and Sutton farms, everything about the town is of the modern date of the present season. The muddy streets are crowded with travelers from all quarters, and no one who has not visited Modoc can be said to have "done" the specimen oil town of the period.

ANGELICA.

This village or town, named after C. D. Angell, "the lord of the soil," has an existence of only a few months, and yet it has attained considerable proportions. Active developments are being pushed on in the neighborhood, and no doubt this village will soon aspire to the title of city.

Such is a brief description of the principal cities and towns, large and small, located in the Pennsylvania Oil Region. Before the oil discovery by Col. DRAKE, in 1859, the older towns were generally farming or lumbering centres, and owe their growth to the Petroleum development. The number of cities projected by sanguine speculators in the flush times of 1864 and 1865, were legion. Beyond a few board shanties, in hopeless decay, nothing now remains of them.

Library of
Early American Business And Industry